AF888283

SPRINGER PROTOCOLS HANDBOOKS™

For further volumes:
http://www.springer.com/series/8623

Methods and Techniques in Ethnobiology and Ethnoecology

Edited by

Ulysses Paulino Albuquerque

Departamento de Biologia, Universidade Federal Rural de Pernambuco, Recife, Pernambuco, Brazil

Luiz Vital Fernandes Cruz da Cunha

Departmento de Ciências Biológicas, Universidade Católica de Pernambuco, Recife, Pernambuco, Brazil

Reinaldo Farias Paiva de Lucena

Departamento de Fitotecnia e Ciências Ambientais, Universidade Federal da Paraíba, Centro de Ciências Agrárias, Paraíba, Areia, Brazil

Rômulo Romeu Nóbrega Alves

Departamento de Biologia, Universidade Estadual da Paraíba, Campina Grande, Paraíba, Brazil

Humana Press

Editors
Ulysses Paulino Albuquerque
Departamento de Biologia
Universidade Federal Rural de Pernambuco
Recife, Pernambuco, Brazil

Luiz Vital Fernandes Cruz da Cunha
Departmento de Ciências Biológicas
Universidade Católica de Pernambuco
Recife, Pernambuco, Brazil

Reinaldo Farias Paiva de Lucena
Departamento de Fitotecnia e
Ciências Ambientais
Universidade Federal da Paraíba
Centro de Ciências Agrárias
Paraíba, Areia, Brazil

Rômulo Romeu Nóbrega Alves
Departamento de Biologia
Universidade Estadual da Paraíba
Campina Grande, Paraíba, Brazil

ISSN 1949-2448 ISSN 1949-2456 (electronic)
ISBN 978-1-4614-8635-0 ISBN 978-1-4614-8636-7 (eBook)
DOI 10.1007/978-1-4614-8636-7
Springer New York Heidelberg Dordrecht London

Library of Congress Control Number: 2013951153

© Springer Science+Business Media New York 2014
This work is subject to copyright. All rights are reserved by the Publisher, whether the whole or part of the material is concerned, specifically the rights of translation, reprinting, reuse of illustrations, recitation, broadcasting, reproduction on microfilms or in any other physical way, and transmission or information storage and retrieval, electronic adaptation, computer software, or by similar or dissimilar methodology now known or hereafter developed. Exempted from this legal reservation are brief excerpts in connection with reviews or scholarly analysis or material supplied specifically for the purpose of being entered and executed on a computer system, for exclusive use by the purchaser of the work. Duplication of this publication or parts thereof is permitted only under the provisions of the Copyright Law of the Publisher's location, in its current version, and permission for use must always be obtained from Springer. Permissions for use may be obtained through RightsLink at the Copyright Clearance Center. Violations are liable to prosecution under the respective Copyright Law.
The use of general descriptive names, registered names, trademarks, service marks, etc. in this publication does not imply, even in the absence of a specific statement, that such names are exempt from the relevant protective laws and regulations and therefore free for general use.
While the advice and information in this book are believed to be true and accurate at the date of publication, neither the authors nor the editors nor the publisher can accept any legal responsibility for any errors or omissions that may be made. The publisher makes no warranty, express or implied, with respect to the material contained herein.

Printed on acid-free paper

Humana Press is a brand of Springer
Springer is part of Springer Science+Business Media (www.springer.com)

Preface

Humans represent just one of many species that constitute the planet's biodiversity. Nevertheless, as the dominant species, humans have been the primary agent of the transformation of natural spaces, which can be seen as positive or negative. Therefore, the study of human interactions, biodiversity, and the environment that surrounds them is a basic tool for understanding the factors that bind human societies to natural resources. Within this context, ethnobiology appears to be a promising discipline that can play a key role as a mediator of dialogue between different academic disciplines and traditional knowledge, a union essential to enable contextualized and sustainable alternatives to exploitative practices and biodiversity management. Thus, this book comes with the original proposal of realizing the main methods that allow us to assess this diverse and complex relationship between people and nature.

This book represents the evolution of the book *Methods and Techniques in Ethnobotany Research* published in Brazil in 2004. We realized that this book was consulted not only by researchers and students interested in ethnobotany but also by people in different fields. Thus, this book, *Methods and Techniques in Ethnobiology and Ethnoecology*, is an improved and expanded edition of the third edition of the Brazilian text (2010). This book was written to guide people in the following methods: think about the data you collect, choose the most appropriate procedures for collecting such data, and reflect on your practice and scientific work. Over the years, we have continued to improve the text, engage new authors, and expand chapters to offer a broader view of the possible range of tools that can be used in this type of research.

Certainly, the text is not without flaws, but as with the Brazilian editions, we are also open to suggestions and comments from readers regarding ways of perfecting each new edition. In this book, the reader will find 28 different chapters covering different qualitative and/or quantitative aspects. We have several works on research methods, and each year texts that are increasingly practical and complete enter the book market. However, the intent of this book is to be a bedside manual to help readers make decisions. Much of this book is the result of the authors' experience with the themes; so their examples give substance to ideas and procedures that are often abstract. Therefore, we are grateful to the commitment and involvement of all the authors, for they made this work possible.

Because this is a translated and revised edition of the Portuguese language version, many of the references the reader finds in the course of the chapters are written in Portuguese. It has been requested of the authors that, to the extent possible, they favor English language publications in the references, especially texts of wide circulation and/or easy access. Unfortunately, this is not currently possible in many cases, and we apologize to our readers. However, surely this aspect does not limit the scope of the work and/or the reader's approach to it.

This book, then, dear reader, is the result of years of investigation and research by the authors with the sole intention of providing a manual with the main methods that can be useful in any ethnobiological investigation.

Recife, Pernambuco, Brazil — *Ulysses Paulino Albuquerque*
Recife, Pernambuco, Brazil — *Luiz Vital Fernandes Cruz da Cunha*
Areia, Paraíba, Brazil — *Reinaldo Farias Paiva de Lucena*
Campina Grande, Paraíba, Brazil — *Rômulo Romeu Nóbrega Alves*

Acknowledgements

A debt of gratitude is owed to many people who helped make this book possible. Thanks are due to Eric Stannard, editor at Springer, for supporting the proposal and Jacob Gallay, who later assisted in the process of producing the book. We also thank all the authors who accepted our invitation to review the texts originally published in Brazil, with a subsequent version produced in the English language, and who understood our requirement to have this version edited by international services specializing in translating and editing scientific texts. We are also grateful to the authors who produced texts specifically for this edition of Springer. We are especially grateful to the National Council for Scientific and Technological Development (CNPq) for financial support in the form of scholarships for scientific productivity given to RRNA and UPA.

Contents

Contributors

ULYSSES PAULINO ALBUQUERQUE • *Laboratory of Applied and Theoretical Ethnobiology, Department of Biology, Federal Rural University of Pernambuco, Recife, Pernambuco, Brazil*

NÉLSON LEAL ALENCAR • *Universidade Federal do Piauí, Floriano, Piauí, Brazil*

ALYSON LUIZ SANTOS DE ALMEIDA • *Laboratory of Applied and Theoretical Ethnobiology, Department of Biology, Federal Rural University of Pernambuco, Recife, Pernambuco, Brazil*

CECÍLIA DE FÁTIMA CASTELO BRANCO RANGEL DE ALMEIDA • *Departamento de Biologia, Centro de Ensino Superior do Vale do São Francisco, Belém do São Francisco, Pernambuco, Brazil*

RÔMULO ROMEU NÓBREGA ALVES • *Departamento de Biologia, Universidade Estadual da Paraíba, Campina Grande, Paraíba, Brazil*

ELBA LÚCIA CAVALCANTI DE AMORIM • *Laboratório de Produtos Naturais, Departamento de Ciências Farmacêuticas, Universidade Federal de Pernambuco, Recife, Pernambuco, Brazil*

HELDER FARIAS PEREIRA ARAUJO • *Agrarian Science Center, Department of Biological Sciences, UFPB, Areia, Paraíba, Brazil*

ELCIDA DE LIMA ARAÚJO • *Departamento de Biologia, Universidade Federal Rural de Pernambuco, Recife, Pernambuco, Brazil*

THIAGO ANTÔNIO DE SOUSA ARAÚJO • *Laboratory of Applied and Theoretical Ethnobiology, Department of Biology, Federal Rural University of Pernambuco, Recife, Pernambuco, Brazil*

ISANETE G.C. BIESKI • *Departamento de Estatística, Instituto de Ciências Exatas e da Terra, Universidade Federal de Mato Grosso, Cuiabá, Mato Grosso, Brazil*

LETÍCIA ZENÓBIA DE OLIVEIRA CAMPOS • *Laboratory of Applied and Theoretical Ethnobiology, Department of Biology, Federal Rural University of Pernambuco, Recife, Pernambuco, Brazil*

ROSSANA DE AGUIAR CORDEIRO • *Department of Pathology and Legal Medicine, Federal University of Ceará, Fortaleza, CE, Brazil*

MARGARITA PALOMA CRUZ • *Laboratory of Applied and Theoretical Ethnobiology, Department of Biology, Federal Rural University of Pernambuco, Recife, Pernambuco, Brazil*

MARIANO MARTINEZ ESPINOSA • *Departamento de Estatística, Instituto de Ciências Exatas e da Terra, Universidade Federal de Mato Grosso, Cuiabá, Mato Grosso, Brazil*

LUCIA REGINA RANGEL MORAES VALENTE FERNANDES • *Academia da Inovação e Propriedade Intelectual, Instituto Nacional da Propriedade Intelectual, Rio de Janeiro, Rio de Janeiro, Brazil*

ELBA MARIA NOGUEIRA FERRAZ • *Ciência e Tecnologia de Pernambuco, Instituto Federal de Educação, Recife, Pernambuco, Brazil*

Washington Soares Ferreira Jr. • *Laboratory of Applied and Theoretical Ethnobiology, Department of Biology, Federal Rural University of Pernambuco, Recife, Pernambuco, Brazil*

Fábio de Oliveira Freitas • *Parque Estação Biologia, Embrapa Genetic Resources and Biotechnology, Brasília, Federal District, Brazil*

Laura Jane Gomes • *Department of Forest Science, Federal University of Sergipe, São Cristóvão, Sergipe, Brazil*

Natalia Hanazaki • *Department of Ecology and Zoology, Universidade Federal de Santa Catarina, Florianópolis, Santa Catarina, Brazil*

Julio Alberto Hurrell • *Laboratorio de Etnobotánica y Botánica Aplicada (LEBA), Facultad de Ciencias Naturales y Museo, Universidad Nacional de La Plata, La Plata, Buenos Aires, Argentina; Consejo Nacional de Investigaciones Científicas y Técnicas, Buenos Aires, Argentina*

Ana Ladio • *Instituto de Investigaciones en Biodiversidad y Medio Ambiente, CONICET-UNCO, S.C. de Bariloche, Río Negro, Argentina*

José Vitor Lima-Filho • *Laboratory of Microbiology and Immunology, Department of Biology, Federal Rural University of Pernambuco, Recife, Pernambuco, Brazil*

Ernani Machado de Freitas Lins Neto • *Departamento de Biologia, Universidade Federal do Piauí, Bom Jesus, Piauí, Brazil*

Reinaldo Farias Paiva de Lucena • *Departamento de Fitotecnia e Ciências Ambientais, Universidade Federal da Paraíba, Centro de Ciências Agrárias, Paraíba, Areia, Brazil*

Domingos T.O. Martins • *Faculdade de Medicina, Departamento de Ciências Básicas em Saúde, Área de Farmacologia, Universidade Federal de Mato Grosso, Cuiabá, Mato Grosso, Brazil*

Maria Franco Trindade Medeiros • *Health and Education Center, Federal University of Campina Grande, Cuité, Paraíba, Brazil*

Patrícia Muniz de Medeiros • *Laboratory of Applied and Theoretical Ethnobiology, Department of Biology, Federal Rural University of Pernambuco, Recife, Pernambuco, Brazil*

Joabe Gomes de Melo • *Laboratory of Applied and Theoretical Ethnobiology, Department of Biology, Federal Rural University of Pernambuco, Recife, Pernambuco, Brazil*

Soledad Molares • *Instituto de Investigaciones en Biodiversidad y Medio Ambiente, CONICET-UNCO, S.C. de Bariloche, Río Negro, Argentina*

Júlio Marcelino Monteiro • *Departamento de Biologia, Universidade Federal do Piauí, Bom Jesus, Piauí, Brazil*

Luciana Gomes de Sousa Nascimento • *Laboratory of Applied and Theoretical Ethnobiology, Department of Biology, Federal Rural University of Pernambuco, Recife, Pernambuco, Brazil*

Viviany Teixeira do Nascimento • *Laboratory of Applied and Theoretical Ethnobiology, Department of Biology, Federal Rural University of Pernambuco, Recife, Pernambuco, Brazil*

Tadeu José da Silva Peixoto Sobrinho • *Laboratório de Produtos Naturais, Departamento de Ciências Farmacêuticas, Universidade Federal de Pernambuco, Recife, Pernambuco, Brazil*

Nivaldo Peroni • *Campus Universitário, Biological Science Center, Department of Ecology and Zoology, UFSC, Florianópolis, Santa Catarina, Brazil*

María Lelia Pochettino • *Laboratorio de Etnobotánica y Botánica Aplicada (LEBA), Facultad de Ciencias Naturales y Museo, Universidad Nacional de La Plata, La Plata, Buenos Aires, Argentina; Consejo Nacional de Investigaciones Científicas y Técnicas, Buenos Aires, Argentina*

Marcelo Alves Ramos • *Departamento de Ciências Biológicas, Universidade de Pernambuco, Nazaré da Mata, Pernambuco, Brazil*

Alessandro Rapini • *Departamento de Ciências Biológicas, Universidade Estadual de Feira de Santana, Feira de Santana, Bahia, Brazil*

Leidiana Lima dos Santos • *Laboratory of Taxonomy, Rural Federal University of Pernambuco, Recife, Pernambuco, Brazil*

Lucilene Lima dos Santos • *Federal Institute of Education, Science and Technology of Pernambuco, Belo Jardim, Pernambuco, Brazil*

Maria Luiza Schwarz • *Federal University of Campina Grande, Cajazeiras, Brazil*

Shana Sampaio Sieber • *Program in Social Sciences, Federal University of Campina Grande, Cajazeiras, Paraíba, Brazil*

Ana Carolina Oliveira da Silva • *Laboratory of Applied and Theoretical Ethnobiology, Department of Biology, Federal Rural University of Pernambuco, Recife, Pernambuco, Brazil*

Rafael Ricardo Vasconcelos Silva • *Laboratory of Applied Ethnobotany, Department of Biology, Federal Rural University of Pernambuco, Recife, Pernambuco, Brazil*

Rafael Ricardo Vasconcelos da Silva • *Universidade Federal Rural de Pernambuco, Recife, Pernambuco, Brazil*

Taline Cristina da Silva • *Laboratory of Applied and Theoretical Ethnobiology, Department of Biology, Federal Rural University of Pernambuco, Recife, Pernambuco, Brazil*

Valdeline Atanazio da Silva • *Botânica—Unidade Acadêmica de Serra Talhada, Universidade Federal Rural de Pernambuco, Serra Talhada, Pernambuco, Brazil*

Gustavo Taboada Soldati • *Laboratory of Applied and Theoretical Ethnobiology, Department of Biology, Federal Rural University of Pernambuco, Recife, Pernambuco, Brazil*

Gardene Maria de Sousa • *Department of Biology, Federal University of Piauí, Teresina, Piauí, Brazil*

Rosemary da Silva Sousa • *Laboratory of Applied and Theoretical Ethnobiology, Department of Biology, Federal Rural University of Pernambuco, Recife, Pernambuco, Brazil*

Francisco José Bezerra Souto • *Área de Botânica, Laboratório de Etnobotânica Aplicada (LEA), Departamento de Biologia, Universidade Federal Rural de Pernambuco, Dois Irmãos, Recife, Brazil*

Fábio José Vieira • *Department of Biology, State University of Piauí, Picos, Piaui, Brazil; Laboratory of Applied Ethnobotany, Rural Federal University of Pernambuco, Recife, Pernambuco, Brazil*

Kleber Silva Vieira • *Laboratório de Ecofisiologia Animal, Departamento de Sistemática e Ecologia, Universidade Federal da Paraíba, João Pessoa, Paraíba, Brazil*

Washington Luis Silva Vieira • *Laboratório de Ecofisiologia Animal, Departamento de Sistemática e Ecologia, Universidade Federal da Paraíba, João Pessoa, Paraíba, Brazil*

Sofia Zank • *Laboratory of Human Ecology and Ethnobotany, Federal University of Santa Catarina, Florianópolis, Brazil*

Chapter 1

Selection of Research Participants

Ulysses Paulino Albuquerque, Reinaldo Farias Paiva de Lucena, and Ernani Machado de Freitas Lins Neto

Abstract

In this chapter the reader will have the opportunity of a brief explanation of the terms most commonly used by ethnobiologist and ethnoecologist as well as access to information about procedures for contact with the informant, theoretical and practical to check the validity and reliability of information obtained, and which the main measures to be adopted in resolving the problem of sampling.

Key words Sampling in ethnobiology, Selection of informants

1 Selecting the Informant/Interviewee

The selection of informants or interviewees for ethnobiological or ethnoecological surveys is an important procedure. In this discussion, we make the distinction between the terms "informants" and "interviewees" because depending on the researcher's theoretical training or methodological choices, the words can have different connotations. Moreover, in the context of research of a participatory nature, it does not make sense to use these terms. Many authors prefer to use the term "informant" to designate people with whom they have developed a lasting relationship, that is, individuals that have participated in more than one interview. Regardless, we believe that the word "informant" merits a distinction that is more than merely conceptual, as explained previously.

The terms *main informant* and *key informant* appear in many studies. A main or key informant is selected from among all the informants to collaborate more actively in the research and is chosen using criteria established by the researcher. Although the researcher must determine the aspects and criteria that should be

Ulysses Paulino Albuquerque et al. (eds.), *Methods and Techniques in Ethnobiology and Ethnoecology*,
Springer Protocols Handbooks, DOI 10.1007/978-1-4614-8636-7_1, © Springer Science+Business Media New York 2014

considered in the choice of informants for the study, the following suggestions may be useful:

(a) Ideally, the selection of the informant should not be based solely on the quantitative aspects of the interview (for example, the number of plants cited) but also on the subtleties of the information obtained.

(b) Regardless of the focus of the study, the researcher should consider the peculiarities and originality of the information.

(c) Finally, the researcher cannot afford to lose the information contained in the subtext of the conversations, whether these conversations are formal or informal. Attention to these details can produce a substantial increase in the information collected and clarify the subsequent process of choosing the informant.

Depending on the research objectives, the researcher must consider whether to work with local experts (people who are recognized in their community as specialists on the subject of the plants and/or animals of the region) or with the community in general (generalists). Many studies have used experienced local experts to evaluate how certain types of resources are used. Given this methodology, for the data to reflect the experience of the whole population, 100% of the informants would ideally be involved. Because of its size, however, a study using the community in general may be preceded by the selection of a representative sample. The issue of sampling is discussed in more detail below.

The researcher is faced with many problems in the field, including contradictions between the information provided and an individual's behavior. In addition, the validity and reliability of the information must be considered. Validity affects the accuracy of the informant, and reliability affects the consistency of the information obtained between and among the informants. However, these definitions for the terms "validity" and "accuracy" are not always the definitions used by researchers [1]. Furthermore, other studies have suggested that because the information obtained can be contradictory, the researcher should diversify the informants to reflect accurately the cultural world of the community studied [2]. A number of suggested methods for evaluating the collected data are outlined below.

Cross-referenced information: This method involves presenting the information provided by a given informant [3] to the other informants, prompting a confrontation in which the information can be refuted or confirmed. This method should be handled carefully because the informants may be offended and believe that the researcher does not have confidence in the information they have provided, which can create problems between the informant and the researcher, compromising the entire study.

Repeated information: The repeated information method is used as a control and considers the time factor. In the repeated information method, the informant is asked the questions, and after a period, the questions are put to the informant again. Once again, this method should be used carefully so that the repeated presence of the researcher does not overwhelm or pressurize the informant.

Historically, especially over the past two decades, several authors (e.g., [4, 5]) have proposed quantitative techniques based on a term known as informant consensus. Undoubtedly, these techniques are based on the idea that culture is shared knowledge, which emerged from the cultural consensus theory developed by cognitive anthropologists in the 1980s.

The cultural consensus theory assumes that *"(a) there is a culturally correct answer to a given question, (b) each informant responds independently of another informant, and (c) the probability that an informant will answer a question correctly in a domain of knowledge reflects the informant's competence in the field"* [6]. Amorozo [7] states that any person in a given culture can be a valid informant because of his "cultural competence." In turn, Reyes-Garcia et al. [6] defined competence, from the cultural consensus perspective, as "the proportion of correct answers given by the informant."

In light of these considerations, let us analyze some of the elements that can facilitate the contact with the informants and fieldwork, such as rapport. The purpose of rapport is the acquisition of trust, which is essential for obtaining information. For example, people who live near protected forest fragments likely make use of the forest's plant and animal resources, even when access is forbidden. Earning the trust of the population will facilitate information gathering because people tend not to tell the truth for fear of being reprimanded.

The researchers can gain trust by following certain procedures, including the following: delimiting the study area; visiting all the households to explain the importance of the work and its purpose; identifying themselves as researchers; and organizing a general meeting to explain the study and the process that will be performed in the community. This meeting can be held in a school or on the premises of a local association. If the goal of the research involves local experts, for example, the researcher can achieve rapport by going from house to house. At this point, it is important to collect the informed consent signatures (required by the research ethics committee at least in Brazil) from all the community members interested in participating in the study, thereby confirming the researcher's ethical commitment to the community and/or the people involved in the research (see Chap. 27).

Because ethnobiological and ethnoecological studies examine different cultural contexts, it is important to consider the behavioral subtleties (physical and oral) of each community.

The physical aspects are described here as nonverbal language, which depending on the population studied, can be of fundamental importance for the understanding of life and the daily routine of the community. The misinterpretation of a simple gesture can compromise the study (e.g., [8, 9]).

According to Viertler [9], nonverbal language is common among the indigenous population of the Bororo of Mato Grosso in Brazil for whom simply keeping quiet or crossing the arms can represent disapproval. Viertler [9] also noted that nonverbal language could have various meanings, from "timidity and humility to disagreement or disapproval." Therefore, before first making contact with the community, the researcher should learn the local customs and practices. Ideally, a local mediator should be used, such as a community member or an individual with substantial access to the community who can provide valuable socio-cultural information and will introduce and guide the researcher in the community, facilitating the development of the study considerably. Table 1 demonstrates a number of characteristics and desirable guidelines for performing fieldwork.

2 Sampling Procedures

In reality, many of the weaknesses observed in ethnobiological and ethnoecological studies relate to the criteria used to choose the survey participants (informants, interviewees, etc.) and the sample size (Boxes 1 and 2). Often, studies reach inadequate conclusions because the sample size is too small or insignificant for the problem in question. The representativeness of a sample is intrinsically related to the selection criteria set by the researcher, which understandably vary. Remember that different factors can interfere with the local knowledge of plants and animals, such as gender, age, occupation, and ethnicity. Therefore, these variables should be considered when designing a research project.

For example, in a study performed by Gomez-Beloz [11], because individuals over the age of 15 were considered adults in their community and were able to form a family, all of the young people over the age of 15 were included as informants. Obviously, this criterion is not useful for all communities. Therefore, it is necessary, at least at this point, to remember the goals of the study so that a truly representative sample can be defined, and any changes that have direct or indirect effects on the representativeness of the sample must be considered. Remember, if the sample is not representative, the strength of the generalizations made in the study is strongly affected and the research may be invalidated.

In studies where there are numerous elements in the selected population, it is necessary to use sampling techniques. An increasingly popular sampling and informant selection technique is the

Table 1
Researcher orientations for maintaining high ethical and scientific standards

Characteristics of good researchers	*Characteristics of bad researchers*
Always truthful	Often deceptive
Openly sharing	Selfish
Considers their own values or religion	Doesn't have values or religion
Respects others' values or religions	Disrespects others' values or religions
Follows customs	Ignores customs
Respectful of elders	Disrespectful of elders
Good researcher orientations	*Bad researcher orientations*
Focus on learning from/with local people	Focus on teaching/convincing local people
Invest time to get to know people	Get work done fast at any cost
Intellectual and real property rights matter	Intellectual or real property rights do not matter
Invest more in the community than is taken	Takes more from the community than is given
Find ways to avoid bribing people	Give bribes to get their way
Good research practices	*Bad research practices*
Embrace collaborations	Avoid collaborations
Listen to people and consider their ideas (use ears more than mouth)	Hear people and tell them what to think (use mouth more than ears)
Contribute to the community before conducting research	Complete research first then, if time is left, contribute to the community
Use informed consent	Do not use informed consent
Ask permission	Assume a right to do research
Use reproducible methods	Work without methods
Test hypotheses	Gather information without a plan
Politely ask appropriate questions	Rudely, asking nosy questions
Respect secrets	Trick people into giving information
Pay workers fairly	Pay workers as little as possible
Collect specimens even when known	Don't collect specimens if already "known"
Offer research results to communities as choices	Present research results to communities as mandates
Clearly present ideas as choices	Use deception to convince people of ideas
Share research results as publications	Keep research results a secret

Bennett's golden rule for ethnobotany field work reproduced from [10]

Box 1 Hypothetical Examples of the Types of Sampling

Example 1. A researcher wants to understand the processes and techniques of healing performed by healers of a particular indigenous group. To meet this goal, we selected a sample of people with more experience in healing and based on indications of others. Opted participant observation and semi-structured interviews that allowed greater freedom of expression on the part of respondents.

This case illustrates a non-probability sampling. The research is primarily exploratory, and its results cannot be extrapolated to the entire community in which they live healers, or even for all healers. Thus, by definition, a sample is not probabilistic *"a subgroup of the population in which the choice of the elements does not depend on chance, but the characteristics of research"* [20].

Example 2. A researcher wants to know what factors affect a community knowledge about the medicinal use of animals, for example, socioeconomic variables (age, sex, education, access to goods, and services), could explain different patterns of knowledge. The researcher conducted a probability sample of 260 people (800 people totaling community) to collect data. Individuals were sampled by means of a draw. This case illustrates a simplified, using probability sampling. The survey was primarily designed to find the association between variables, allowing the results to be extrapolated to the entire universe (community). Thus, by definition, represents a random sample *"subgroup of the population in which all elements have the same probability of being selected"* [20].

> *Choose among a random sample or a non-probability depends on the objectives of the study, the type of research and the contribution thought to do with it* [20].

Box 2 Sample Size

When making a probabilistic sample (remember that we are with the quantitative approach), we must ask: given that a population is N, which is the lowest number of sample units (people, organizations, chapters of novels, etc.) that it is necessary to compose a sample (n). I ensures a standard error of less than 0.01? The answer to this question seeks to find the probability of ye my estimate y of Υ approaches, the actual value of the population. If we set the standard error and fix 0.01, suggest that this fluctuation of our average estimate y with respect to the actual values of the population is not $\Upsilon > 0.01$, i.e., that of 100 cases, 99 times my prediction is correct and that the value of y is at a confidence interval that includes the value of Υ. In summary, for a given variance (V) y, how large must be my sample? This is determined in two steps:

1. $n' = s^2 / V^2$ = size of sample = provisional sample variance/population variance $n' = n' / (1 + n'/N)$

Consider the following case. (…) Delimit a population saying that for a study of CEOs, we consider all those CEOs of industrial and commercial enterprises which, in 1983, has a registered capital exceeding 30 million pesos, with sales up to 100 million pesos or more than 300 employees.

(continued)

Box 2 (continued)

With these characteristics, it was determined that the population was N=1.176 CEOs since 1.176 companies met the above characteristics. So what is the number of CEOs that n must be interviewed, in order to have a standard error of less than 0.015, and since the total population is 1.176?

N=population size of 1.176 companies.

y = average value of a variable = 1, a general manager for the company. se = standard error = 0.015, determined by ourselves.

V^2= the population variance. Its definition (If): square of the standard error.

s^2 = variance of the sample expressed as the probability of y.

n' = sample size without adjustment.

n = sample size.

Substituting, we have:

$$n' = \frac{s^2}{v^2}$$

$$s^2 = p(1-p) = 0.9(1-0.9) = 0.09$$

$$V = (0.015)^2 = 0.000225$$

$$n' = \frac{0.09}{0.000225} = 400$$

$$n' = \frac{n'}{1+n'/N} = \frac{400}{1+400/1.176} = 298$$

$$n' = 298$$

Therefore, for our research, we will need a sample of 298 CEOs. This is the first procedure to get a random sample: determine your size based on population estimates. The second procedure is based on how and where to select these 298 individuals.

Reproduced from Sampieri et al. [20].

"snowball" sampling technique [12]. This technique is used for the intentional selection of informants. For example, to explore the local healing systems in a community, one could choose to work only with "local experts" representing the set of individuals who are socially legitimized and recognized as holders of a particular knowledge. From the initial contact with the community, the first expert is identified, and this expert indicates another expert, and so on, until all the community experts are involved. It may happen that, for various reasons, a certain expert does not indicate another name, leaving the researcher to take stock of the situation and perhaps restart the process. It should be noted that the researcher can and should adapt the method to the particularities of the population under study.

For example, Lins Neto et al. [13], studying the diversity of knowledge and the use of a fruit native to the caatinga (*Spondias tuberosa* Arruda), selected eight main informants using information from field observations and previous studies performed in the region. Subsequently, the authors applied the snowball technique to the eight informants and obtained eight new names, demonstrating that for a specific situation, the snowball method can be adjusted to suit the reality of a situation. In this case, if the snowball technique had been used with only one informant, it is possible that only a small number of cited experts would have been obtained. This situation could arise if the informant is uncomfortable providing the names of other experts, possibly because of vanity (the informant wants to be regarded as the primary holder of the knowledge about the region's plants) or personal reasons, such as relationship problems.

Several books on sampling techniques have been published (for example, [1]). Next, we will examine the most commonly used techniques and those that are of direct interest to ethnobiologists and ethnoecologists. Generally, the two procedures for Box 1 illustrates a number of examples of sampling types.

2.1 Probabilistic Sampling

In this procedure, the elements of the sample are selected randomly, which means that from all of the existing elements, each element has the same chance of being selected. In other words, when the objective is to describe or draw conclusions regarding the characteristics of the population, probabilistic sampling techniques are used [14]. Typically, a draw is conducted based on the random number tables that can be found in good statistical texts. Remember that in any sampling procedure, the error is reduced as the sample size is increased and there is greater homogeneity.

2.1.1 Simple Random Sample

This technique is the most basic form of sampling but has a somewhat complex mathematical reasoning. The technique consists of enumerating all the elements and using a table of random numbers to select the elements that will be part of the sample. This approach allows each member of the population to have an equal chance of being included in the sample, i.e., to be chosen at random [15]. Although this procedure is more appropriate for probabilistic precepts, it is not always easy to apply because it requires each element of the population to be identified; i.e., the researcher has a list of all the elements that make up the population. Sometimes, this type of sampling does not reflect an accurate representation of the population. For example, if 200 individuals are selected according to simple random sampling criteria from a population in which the ethnic distribution is 75 % white, 10 % African descent, 10 %

Hispanic, and 5 % Asian, the Asian population would have a very low probability of participating in the survey, which would affect the representativeness of the sample [14].

2.1.2 Stratified Sample

This technique is used when it is more difficult to obtain a complete list of the elements that make up the population, as mentioned above. The first step is to stratify the population into categories based on the criteria established by the researcher, such as gender, occupation, ethnicity, and region. Stratification ensures the representativeness of the sample, allowing at the same time an awareness of the error. Simple random samples can be obtained from each stratum or category, which when combined, comprises the final sample. This sampling has the main advantage of ensuring representativeness regarding the characteristics of the population [15]. Returning to the example of Spata [14], the adoption of stratified sampling would enable the 200 individuals to be selected proportionally among the ethnic groups, i.e., 150 whites, 20 of African descent, 20 Hispanics, and 10 Asians, thus ensuring the representativeness and homogeneity of the sample.

2.1.3 Cluster Sampling

The procedures listed above are in some ways simple and practical. There are occasions, however, when the researcher does not have a list of the elements that comprise the population and has to address broader categories (clusters or groups), such as the population of a city and an indigenous community. The cluster sampling procedure involves two steps: listing and sampling. For example, to perform an ethnobotanical study in a rural community formed by several households, one might observe how the community is structured, whether in blocks, zones, etc. Each subset (for example, blocks) would be termed a cluster. A refinement must then be made using a stratification process, resulting in the formation of subunits. The listing of each of these elements and the selection of the elements forms the sample, in a similar manner to simple random sampling. It is important that the entire process be performed randomly. Cluster sampling is advantageous over random sampling in which the costs (time and resources) of preparing a list of the entire population are high [15]. However, according to Cordeiro [16], the disadvantages of cluster sampling are "a greater complexity and difficulty in statistical analysis and, generally, an increase in the variance of the estimators used, with a consequent decrease in the accuracy of the study."

2.1.4 Area Sampling

Area sampling is commonly used in community studies and/or when the elements of the population are entirely unknown. The technique makes use of maps or aerial photos for the subdivision of the area into smaller units, and a draw is performed to select the sample areas. After drawing the areas, the researcher can research all

Table 2
Sizes of samples required for various sizes of people, considering a confidence interval of 5 %

Population size	Sample size
50	44
100	80
150	108
200	132
250	152
300	169
400	196
500	217
800	260
1,000	278
1,500	306
2,000	322
3,000	341
4,000	351
5,000	357
10,000	370
50,000	381
100,000	384

Reproduced from Bernard [17]

the elements in the draw area or make new draws, adopting a procedure similar to that described for the cluster sampling technique.

2.2 Sample Size

The researcher always has one question in mind: what is the optimal size of a sample? The sample size can be estimated simply by selecting a proportion of the elements of each cluster or of the total population (such as in simple random sampling) (Table 2). A more detailed discussion of this technique can be found in Babbie [1]. We emphasize that for a sample to be representative of a population, it must be composed of a sufficient number of cases (as outlined above). The number of cases, in turn, depends on the following factors: the size of the sampling population, the established level of confidence, the maximum allowable error, and the percentage in which the phenomenon occurs [15]. The following table, reproduced from Bernard [17], may be helpful; however, the reader should be careful when using this table and remember that sampling methods depend on the research objectives, the contribution of the study, and the feasibility of the fieldwork.

2.3 Non-probabilistic Sampling

As we have demonstrated, random samples are formed by entirely random processes, and despite their robustness and superiority, it is sometimes necessary to choose a sample in which the elements are chosen intentionally, depending on the characteristics of the research to be developed. The weakness of this approach is that it is limited when generalizing. However, if the intention is to describe only a phenomenon and to generate data and hypotheses, it may not be advantageous to generalize the data to all members of the population. This situation calls for the use of non-probabilistic sampling techniques [14]. The "snowball" technique described above is a good intentional selection strategy. The reader should refer to Part 2 of the book published by Flick [18], which discusses the problem of sampling in the context of qualitative research.

2.3.1 Intentional or Judgment Sampling, or Rational Selection

In this technique, the researcher focuses on specific groups based on their experience or knowledge of the population, for example, priests from Afro-Brazilian cults who have in-depth knowledge about the ritual and medicinal uses of plants. The "snowball" method is one example; however, the method only allows generalizations about the specific group studied and not the community as a whole. Almeida and Albuquerque [19] studied the knowledge of "herbalists" at a major street fair regarding medicinal plants and animals. Because the main goal of the research was to understand the system of local knowledge, the resources that were offered to the population and whether it was shared by all elements, the researchers chose to interview all the fair vendors who sold plants and animals for therapeutic purposes. The primary advantage of this method was the low cost involved, whether in terms of time or preparation; however, this method requires detailed knowledge of the population [15]. Box 3 presents some tips for using this type of sampling.

2.3.2 Accidental Sampling

Accidental sampling is an interesting procedure in exploratory studies, where the researcher is not well right questions of your research. It can be done as an initial procedure for guide. The researcher listening to people as they come up in the course of his fieldwork, so entirely free and accidental, until the completion of the stated size for the sample. It is rare to find this procedure in ethnobotanical research. The researcher can adopt this procedure to learn a little community, the universe conceptual and practical, that will be faced. Example: interviewing people about the use and knowledge of medicinal plants in health centers or hospitals while awaiting medical care.

Box 3 Steps in purposive sampling. Reproduced from Tongco [21]

1. Decide on the research problem.
2. Determine the type of information needed.
   Information from every individual in the community is potentially valuable > use random sampling.
   Time and resources are too limited for random sampling > use purposive sampling with caution
   Information is held by only certain members of the community > use purposive sampling.
   Information needs a high degree of interpretation regarding cultural significance > use key informants
3. Define the qualities the informant(s) should or should not have.
4. Find your informants based on defined qualities.
   Research about the area and community.
   Ask for help before going to the site and upon arrival at the site.
   Realize finding informants may be a trial and error process. Be patient and persistent!
5. Keep in mind the importance of reliability and competency in assessing potential informants.
6. Use appropriate data gathering techniques.
7. In analyzing data and interpreting results, remember that purposive sampling is an inherently biased method.
   Document the bias.
   Do not apply interpretations beyond the sampled population.

3 Final Considerations

The selection of survey participants is crucial in ethnobiological and ethnoecological studies, and special attention should be paid to this process because the success and proper execution of a study depends on this step. It is essential that the researcher carefully and thoroughly evaluates the techniques employed for the selection of informants, always paying attention to the peculiarities of the cultural group being studied as well as the goals of the research. The primary recommendation of this chapter, which will be reinforced in subsequent chapters, is that each researcher must choose the technique that best meets the objectives of the study being developed, ensuring the proper execution of the research and reliable results.

References

1. Babbie E (2003) Métodos de pesquisas. Editora UFMG, Belo Horizonte
2. Combessie JC (2004) Métodos em sociologia: o que é, como faz. Editora Loyola, São Paulo
3. Montenegro SCS (2001) A conexão homem/camarão (*Macrobrachium carcinus e M. acanthurus*) no baixo São Francisco alagoano: uma abordagem etnoecológica. Tese de

Doutorado—Universidade Federal de São Carlos, São Paulo
4. Phillips O, Gentry AH (1993) The useful plants of Tambopata, Peru: I. Statistical hypothesis tests with a new quantitative technique. Econ Bot 47:15–32
5. Phillips O, Gentry AH (1993) The useful plants of Tambopata, Peru: II. Additional hypothesis testing in quantitative ethnobotany. Econ Bot 47:33–43
6. Reyes-Garcia V, Byron E, Vadez V, Godoy R, Apaza L, Limache E, Leonard W, Wilkie D (2004) Measuring culture as shared knowledge: do data collection formats matter? Cultural knowledge of plant uses among Tsimane' Amerindians, Bolivia. Field Methods 16(2):135–156
7. Amorozo MCM (1996) A abordagem etnobotânica na pesquisa de plantas medicinais. In: Di Stasi LC (org.) Plantas medicinais: Arte e Ciência. Editora Unesp, São Paulo
8. Posey D (1987) Introdução—etnobiologia: teoria e prática. In: Ribeiro B (ed) Suma etnológica Brasileira, vol 1, Etnobiologia. Ed. Vozes, Rio de Janeiro
9. Viertler RB (2002) Métodos antropológicos como ferramenta para estudos em etnobiologia e etnoecologia. In: Amorozo MC, Ming LC, Silva SP (eds) Métodos de coleta e análise de dados em etnobiologia, etnoecologia e disciplinas correlatas. Editora UNESP, Rio Claro
10. McClatchey W, Gollin LX (2005) An ethnobotany research training workshop in Madagascar. Ethnobot Res Appl 3:309–327
11. Gomez-Beloz A (2002) Plant use knowledge of the Winikina Warao: the case for questionnaires in ethnobotany. Econ Bot 56(3): 231–241
12. Bailey K (1994) Methods of social research, 4th edn. The Free Press, New York
13. Lins Neto EMF, Peroni N, Albuquerque UP (2010) Traditional knowledge and management of umbu (*Spondias tuberosa*, Anacardiaceae): an endemic species from the semi-arid region of Northeastern Brazil. Econ Bot 64(1):11–21
14. Spata AV (2005) Métodos de pesquisa: ciência do comportamento e diversidade humana. Editora LTC, Petrópolis
15. Gil AC (1999) Métodos e técnicas de pesquisa social, 5th edn. Editora Atlas S.A., São Paulo
16. Cordeiro R (2001) Efeito do desenho em amostragem de conglomerado para estimar a distribuição de ocupações entre trabalhadores. Rev Saude Publica 35(1):10–15
17. Bernard HR (1988) Research methods in cultural anthropology. Sage, Newbury Park
18. Flick U (2004) Uma introdução à pesquisa qualitativa. Editora Bookman, Porto Alegre
19. Almeida CFCBR, Albuquerque UP (2002) Uso e conservação de plantas e animais medicinais no Estado de Pernambuco (Nordeste do Brasil): Um estudo de caso. Interciencia 27(6): 276–285
20. Sampieri RH, Collado CF, Lucio PB (2006) Metodologia de pesquisa. Editora McGraw-Hill, São Paulo
21. Tongco MAC (2007) Purposive sampling as a tool for informant selection. Ethnobot Res Appl 5:147–158

Chapter 2

Methods and Techniques Used to Collect Ethnobiological Data

Ulysses Paulino Albuquerque, Marcelo Alves Ramos, Reinaldo Farias Paiva de Lucena, and Nélson Leal Alencar

Abstract

The aim of this chapter is to discuss the methods and techniques most widely used to elicit information from informants or interviewees, many of which are derived from anthropology. We will thoroughly describe such methods and critically appraise their advantages and limitations based on our personal experience and the published literature. Interviews are the main tool that researchers have to elicit information from study populations. Many techniques are available to conduct interviews, ranging from individual approaches to research conducted with groups of people.

Key words Ethnobiology, Traditional knowledge, Local knowledge, Methods in Ethnobiology

1 Introduction

We are aware of many handbooks that address methodological issues related to the study of the relationship between human beings and the environment, particularly within the scope of the social sciences. However, in the present chapter, we present a different approach by focusing on the advantages and disadvantages of the various techniques used for data collection, thereby guiding the reader in the choice of the most appropriate tool for their studies.

As in any ethnobiological study, the stage of data collection involves a combination of theoretical and methodological competences arising from various disciplines. We chose to select methods that are frequently used by investigators in various parts of the world. Our purpose is not to restart the old debate opposing qualitative to quantitative research but to propose a strategy based on their mutual complementarity, thus paving the way for a multimethod approach. The text transcribed in Box 1 is intended to help the reader to reflect on this issue.

Ulysses Paulino Albuquerque et al. (eds.), *Methods and Techniques in Ethnobiology and Ethnoecology*,
Springer Protocols Handbooks, DOI 10.1007/978-1-4614-8636-7_2, © Springer Science+Business Media New York 2014

Box 1 Quantitative-Qualitative: Opposition or Complementarity?

(...) there is no contradiction, just as there is no continuity, between quantitative and qualitative research. Their natures are different.

The former acts at levels of reality, where the data appear to the senses: "ecologic and morphologic levels" (...)

The latter operates with values, beliefs, representations, habits, attitudes, and opinions.

The field of practices and goals of the former is meant to bring observable data, indicators, and tendencies to light. From the social perspective, it must be used in the approach to large agglomerates, demographic sets, for instance, to classify them and make them intelligible by means of variables.

The latter is fit to deepen into the complexity of particular phenomena, facts, and processes specific to groups that are more or less defined as to their extension and liable to a more intensive delimitation.

From the epistemological perspective, none of these approaches is more scientific than the other. What good might the use of highly sophisticated measurement tools bring to a researcher when such tools are not adequate for the understanding of the data or do not answer the fundamental questions? This is to say, the mere fact that a study is quantitative does not make it "objective" and "better", even though it makes sophisticated use of analytic instruments, when it distorts or ignores important aspects of the investigated phenomena or social processes. Similarly, a qualitative approach does not ensure deep insights by default.

This observation is needed to refute the hypothesis by several scholars that the quantitative approach is the most perfect one from the scientific point of view, whereas the qualitative studies are characterized as mere "subjectivism", "impressions", or "exploratory activities" in the best of cases.

This study is not the place to discuss this subject at length; however, in both the quantitative and qualitative perspectives, all the stock of methods and techniques that both approaches developed to be considered scientific must be used.

Nevertheless, although the relationship between the quantitative and the qualitative, objectivity and subjectivity cannot be reduced to a continuum, it cannot either be conceived of as a contradictory opposition. On the very opposite, it is desirable for the social relationships to be analysed in their more "ecologic" and "concrete" features, and explored in-depth as to their more essential meanings. Thus, the quantitative approach might raise issues to be explored in-depth and qualitatively, and vice versa.

Excerpted from Minayo and Sanches [1].

2 The Interview

Interviews are one of the most basic procedures used to collect data in ethnobiological studies. Although interviews might appear to be simple, they are enmeshed in countless subtle features and details that must be controlled by the investigator.

A poorly planned or conducted interview might result in data with bias. The data resulting from an interview might therefore become contaminated on several grounds:

(a) The interviewer's behavior and appearance: when the investigator behave in an inquisitorial manner and thus constrain the interviewee as a function of, e.g., the image that he or she conveys (for instance, style of dress or language used) or, conversely, when behave in a hesitant manner, raising doubts in the interviewee regarding the honesty of the study.

(b) The manner in which the questions are directed and posed: questions that are imprecise regarding their goals might confound the interviewee, or conversely, if too direct, they might suggest probable answers. It is worth remembering that there is an inherent constraint between interviewee and interviewer, and a well-known cultural trait makes us refrain from disagreeing with the opinions of people with whom we are not acquainted. For that reason, interviewees might feel compelled to agree with opinions subtended during an interview.

(c) Cultural interference: on the investigator's side, distortions might be possibly introduced in the data resulting from a mistaken interpretation of the culture of the interviewee. On the interviewees' side, their manner of reacting and expressing themselves in the presence of an unfamiliar person might compromise the quality of the collected data.

(d) The interview setting: the interview might be conducted in the interviewees' primary environment (e.g., their homes), a secondary territory (e.g., close to their homes, at school, or in the workplace), or a tertiary territory (corresponding to public areas, such as squares, bars, commercial areas, or restaurants) [2]. The place where the interview will be conducted is of paramount importance because human behavior can be deeply influenced by the environmental context. In this regard, primary and secondary territories are most often recommended because interviewees know and master these places best. Miranda et al. [3], for instance, noticed that the responses in free list-based interviews were influenced by the place in which they were conducted.

Closely connected with the interview setting, attention must be paid to the presence of third parties (e.g., an interviewee's relative). Several authors have already shown that the presence of a third party during an interview might elicit changes in the interviewees' answers, although very few studies have approached this issue directly (e.g., Boeije [4]). Although it might appear insignificant, the time of the day during which the interviews are conducted must also be taken into consideration because some interviewees do not want to be bothered at particular times of the day; thus, it is advisable to schedule the interviews beforehand at a time that is convenient to the interviewee.

Box 2 Five Steps for Achieving Success in the Use of Interviews as a Method to Collect Ethnobiological Data. Based on Richardson [6] and Babbie [7]

1. Establish a friendly and trustworthy atmosphere for the interviewee. In this regard, it is worth insisting on the interviewer's appearance and behavior: (a) dress in a style similar to that of the interviewees, that is, simple and modest; (b) behave in a pleasant manner and try to act like the type of people with whom the interviewees feel most comfortable.
2. Acquaint yourself with the questionnaire that will be applied in the interview, which requires thorough, item-by-item study. Practice reading the questionnaire aloud and, when applying the questionnaire to the informant, read the questions naturally using a conversational tone. Avoid posing the questions in your own words, as this increases the risk of distortion.
3. After having posed the questions, allow time for the interviewee to answer at length, and show interest in the subject that is being addressed, even when you already obtained the intended answer. Do not hurry the interviewee!
4. Do not argue with the interviewee, do not offer advice, nor make moralistic comments. Avoid appearing as the "protagonist" of the interview, and avoid any authoritarian attitude.
5. Do not restrict your attention to what the informant is able to report but also concentrate on what he or she cannot or might not express without your help.

Although biases always exist, formulating a specific design for an intended study can minimize its effects [5]. In Box 2, we describe five general recommendations that must be taken into consideration during the planning and execution of an interview, which complement the considerations stated above.

Interviews are often conducted with informants who state that they do not know enough about the subject to participate usefully in the study. In such cases, the interviewer must make clear that the interviewee's participation is crucial and that he or she is truly interested in what the interviewee has to say. For this purpose, some techniques can facilitate interaction with the interviewees [8]. In addition, asking the interviewee's consent to record the interview and make notes is always important.

Universal questions that haunt every investigator, especially beginners, while planning or at the beginning of an interview include what to ask and how to pose the questions.

According to Weller and Romney [9], the first step in collecting information during studies is to decide on the cultural domain of interest that represents the body of the study, e.g., nutritional plants, medicinal animals, or water resources. Finding the best way to pose questions on the domain of interest is of paramount importance; thus, ideally, every question must be previously tested with other people, e.g., other colleagues at the laboratory. However, at a later time, the questions must unavoidably be pretested with members of the community in which the study will be conducted to identify cultural biases that might compromise the collection of data. A well-formulated and well-written question might elicit much valuable information and eventually contribute to the satisfactory unfolding of the interview itself. Unfortunately, a portion of informants offer little information, indicating the need to reformulate a question or apply probing methods [9]. For instance, "You said that plants *x* and *y* are used to treat inflammation. What other plants are also used to treat it?" or, "You said that animals *x* and *y* are hunted in this area, what other animals are also hunted?"

From a formal point of view, interviews might be classified according to the technique applied to elicit information as structured, unstructured, semi-structured, and informal.

2.1 Structured Interviews

In structured interviews, the stimulus applied to each interviewee must be the same, i.e., previously devised questions are posed to each informant independent of previous contacts with the target population. This approach requires that the investigators have perfect mastery of the most relevant issues that will be explored. This type of interview imposes limits on the interviewee's (or respondent's) answers; however, advantageously, it facilitates the codification/categorization of the answers and allows faster production of materials for analysis.

In this type of interview, the meetings between the interviewer and each interviewee must be as similar as possible. That is, biases introduced by, for instance, the setting, context, and schedule must be the same for all the interviewees; conversely, different situations will be interpreted as being the same phenomenon.

For this reason, questionnaires and/or forms are extremely useful tools to conduct structured or semi-structured interviews. According to some authors, the use of questionnaires defines structuration by default; thus, an interview is structured whenever they are used. Nevertheless, a distinction is sometimes made between the use of questionnaires and forms, as a function of the manner in how the data are collected. When the data are collected during direct and personal interviews, where the interviewer records the data, the product is designated as a form; conversely, when the informant fills in the data by him/herself, the product is designated as a questionnaire. Why do we hold that an explicit indication of the tool used is important? Both formats have their advantages and disadvantages; however, we tend to use forms because they allow the informants to elucidate certain issues and provide more insightful information. Nonetheless, the use of forms demands more time and is more expensive.

The limitations of these instruments are more or less accentuated as a function of the type of questions asked or the manner in which they are posed. For example, in Brazil, one obtains more reliable information when the interviewee is asked for his or her date of birth than for his or her age because there is a cultural tendency to describe age using imprecise terms.

The questions included in the instruments might be classified as open-ended or closed-ended, with each type having its own advantages and disadvantages, mainly with respect to statistical analysis. Open-ended questions give interviewees more freedom to answer because the interviewer does not provide response options; e.g., "Tell me about the harvesting of plants for firewood". Closed-ended questions, in turn, are highly useful because they elicit uniform answers that might be (a) dichotomous (yes or no, or true or false; e.g., "Do you use plants for firewood?") or (b) multiple choice [e.g., "Which of the following materials are used for cooking in your home? Firewood (), Gas (), Coal (), Bagasse (), or Manure ()"]. According to several authors, there is also the semi-open-ended questions, used to elicit short answers, e.g., "How often do you gather firewood?" Thus, as a function of the questions they include, questionnaires might represent an important source of research bias."

As mentioned above in Box 3, questionnaires and forms should be tested before their actual application; for that purpose, we recommend performing an initial, rapid pretest or pilot study with a small group of people to ensure the instrument's quality, clarity, validity, and reliability [10].

We now discuss how Bernard [8] classifies and explains the use of these instruments, as well as their advantages and disadvantages.

(a) Face-to-face encounter forms are applied in direct encounters between the investigator and interviewee in which the former poses questions and records the answers on a form, elucidates doubts, and makes comments when they are pertinent. The advantage of these forms is that they elicit information from people who might not provide it otherwise (e.g., illiterate, blind, or elderly people) in addition to ensuring that all of the questions are answered. Considerable skill is needed to apply this type of form, particularly in the case of intrusive or highly reaction-inducing questions. However, personal interviews are money- and time-consuming and almost always demand representative samples (see Chap. 1). Forms are widely used in

Box 3 Writing Questions [6]

1. "Never include a question without a clear idea of how the information it provides might be used and how much it will contribute to the study goals".
2. "Use precise vocabulary to ask what you really want to know".
3. "Avoid posing two different questions in one single sentence".
4. "Avoid the use of terms that the interviewees might ignore or not master".
5. "The questions must be adjusted to the interviewees' answering possibilities".
6. "The use of short items is preferable".
7. "Avoid negative questions. That type of question usually and easily leads to error".
8. "The questions must not follow a definite direction nor reflect the investigator's position on a given subject. They must be formulated in an objective manner for the interviewees not to feel pressed to give the answer that they believe agrees with the investigator's opinion".
9. "Finally, the investigator must be careful in interpreting the interviewees' answers".

ethnobiological research, albeit often misnamed as "questionnaires".

(b) Questionnaires (self-applied interviews) might be sent by mail to be delivered at the respondent's home, school or workplace because they are filled in without the presence of an interviewer. Although questionnaires facilitate the collection of data, they are limited by the respondents' potential lack of adherence or response. All interviewees are posed the same questions; thus, misinterpretation by the investigator is avoided. In addition, more complex questions might be posed, as well as questions that require relatively longer times to answer. The use of questionnaires allows bypassing the interviewee's inhibition caused by the presence of the interviewer or if delicate or intrusive topics are addressed (e.g., the collection of resources in illegal areas and religious or sexual subjects). However, when questionnaires are used, the interpretations of the questions by the interviewees cannot be controlled, nor can the order in which they answer the questions. Therefore, a given question might be read before the preceding questions have been answered, thereby influencing the answers to the previous questions. Few researchers appear to use questionnaires in ethnobiological research.

2.2 Unstructured Interviews

This type of interview does not include preestablished questions but unfolds as openly as possible and might be conducted at any time or place (e.g., at home, walking along the road, or while weeding the fields). A striking example of this type of interview is the data-generating method [11], in which the informant is requested to speak about some topic by merely stating, "Speak about this". Thus, unstructured interviews contribute not only to collecting information relative to the informant's knowledge but also to recording his or her behavior, which is hardly possible using the closed-ended questions of structured interviews.

Contrary to widespread perceptions, the use of unstructured interviews does not involve letting the interviewee speak freely because their focus (i.e., the core subject of the study) must be explained to the interviewee [12]. Therefore, the investigator must guide the course of the interview by following an order in describing the phenomena of interest for his/her study. For this reason, the interviewer should keep the conversation focused on a topic, providing space for the informant to talk about the content of the discussion [8].

2.3 Semi-Structured Interviews

As their name indicates, semi-structured interviews are intermediate between the two abovementioned types. In this case, although the questions are partially established by the investigator before conducting actual field research, they are largely flexible and allow more attention to be paid to issues that might arise during the interview because the investigator may announce the topics beforehand and has a guide available to orient the interview. If it is not possible to interview the same informant twice, semi-structured interviews are the best tool to use.

The "open" side of this technique allows for the interviewees to answer the questions according to their own conceptions; however, this approach does not mean that the investigator should allow them speak freely. The interviewer should not allow the focus of the study to be lost, but neither must he/she interrupt the informants abruptly when they deviate from the main topic of the interview. According to Aguiar and Medeiros [12], when this technique is used, the investigator must conduct the interview by him/herself, rather than using other interviewers, because a thorough knowledge of the subject is required.

The interview guide (similarly to an observation plan—see participant observation) must be elaborated beforehand and comprise a list of the topics (subjects, issues, and doubts) that will be approached. Therefore, it must be easy and quick to check the guide [10] because it sets the agenda for the interview and will guide the interviewer to avoid over-focusing on certain subjects until exhaustion while others are merely skimmed over.

2.4 Informal Interviews

This type of interview is similar to an unstructured interview but differs in that it is totally beyond the investigator's control. In this instance, the investigator must use a field diary to record all of the events observed or heard during the observation process. Bernard [13] observed that this method is often used during the early stages of participant observation but can also be used throughout the study to strengthen the affective bonds with the investigated population and to identify new facts that might be of interest.

3 Free Listing

Free listing is a technique that is commonly used in the social sciences to extract information [9, 13] and is considered by some authors to be a variant of the structured interview. Several researchers use this technique to identify items in cultural domains and calculate their cultural salience (prominence, familiarity, and representativeness). According to Quinlan [14], free listing exhibits three basic characteristics that must be considered: (1) The interviewees tend to list terms according to their order of familiarity; (2) Individuals who know more about a given domain tend to cite more terms than individuals who know less; (3) The most cited terms denote more locally prominent or salient items.

Free listing seeks to identify specific information on a given cultural domain of the investigated community. The participants are requested to list, for instance, medicinal plants that they know or animals that have been observed in the area. The basic premise is that the most culturally important items will be cited in many lists and in their corresponding order of importance. This technique is useful, and the lists produced by one or several communities might be compared using software, such as ANTHROPAC®.

Free listing might also be useful to identify individuals in local communities who are specialists in a given cultural domain and to investigate the intracultural variation of that domain [14]. However, free listing might be limited when questions are posed about a highly specific domain with which the interviewees are not familiar. According to Quinlan [14], another limiting factor is the fact that interviewees cite terms from their active vocabulary and thus might omit information regarding items that no longer belong to their everyday experience. This phenomenon might be observed, for example, with famine foods that are normally used under adverse and/or low availability circumstances when the preferred foods are less/not available. In studies examining animal species that have become rare in the investigated area, these animals might not occupy an important place in free listing, especially when their scarcity is so severe as to reduce the odds for the informant to have actual contact with them. Therefore, when knowledge regarding such animals is of interest, the interviewer must supply additional stimuli for the informants to cite species that are no longer used/consumed.

Other problems exhibited by free listing include the fact that informants might not remember all of the items that they know. For this reason, Brewer [15] suggests the use of some complementary techniques that might expressively enrich the initial lists, including the following:

- Nonspecific prompting—This technique involves continuing to pose questions to informants even after they have declared that they cannot remember further items. For instance, in a study on plants known to be medicinal, the interviewer should ask the interviewee, "What other plants might be used for health problems?" Prompting must comprise positive assertions and not induce "yes or no" answers.
- Reading back—This technique might follow the technique described above when the informant once again declares that they cannot remember further items. This method involves slowly reading back all items cited by the informant, thus facilitating the addition of items that were not previously listed.
- Semantic cues—In this case, free listing is used as a semantic cue. Brewer [15] holds this approach to be a powerful technique to enrich lists; this technique is based on the process of natural association between items. This method might be used following the techniques mentioned above and involves questioning the informant on additional items belonging to the domain that are similar to those that he or she has already mentioned. For instance, if oranges are mentioned on an initial list of known fruits, then the informant might further suggest tangerines, lemons, and limes following this semantic cue.

As with any other type of interview, free listing might also be influenced by the setting where the interview is conducted or the activity that the interviewee is performing at the time he or she is interviewed. Because this technique assumes that the most frequently and first cited items are the most important from the cultural perspective, these variables must be controlled during the interview. For instance, an informant approached in front of or inside a vegetated area might use it as a source for consultation while citing plants, whereas an informant interviewed at home will lack that source of stimulation. An informant interviewed while preparing food might cite more plants involved in that process (e.g., nutritional plants and plants used as firewood or seasonings) than an informant interviewed in a different setting.

Therefore, every study design must consider the settings included in the investigated area and how much they might contribute to eliciting information relative to each investigated cultural domain.

4 Field Journal: The Use of Audio and Video Recorders

Field journals are used by investigators to record their observations and may be combined or not with other techniques for data collection. In this way, investigators perform their first reading of cultural systems, which allows reading the cultural facts as has been described [16]. In a field journal, all of the events that occur during one working day must be recorded together with the researcher's perceptions and conclusions regarding the people with whom he/she was in contact. The researcher might choose to organize their diary as he/she deems best and either follow a chronological order or classify the data according to the main subjects. Finally, even when the latter criterion is selected, a temporal order may be adopted. For instance, the diary might be subdivided into chapters, such as medicinal plants, medicinal animals, hunting procedures, plant cultivation, and harvesting, and the events might subsequently be described in the order in which they occur.

Importantly, in addition to field diaries, personal journals might be used, which afford a valuable tool in studies on the relationship between people and the environment. The difference between both types of journals is that personal journals are written by the informants themselves upon the investigator's request; thus, the narrative and reflections are fully determined by the informant based on his/her personal experience [17].

During interviews, and especially when they are guided using forms with open-ended questions, the use of tape recorders is indispensable to register the entire dialogue. If the conversation were to be periodically interrupted to record the information, it would be difficult and complicated to record the entire information provided because the informant would lose his/her line of thought, and important information could be lost. Conversely, the use of a tape recorder allows the conversation to flow freely without any concern. The interviewees' permission should always be sought before a tape recorder is used, and the researcher should determine whether its presence imposes any constraint on the informants. If the latter occurs, the tape recorder should be switched off to allow the interviewee to feel at ease while discussing embarrassing or polemical subjects off the record. In addition, it is important to reassure the interviewees regarding the confidentiality of the recorded material and to explain the use that will be made of the recorded material. It is also important to minimize the time between an interview and its transcription because long delays might result in a loss of information; sometimes the tape is contaminated by external noise that was also recorded. It should be remembered that the process of interview transcription can take a long time and demands considerable attention; depending on the amount and quality of the data, 45 min of recording might require 8 h of transcription [18]. Richardson [6] estimates that the time needed for transcription is twice that devoted to the actual interview.

Video recording might represent a valuable tool to record ethnobiological data because it allows a given practice to be captured in its minute and subtle details. However, the actual use of this technique demands expertise and experience to produce materials with high scientific and artistic quality. Among several suggestions, Fuller [19] recommends the following: paying attention to the ethical and practical issues involved in the planning of documentation regarding practices that involve the use of plants; discussing the ethical issues relative to ethnobiological research with the involved community, including obtaining informed consent; and assessing commercial implications in case the video is exhibited to large audiences, including the discussion of the intellectual property rights.

5 Guided Tour

This field research method usually requires the participation of a member of the community skilled in moving across the local vegetation who is knowledgeable regarding the local flora and fauna or regarding certain main informants selected in the investigated communities An example of the application of this technique is the identification of the animals and/or plants cited by their vernacular names in the interviews in anthropogenic areas or forests. That is, a guided tour consists of giving foundations and validating the names of cited plants and animals because the vernacular name of a given species might exhibit considerable variation among different areas and even among the individuals of a single community. Such differences might be solved by observing a plant that was given one or more vernacular names, which thus become synonyms [20]. This method is also referred to by some authors as the field informant technique or a "walk-in-the-woods". We further recommend recording this activity as hours/informant. It is worth noting that the names cited by an interviewee during a guided tour might not reflect consensus use within his or her own community. For that reason, tours with other members or other activities involving a larger number of community members ought to be performed, such as participant workshops or group dynamics.

6 Participant Observation

We noted above that when information is sought regarding the experiences and opinions of people regarding a given subject, they can be asked directly using questionnaires and interviews. However, Pinheiro et al. [17] observe that when the subject of investigation is the interaction between people and the environment, many of the everyday activities that people perform pass unnoticed by them. Thus, even when these people are willing to report on such activities, they are often unable to provide reliable information on some topics. Within this context, observation-based studies play an important role.

According to Combessie [10], participant observation is an appropriate technique to establish initial contact with a community because it allows room to explore its reality. Participant observation allows the gathering of information on the everyday life of the investigated community, for which purpose the investigator must be able to memorize and remember the events observed and heard and later write them down in a certain order, preferentially a chronological one [8, 13, 20]. Many of the collected data are qualitative, namely, field notes relating to situations, photographs, or dialogue records. Participant observation should not be confounded with participatory research because these methods pursue different goals (Table 1).

Although participant observation affords the investigator a deeper experience of the entire investigated phenomenon, it is limited because no one is able to observe absolutely everything in a trustworthy manner; participants, facts, and situations must be selected. Babbie [7] observes that such selection must be controlled to avoid the risk of constructing a biased set of data.

According to Montenegro [20], participant observation might be complete or characterized by full detachment (the latter is discussed in the next section). Complete participation implies the full involvement of the investigator with the local community and its costumes and practices.

"That method further allows for an analysis of the observed reality 'from within' and thus enables one to perceive how a cultural element, anchored in the common sense, in that which is collectively accepted and legitimated, develops the knowledge on the local plants" [21].

Research originally based on direct observation or participant observation unfolds along three stages, with each one being sequentially dependent on the previous one [22]:

1. The selection and definition of problems—"In this stage, the observer looks for the problems and notions able to afford the widest comprehension of the organisation he is studying" [22]. The investigator must always consider the credibility of

Table 1 Differences between participant observation and participatory research

Participant observation	Participatory (action) research
Aims to learn about and understanding reality	Aims to transform reality
The definition of objects and techniques, and the analysis of data are centered on the investigator	The members of the investigated group participate throughout the study process
	Involves a process of research, education, and action

Source: Víctora et al. [23]

informants and the nature of their statements, i.e., whether they are directed or spontaneous. In such cases, all singularities relating to the observed fact or information supplied ought to be recorded in the field diary for later critical appraisal.

2. Control of the frequency and distribution of the phenomena—"Having a grasp of many problems, notions, and transient indicators, the observer now wants to know which among them are worthy of pursuing as the main focus of his study" [22].
3. Model construction—"The final stage of field analysis involves incorporating individual discoveries into the generalised model of the social systems or organisation under study, or into a part of that organisation" [22]. In our specific case, this *stage seeks to attribute meaning to the set of data collected on the relationship between people and plants. The observer might begin by constructing and specifying the relationships that exist within the system, and then* systematize the phenomena by means of an analytical operation.

Notably few ethnobiological studies use observation as the sole methodological strategy used. As a rule, investigators need to harmonize data collected through participation with data obtained using other methodological tools. Contradictory elements appear quite often and demand intense reflection and the formulation of questions in an attempt to understand what might prove to be merely an apparent contradiction.

7 Nonparticipant Observation

Also known as direct observation, this technique might be considered a variety of participant observation because it involves greater contact with the community without, however, exhibiting the strong degree of involvement demanded by the latter. From the methodological point of view, non-participant observation consists of a free observation and recording of the phenomena observed in the field, where the investigator behaves as an attentive spectator.

This technique is also known as participant observation with total detachment. As a rule, a large amount of information is obtained, which thus demands a strong sense of organization and the systematization of ideas by the investigator.

8 Focus Groups

Focus groups are employed as a data collection strategy that prioritises interaction and discussion between a group and the investigator. The success of this technique depends on the definition of a "focus" subject to be explored by the group with the investigator's mediation. This technique might be applied to the collection of ethnobiological data for the following purposes: (1) to generate hypotheses based on the discussion by informants of an issue raised by the investigator; (2) to obtain the group's interpretations of phenomena, facts, or categories identified during the study; and (3) to assess strategies for data collection because the use of focus groups might be complementary to other methods. For instance, a focus group might include individuals from the community who are involved in healing practices to establish the meaning of disease categories. The investigator must practice great discernment when choosing this strategy and must consider the following recommendations: (1) the groups should preferentially comprise unacquainted individuals bearing no family relationships; (2) the data should be documented appropriately to allow the identification of each participant's discourse and mutually convergent discourses; and (3) the working dynamics should be controlled because the more expansive individuals tend to inhibit others or centralize speech. This technique might be applied in three phases (Box 4).

9 The Use of Visual Stimuli

The investigator might resort to several visual stimuli to elicit information from people. Such stimuli might include fresh plants, dried pressed plants, photographs, artefacts made from plants, and even in situ individuals. The literature includes a wide variety of terms to designate methods based on such stimuli. For instance, the artefact inventory method, also known as artefact interview, involves the collection of information on plants; their parts used to construct artefacts, which are shown to informants to elicit specific information. Similarly to the previous method, checklists involve presenting visual stimuli to people to obtain information on plants. Due to the variation in such terms, Medeiros et al., in Chap. 5, suggest the following standardization: (a) *Inventory/Interview*—studies involving a vegetation inventory associated with the collection of ethnobiological data from informants at the field; (b) *Walk-in-the-woods*—studies involving tours to collect and

Box 4 Application Phases of the Focus Group Technique. Transcribed from Víctora et al. [23]

Phase I

(a) Potential participants are invited: participants are selected based on preestablished criteria; the meeting's nature is explained; and the date, time, place, and duration of meetings are defined.

(b) Meeting preparation: a guide is elaborated using open-ended (e.g., "What do people here do when...?") or hypothetical (e.g., "What would people here do under *x* conditions?") questions; materials are organized (e.g., tape recorder and tapes); a meeting place is prepared. Due to ethical reasons, it is advisable to avoid directly questioning the participants regarding their personal lives. Nevertheless, some participants might possibly share personal experiences with the other participants.

Phase II

(a) The meeting: the facilitator must maintain leadership over the group, listen to the participants attentively, show involvement, encourage participants, be receptive to changes, suggestions, and interruptions, and ensure that all participants speak, rather than only the most vocal ones. The presence of an assistant is advisable to take care of other tasks, e.g., recording the events, observation of nonverbal language, practical actions such as changing the recorder tapes, and ensuring that the external environment does not alter the internal environment.

(b) Mapping of the participants: performing a short interview with each participant in the group is recommended to collect general data such as age, gender, educational level, occupation, marital status, number of children, and certain specific issues relative to the study subject that cannot be raised with the group for ethical reasons.

Phase III

Transcription of the tapes and compilation of the data, always maintaining the facilitator's attitude and influence relative to the obtained answers.

identify plants, and obtain data from informants unrelated with the vegetation inventory; (c) *Checklist/Interview*—studies based on the use of stimuli detached from their original context (parts of fresh or dried plants, photographs, and drawings); (d) *Artefact/Inventory*—studies using plant-derived products (e.g., tools, building materials, and toys).

10 Ranking

Ranking analysis might derive directly from free lists elaborated by the investigator, whereby the informants are invited to rank the plants as they deem fit. These data are used to construct a matrix, which allows attributing a score to each species and the subsequent

Table 2 An example of the organization and analysis of data using the ranking technique

Plant	Informant code								
	A_1	A_2	A_3	A_4	A_5	A_6	Total	Mean	Rank
Coconut	3	6	3	6	5	1	24	4.00	5
Guava	6	1	4	1	4	6	22	3.67	4
Mombin plum	2	4	6	5	1	3	21	3.50	3
Brazil plum	5	2	1	4	3	5	20	3.33	2
Manga	4	3	5	2	2	4	20	3.33	2
Avocado	1	5	2	3	6	2	19	3.17	1

calculation of an index. Ranking might be extremely useful in studies of local preferences. The reader is referred to the example given in Table 2, in which the numbers represent the order according to which six informants cited plants (A_1–A_6). The scores for each plant are first added and later divided by the number of informants. In this example, avocado might be considered as exceptional among this particular set of interviewees as a function of its ranking.

10.1 The Ordering of Photographs

The ordering of photographs (a type of projective interview) consists of the classification of photographs by informants according to a criterion established by the investigator. This technique might be useful to identify ecological areas that are preferred for the collection of plant resources or to characterize animals or landscapes according to the importance attributed to them by informants among other uses. This technique allows the investigator to perform analyses similar to the ranking described in the section above. Notably, several factors might affect the quality of the data thus obtained, including the quality of the photographs or illustrations that are used and the inadequacy of the instruments used to the formulated research question. Víctora et al. [23] recommend the investigator to also record the impressions and commentaries that might arise when this technique is applied to understand the informants' representations of the investigated subject. Chapter 5 provides several notes on the use of photographs in ethnobiological research.

11 Twenty-Four Hour Recall

This technique consists of stimulating the interviewees to list the items (e.g., plants, medicines, or animals) with which they had contact during the previous 24 h. This approach is usually employed in studies that seek to gather information regarding diets, nutritional habits, and hunting behavior. This method is a variant of free listing in that it involves consecutive or periodic visits to the interviewees during which they are inquired about the contents of the meals they had in the previous 24 h. The relative importance of dietary components might be calculated based on the same criteria as those used in the ranking technique.

One of the limitations of this technique derives from the fact that the informants might present the investigator with an underestimated list of species, mainly because it depends on their ability to remember the items that they used and on their interpretation of the question posed by the interviewer. For instance, a given investigator might want to investigate the diversity of plant items used for human nutrition in a given community and ask his informants about the plants that they consumed during the previous 24 h. The informants might have a different understanding/interpretation of the meaning of the word "plant" and might therefore not cite plant items that should have been listed. In addition, it worth noting that informants might feel embarrassed or intimidated by having to mention certain of the items that they use in everyday life. This reluctance might occur due to cultural factors, according to which the consumption of some resources might be considered a synonym for poverty, or for safety reasons, especially in communities that reside in protected areas where the collection or hunting of some resources might be illegal.

12 Community Mapping

Community mapping is used to gather information on natural resources and the local perception in a given geographical context, including the mapping of resource areas. The collaboration of local informants is sought; e.g., informants who are requested to draw the local landscape. This procedure is useful in ethnobiological studies for recording local knowledge of ecological areas, the distribution of vegetation, harvesting and hunting areas, and the distribution of useful resources. The procedure includes the following stages:

1. Each informant draws or prepares a basic map providing the requested information. Because many people might find it difficult to accomplish this task, detailed and careful explanations are often needed.
2. Each informant is requested to name and identify each item he or she has represented on their map.
3. Field trips might be performed with the participants to revise and refine the maps.
4. At the end of the process, a consensual map might be built based on a critical appraisal by all of the involved participants.

5. The investigator might use specific symbols and colors to represent items.

Nevertheless, the construction of maps by informants might be difficult; it might also be inappropriate. This difficulty occurs, for instance, when the informants have trouble developing visual representations of the land relief and the vegetation of the area. In these cases, it is recommended that the investigator employs aerial photographs or even landscape records in which the full area of interest is depicted. Such illustrations are shown and explained to the participants, who are later requested to identify and demarcate the corresponding areas. It is worth remembering that the use of technological resources that are alien to the interviewees' reality might hide limitations. For this reason, one should pay special attention while conducting interviews supported by such resources.

13 Life History

This is an entirely open-ended and subjective approach, where communication occurs in the first person. More usually employed in sociology, the life history or oral history technique might contribute to expanding on certain issues and detach itself from the autobiographical character that it usually bears. The documents thus produced are usually classified as focal and case studies. Such studies are seldom used in ethnobiological research, perhaps because they do not afford the elements needed for understanding more general phenomena. Because this is an extremely open-ended approach, a degree of caution is needed for the intended goals to be successfully accomplished. The most appropriate procedure is for the investigator to lead the informants through the relevant paths of their lives that fit with the intended goals of the study. In ethnobotany, for instance, the subject of study might be the life an individual who has a wide knowledge of local plants. Because such a study seeks to portray several facets of the selected character, it will be limited from the perspective of generalization. Nonetheless, such a study will be able to elucidate some issues, such as the transmission of knowledge between the interviewee and members of their community or the source of his or her learning in regard to definite cultural domains.

In addition to the care needed in the selection of the most appropriate methods/techniques for your study, you must pay attention to the materials/resources needed to achieve success in fieldwork. An investigator needs to have an "ideal basic kit" of tools and materials to facilitate, optimize, and systematize the work during his or her stay at the community. Box 5 describes a checklist that might be useful for this purpose.

14 Data Triangulation

The procedures chosen to approach some problem do not always afford the ideal solution to tackle a given subject. Sometimes the subject of study bears a level of complexity that poses a major challenge to the investigator. Ideally, the methods selected in such studies should be subjected to continuous criticism and reflection such that the investigator is able to make better decisions regarding data interpretation. Recently, several authors assessed the efficiency of various methods (see, e.g., Gaugris and Rooyen [24]).

One of the approaches we suggest for planning research is triangulation. This name designates the combination of several methodological procedures for the study of a single research problem. Several authors have described various types of triangulation, but in this study, we specifically allude to methodological triangulation, which might involve triangulation within a single method or between two different methods (see Flick [25]). "An example of

Box 5 Checklist of Materials Useful and/or Essential for Ethnobiological Field Research

1. Notebook. To be used, for instance, in interviews, to record materials for the database, to perform data analysis in the field.
2. Digital camera. To document, for instance, plants, animals, activities, and landscapes.
3. Digital voice recorder. To document, for instance, interviews and guided tours.
4. Global positioning system (GPS). To reference, for instance, collection sites, resource areas, and plants.
5. Extra batteries for the equipment.
6. A two-way radio. For communication between team members in the field and as a safety measure.
7. To collect botanical materials, the following are useful: old newspapers, collection plastic bags, pruning shears, wooden presses, a field copybook, a field hothouse, pens, porcelain painting pencil (due to its water resistance), graphite pencils, adhesive tape, large knife.
8. Materials for vegetation studies that are dependent on the sampling methods used, e.g., a tape measure and labels; safety materials, such as boots and/or gaiters, long-sleeved shirts, and hats.
9. Plastic or wooden clipboards (for interviews).
10. A field diary.
11. A full first aid kit.
12. Investigator's identification card including: full personal data, blood type, contact telephones and addresses, and drug allergies.

the former strategy is the use of different subscales to measure one item in a questionnaire, and of the latter, the combination of a questionnaire and a semi-structured interview" [25].

As an example, while conducting interviews, investigators might also perform guided tours to collect materials mentioned in the interviews and identify ethnospecies recognized by the community. In this way, overestimating the diversity of species recognized by the community will be avoided, and the classification of the species present in the area can be understood (Box 6).

Box 6 Summary

Procedures for field safety (based on Bridges and McClatchey [26])

- Pay attention to the social rules at the working sites. Avoid traversing the area without knowledge of the potential dangers.
- Explore the landscape in the working area using maps and, more particularly, resources such as Google Earth. Look for maps containing useful information (such as the vegetation cover).
- Acquaint yourself with the main roads, motorways, rivers, mountains, and railways. Develop a habit of recording the main local coordinates (latitude and longitude) of the main research sites.
- Locate the hospitals, transport centers, and the seat of the local government.
- Make a list of contacts at your place of origin (e.g., doctor, dentist, bank, and people that should be notified in case of emergency) and at your destination. Ensure that this list remains always accessible.
- Identify the most probable health problems and consider how you might address or even anticipate them; e.g., poisonous snakes. Always carry an emergency kit with you.
- Establish a means of communication so that people will know about the progress of your fieldwork on a regular basis and whether you are having problems. Name a laboratory researcher or the study coordinator as your contact for this purpose.
- Wear appropriate clothes when performing biological procedures and with regard to your insertion in the community.

Procedures for contact with the community

- Attempt to understand the structure and network of the local social relationships. As a rule, contact with local leaders facilitates the investigators' access to research sites. For this purpose, and as the initial stage of the study, contact the county government, residents' associations, religious leaders, and health agents to obtain information regarding the characteristics of the community and to facilitate the acceptance of the study by the community.
- Whenever possible, organize meetings and/or workshops to explain the project and the adherence to legal regulations, and obtain signed informed consent.

(continued)

Box 6 (continued)

- The experience of our laboratory in establishing "headquarters" at research sites that allow for closer contact to be made with the community proved to be beneficial and profitable. However, each community has its own rules of social coexistence, and every effort should be taken not to transgress them.
- Fieldwork involves many expenses that might be shared (when a project has no funding) by members of the collective project team. In this regard, the expenses must be proportionally divided during the financial planning of the project among the graduate and postgraduate students.
- Attempt to have the researchers working at the community wear a uniform. This makes movement in the area during the study safer and quieter. However, wearing uniforms might impair the study inasmuch as community members might identify the investigators with, e.g., members of environmental protection agencies.
- Attempt to make appointments to interview informants. This makes the interviewees more comfortable and facilitates the research dynamics. To appear unexpectedly at an interviewee's home or workplace might make them uncomfortable, and this discomfort would make the interview more difficult.

General procedures for data collection

- The data collection procedures might be liable to several sources of error arising from various circumstances. For that reason, a pilot study or pretest should be conducted whenever possible to assess the efficiency and quality of the research tools.
- As much as possible, attempt to collect the plants and/or animals cited by each interviewee during the events scheduled for collection of ethnobiological information. Individuals belonging to a single community might attribute identical names to different botanical species or vice versa.
- Attempt to systematically elucidate the meaning of native categories, even when they appear obvious. For instance, the category "haemorrhoids" represents a type of parasitic infection to certain communities in the semiarid area of Pernambuco. Similarly, attempt to be as specific as possible while recording data.
- When needed, discuss the data collection progress with the project's principal investigator at each step to identify possible problems.

General procedures for the collection and identification of botanical material

- Collect at least three samples from each plant. Each sample must be identified using the same collection number.
- The samples must be identified and/or confirmed by specialists.
- Upon collection, characteristics that are lost during herborising must be recorded, including: (a) flower color, (b) plant height, (c) presence or absence of underground systems, and (d) frequency in the area.
- Data pertaining to the collection must be recorded in field notebooks following the model recommended by each research institution.

References

1. Minayo MCS, Sanches O (1993) Quantitativo-qualitativo: oposição ou complementaridade? Cad Saúde Pública 9:237–248
2. Gunther IA (2008) O uso da entrevista na interação pessoa-ambiente. In: Pinheiro JQ, Gunther H (eds) Métodos de pesquisa nos estudos pessoa-ambiente. Casa do psicólogo, São Paulo, pp 53–74
3. Miranda TM, Amorozo MCM, Govone JS et al (2007) The Influence of visual stimuli in ethnobotanical data collection using the listing task method. Field Methods 19:76–86
4. Boeije HR (2004) And then there were three: self-presentational styles and the presence of the partner as a third person in the interview. Field Methods 16(1):3–22
5. Alexiades MN (1996) Selected guidelines for ethnobotanical research: a field manual. The New York Botanical Garden, New York, p 306
6. Richardson RJ (1999) Pesquisa Social: Métodos e Técnicas, 3rd edn. Revista e ampliada. Editora Atlas S. A, São Paulo, p 334
7. Babbie E (2003) Métodos de Pesquisas de Survey. Editora UFMG, Belo Horizonte, p 519
8. Bernard HR (2006) Research methods in cultural anthropology, 4th edn. Sage, Newbury Park, CA
9. Weller SC, Romney KA (1988) Systematic data collection. Sage, Newbury Park, CA
10. Combessie JC (2004) O método em sociologia o que é, como se faz. Ed. Loyola, São Paulo
11. Posey D (1987) Introdução—Etnobiologia: teoria e prática. In: Ribeiro B (ed) Suma Etnológica Brasileira, vol 1, Etnobiologia. Ed. Vozes, Petrópolis, pp 15–25
12. Aguiar VRL, Medeiros CM (2009) Entrevistas na pesquisa social: o relato de um grupo de foco nas licenciaturas. In: Anais do IX Congresso Nacional de Educação, PUCPR, pp 10710–10718
13. Bernard HR (1988) Research methods in cultural anthropology, 2nd edn. Sage, Newbury Park, CA
14. Quinlan M (2005) Considerations for collecting freelists in the field: examples from ethnobotany. Field Methods 17(3):219–234
15. Brewer DD (2002) Supplementary interviewing techniques to maximize output in free listing tasks. Field Methods 14(1):108–118
16. Silva VG (2000) O Antropólogo e sua magia. Ed. Edusp, São Paulo, p 200
17. Pinheiro JQ, Elali GA, Fernandes OS (2008) Observando a interação pessoa-ambiente: vestígios ambientais e mapeamento comportamental. In: Pinheiro JQ, Gunther H (eds) Métodos de pesquisa nos estudos pessoa-ambiente. Casa do psicólogo, São Paulo, pp 75–104
18. Werner O, Schoepfle GM (1987) Systematic fieldwork: foundations of ethnography and interviewing. Sage, Newbury Park, CA, p 416
19. Fuller RJM (2007) Guidelines for using video to document plant practices. Ethnobotany Res Appl 5:219–231
20. Montenegro SCS (2001) A conexão homem/camarão (*Macrobrachium carcinus* e M. *acanthurus* no baixo) São Francisco alagoano: uma abordagem etnoecológica. Tese de Doutorado, Universidade Federal de São Carlos, 209p
21. Albuquerque UP (2005) Introdução à etnobotânica. Interciencia, Rio de Janeiro, p 80
22. Becker HS (1999) Métodos de Pesquisa em Ciências Sociais. 4ª edição. Editora Hucitec, São Paulo, 178p
23. Víctora CG, Knauth DR, Hassen MNA (2000) Pesquisa qualitativa em saúde—uma introdução ao tema. Tomo Editorial, Porto Alegre, p 133
24. Gaugris JY, van Rooyen MW (2006) Questionnaires do not work! A comparison of methods used to evaluate the structure of buildings and wood used in rural households, South Africa. Ethnobotany Res Appl 4:119–131
25. Flick U (2004) Uma introdução à pesquisa qualitativa. Bookman, Porto Alegre, p 312
26. Bridges KW, McClatchey W (2006) Expecting the unexpected: safety considerations for ethnobotany field research. Ethnobotany Res Appl 4:41–50

Chapter 3

Participatory Methods in Ethnobiological and Ethnoecological Research

Shana Sampaio Sieber, Taline Cristina da Silva, Letícia Zenóbia de Oliveira Campos, Sofia Zank, and Ulysses Paulino Albuquerque

Abstract

This chapter introduces a brief discussion of the paths of participatory methods, within a multidisciplinary perspective, exemplifying its use as an instrument for data collection in ethnobiological research, as well as a discussion of some advantages and disadvantages.

Key words Participatory approach, Ethnobiology, Ethnoecology, Ethnobotany

1 Paths to a Participatory Approach: Participatory Methods

Participatory methods (PMs) can be understood as a set of procedures backed by the active participation of social actors [1–4] that provide for the collective construction of solutions and diagnoses to be made by the community itself. In the past, PMs have mainly been used by NGOs and by some innovative government agencies, with the objective of interacting with local populations, based not just on individual analysis but also on collective analysis [5].

One of the main features of PMs is the dynamics of the process, which is developed within a specific community context [3]. Thus, these methodologies allow for a true dialogue with the community [2], so that the researchers may come to appreciate their knowledge, share their experiences, and analyze different alternatives for improving their planning skills and actions [3, 5]. It is worth emphasizing the importance of using collective spaces for this type of study because, in participatory research, the community is not only a source of information, it is part of a construct of social nature; therefore, solutions must be sought through a collective construction process. In this process, the researcher acts more

Ulysses Paulino Albuquerque et al. (eds.), *Methods and Techniques in Ethnobiology and Ethnoecology*, Springer Protocols Handbooks, DOI 10.1007/978-1-4614-8636-7_3, © Springer Science+Business Media New York 2014

as a facilitator, that is, organizing the activities and discussions in an objective way, contributing to the participation and reflection of those involved and facilitating interpersonal communication [3, 6, 7].

In recent years, the use of participatory approaches has grown considerably in research activities worldwide [5, 8, 9], principally because of the previous lack of community participation, which led to the failure of many projects [2]. Research using a participatory approach has surpassed research involving studies based entirely on the researchers' scientific-technological knowledge [10].

The techniques of PRA (participatory rural appraisal) and its tools are very useful when applied to ethnobiological issues and may be understood as a new method for working with a community's reality and its knowledge of natural resources. In this sense, the PRA can be understood as a set of methods aimed at sharing, improving, and analyzing a rural community's knowledge of the environment and of its living conditions [4, 5].

Over time, the PRA has received significant contributions from various fields of research. Among these, the participatory research inspired by Paulo Freire's methods is prominent. In the mid-1960s, his work was based on the importance of dialogue as a way of learning and enhancing the capacity of communities to conduct their own analyses of their living conditions [1]. Furthermore, since popular knowledge became valued and was treated in a horizontal form, PRA techniques have been used and adapted by several research communities, such as the social sciences, agronomics, and natural sciences [5]. Currently, some good examples are the uses of participatory tools in various approaches to research conducted in the fields of anthropology, sociology, agriculture, biology, forestry sciences, business, and economics. Within this context, studies related to ethnosciences, such as ethnobiology and ethnoecology, might utilize these tools, which provide a didactic and participatory process for gathering information.

From this reflection we would like to highlight that the concept of PRA, rapid rural appraisal (RRA) is considered to be one of the precursors of PRA [5] and can be understood as a systematic and semi-structured activity developed in the field by a multidisciplinary team to rapidly acquire new information about rural life [11]. The RRA may use various approaches depending upon the intended objectives, such as exploratory (obtain initial information about a particular topic), thematic (investigate a specific topic), participatory (involve different social actors for any further actions), and monitoring (monitor the progress of actions and experiments, for example) [11]. In this context, participatory methods can be applied to RRAs, but their purpose and processes are based on the collection of local information and on learning from external actors [5]. In contrast, PRAs are created not only to generate information but also to provide mutual learning between social actors through dialogical communication [5, 12].

The application of participatory methods requires the same level of care normally applied by researchers to the analysis and interpretation of data in conventional research. Cross-referencing methods or triangulation methods are important for triangulating the information acquired from various different perspectives [2, 13]. Special care should be taken by the facilitators–researchers to avoid transforming the participatory methods into a persuasion tool, imposing solutions without a commitment to democratic management or real participation and merely giving the appearance of involvement with the community [13]. The facilitator's role is to guide the learning process and to allow decisions to be made by the group involved [14]. Chart 1 demonstrates some of the most obvious advantages and disadvantages of participatory methods.

Chart 1 Advantages and Disadvantages of Using Participatory Methods

Advantages

- Allows moments of self-reflection and sharing of experiences, knowledge, planning, and community actions in a collective form and in a more practical and participatory manner.
- Promotes the formation of groups from different social environments, social-demographic patterns or different ages and genders, enabling intersocial analysis.
- Enables the externalization of an individual's knowledge by providing an opportunity for participants to express their opinions more spontaneously through diverse exercises and by encouraging respect for local knowledge. Motivates people to participate and allows a dynamic work process.
- Allows multidisciplinarity, which is ideal for a visualization of the "big picture."
- Helps communities identify their needs and collectively propose changes.

Disadvantages

- Allows for the voices of those who speak the loudest to prevail, emphasizing the views of the more dominant individuals in the community during participatory meetings and gatherings. In this context, the views of marginalized groups, specifically women, cannot be taken into consideration, which hampers the effective participation of all stakeholders.
- Unfeasible to conduct pretesting of methods, such as those in conventional research. For example, using questionnaires with a small group of informants. The completion of pretests in participatory research implies additional costs or demands excessive time from community members involved in the process.
- Limited power to make generalizations. Here, researchers need to be very attentive to the possible applications of the results obtained.

See [3, 13, 15].

2 Methods That May Be Used in Ethnobiological Research

2.1 From Interviews to Meetings, Workshops, and the Formation of Focus Groups

While the researchers are still in the process of forming the collective spaces for participatory activities, the use of interviews can play a very important role in participatory appraisal because, in many cases, this tool can be used during the early stages of the research as a means of orientation, providing the earliest contacts with the community. Individual interviews conducted early in the research play a strategic role in the process of gaining people's trust in the researcher, and such interviews may establish a true "rapport" with the community, facilitating the PRA. In other cases, interviews may be used secondarily to acquire supplementary information [2, 15].

Likewise, interviews with focus groups can be performed by employing naturally occurring groups, taking into account the social environments, sociodemographic patterns, or differences in age and gender of the participants [2, 16]. Interviews with focus groups promote spaces for discussion and interaction between different social actors, especially when conducted with groups that are knowledgeable about the issues to be addressed [2].

Along the same lines, meetings and workshops are participatory strategies in ethnobiological research that can provide both public and collective moments. At first, meetings work in support of the presentation and explanation of the study objectives, generating moments of reflective discussion with the participants [15, 17]. Throughout the process, dialogue is a key activity that provides a process for the creation of critical knowledge about the actions being taken in the community [2]. Visual stimuli may be employed, such as slides and photos, during the first moment to begin the conversation and facilitate the process of dialogue. Meetings are important for a preliminary survey of the experiences and expectations of the community, as well as to experience the scenario that researchers will encounter.

Participant selection for Rural Appraisal, will depend on the research objectives, the workplace, and the interest and involvement of the community. Assuming that the formation of different groups comprises people who interact and collectively share a common past in the same social environment facilitates the exchange of experiences in participatory workshops [16], participatory exercises consisting of group meetings can be performed in accordance with the goals of each exercise. The latter would facilitate conduct and communication with participants, who were selected because they understand the same social environment. As an example, Sieber et al. [18] used focus groups represented by local experts in medicinal plants. In a complementary fashion, another group was formed, which consisted of relatives of those experts gathered in the homes of the participants, to explore the perceptions of people living in the same environment, that is, the family space. The formation of family groups can facilitate knowledge exchange between local experts and their families, who often do not have moments in their daily routine that facilitate this process.

Below are some participatory procedures that may be useful when collecting ethnobiological data. In using these procedures, it is important that the researcher–facilitator use his or her creativity to change and adapt the participatory procedures for the local context. Some of these methods can be used in strategies intended to "give feedback" to a community. Thus, for example, one can reflect on the problems and possible solutions for future plans relating to the management of the natural resources that are part of the environment and a reality for each community.

To implement these tools, one should take into account that the tools need to be understood by the community. Unlike interviews, participatory techniques need to make information visible to the community and allow for reflection and ownership of the results. It is important to create legible cards with large letters and, in the case of illiterate communities, drawings would be an alternative for representing the information. Chart 2 is an example of an ethnobotanical study that used participatory methods as tools for data collection.

Chart 2 Example of an Ethnobotanical Study That Used Participatory Methods as Tools for Data Collection

Example of a participatory study conducted by CIFOR (Center for International Forestry Research—Science for Forests and People)

Study area

District of Malinau in Kalimantan Oriental, Indonesia.
A pilot study in two indigenous communities with subsequent revisions and final application in five additional communities.

Objectives and study characteristics

- Multidisciplinary research.
- Contribution to the sustainability of forests in the tropics.
- Integration of social, environmental, forestry, and biodiversity objectives for the achievement of long-term, multiple-use management.

Participatory Research

Multidisciplinary and collaborative process used to define and collect useful and decisive information related to environmental impacts and the perspectives of local communities.

Participatory Methodologies Utilized

- Community meeting for initial presentation and explanation of the research. Development of a community map to gather information on natural resources and local perceptions.
- Conducting of scoring exercises, specifically, the rock distribution method (RDM), to quantify the importance of natural resources and territorial units.

See [17].

2.2 Community Mapping

Community mapping is based on building a map that synthesizes relevant information, using the local perspective to represent the natural resources using a base map, a map of the area, and aerial photos or sketches on which participants may designate, appoint, or delimit places where they obtain resources [17, 19]. The first step in this exercise is to obtain from the participants the names used for the identification of places and the flows of major rivers and, later, to include additional reference sites (such as mountains and villages) used to locate specific types of vegetation coverage, resources, and risks (natural or anthropogenic) [17]. Community mapping is an effective tool for developing community management projects, gaining an appreciation of local knowledge, and understanding local ownership, the use of resources, and the demarcation of territories [13, 17, 19].

2.3 Historical Timelines

"Historical timelines" (Fig. 1) is another tool that can be used to analyze significant changes in the community's past that can explain the present, especially in terms of natural areas or specific resources. Participants indicate key events and state their comments in a discussion, beginning with questions pertaining to the time when the community was founded, the people that were the first to arrive and the most important events [2, 17]. Participants are also asked about the changes that have occurred in the landscape over time, when they occurred, the nature of the changes and their causes; the answers should include a local representational approach to the transformation of the landscape and possible threats to

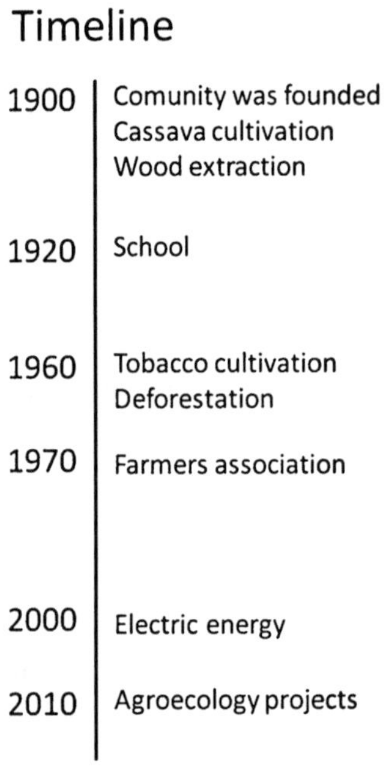

Fig. 1 Representation of participatory exercises: "Historical Timeline"

biodiversity. Insofar as the participants recall events, a vertical line is drawn on a poster board to represent a timeline. Events can be recorded on individual cards and then placed in sequence on the vertical line, with the oldest events located towards the top of the line in the upper section of the line (Fig. 1). All information should be written on a pre-prepared form [2, 17].

2.4 Historical Charts

Complementary to the timeline, the "history chart" [2] or "historical calendar" (Fig. 2) [15] can be used as a tool to graphically depict and quantify the changes that have affected the community as they relate to the researcher's specific interests. The researcher–facilitator might discuss only one element with the community, e.g., forest cover, or he or she can analyze various elements sequentially, forming a matrix with the elements in rows and the time periods in columns (Fig. 2). Such a chart or calendar might involve, for example, an analysis of the vegetation dynamics that operate within the community, representing the historical changes that have affected the landscape in the areas that supply native resources [18]. In this example, the graph visually represents, on illustrated cards, several decades of the community's history, beginning with the present day. By visualizing these graphs, participants indicate the relative numbers of species available in the various resource units during each season and, thereby, provide a representation of the abundance and decline of these species in the environment over time. For each identified species, illustrated notecards with drawings of trees can express the quantities of plants available for each species in each resource unit, which allows for an analysis of how they are represented among the participants.

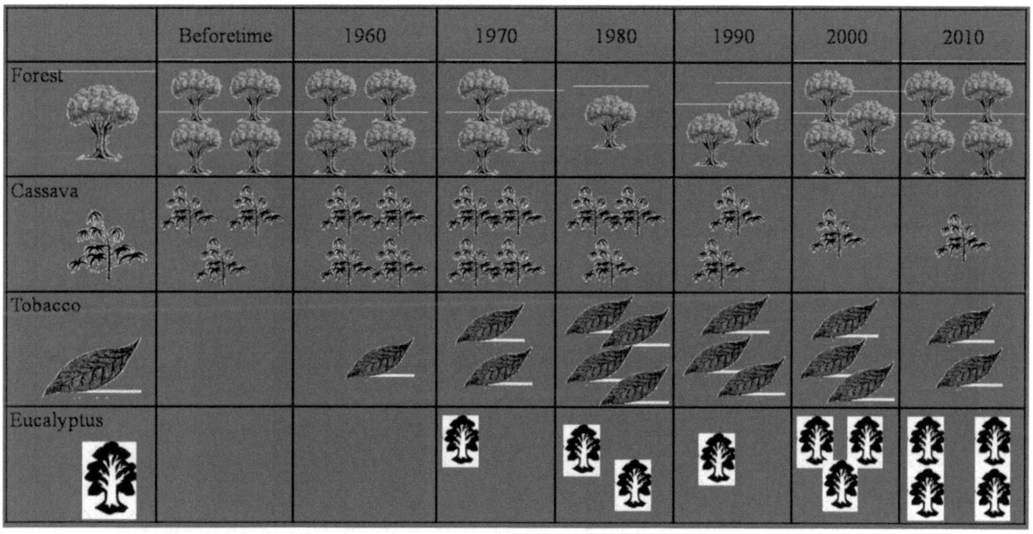

Fig. 2 Schematic of a historical chart analyzing various environmental elements

If the researcher wants to understand the dynamics of change in the landscape, an "environmental historical chart" can be constructed with a "historical succession" of, for example, agricultural production systems and livestock [18]. The graph can visually represent, through drawings, the production units (for cultivation and grazing, for example) and the process for supplying the relevant units of natural resources (native forest) that have historically been transformed; it can also provide a perspective for the future, according to the expectations of the participants. Thus, one can diagnose the changes that have affected the community by analyzing its occupied spaces, i.e., the availability of each resource area's productivity and its relationship with the transformation of the species from the native forest. For example, by researching local perceptions concerning the abundance of species in an area of riparian vegetation in northeastern Brazil, Silva et al. [20] generated a historical chart with a time frame of 3 decades. The data generated from the graph were supplemented with satellite images of the same vegetation during the same time frame. This approach results in more robust information for the preparation of management plans aimed at conserving natural resources.

2.5 Scoring and Ranking Exercises

To cross-reference the information gathered via the different techniques described above, scoring exercises (Fig. 3) can be performed to determine the relative importance of the environment and/or the resources from the participants' representations and seek information about decision-making related to the conservation of local resources [8, 17]. At each stage of the exercise, participants are asked to distribute 100 seeds, for example, among a group of illustrated cards that may represent the areas from which certain resources are obtained or the species that are used in the community, according to their importance. The cards display the

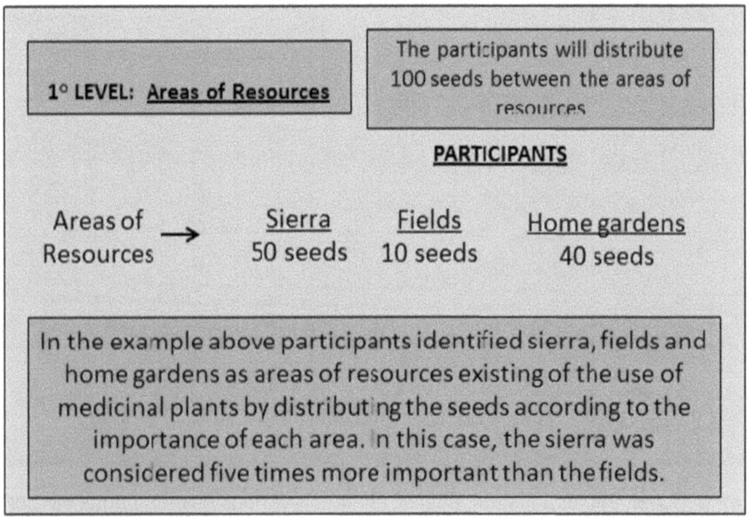

Fig. 3 Scoring exercises

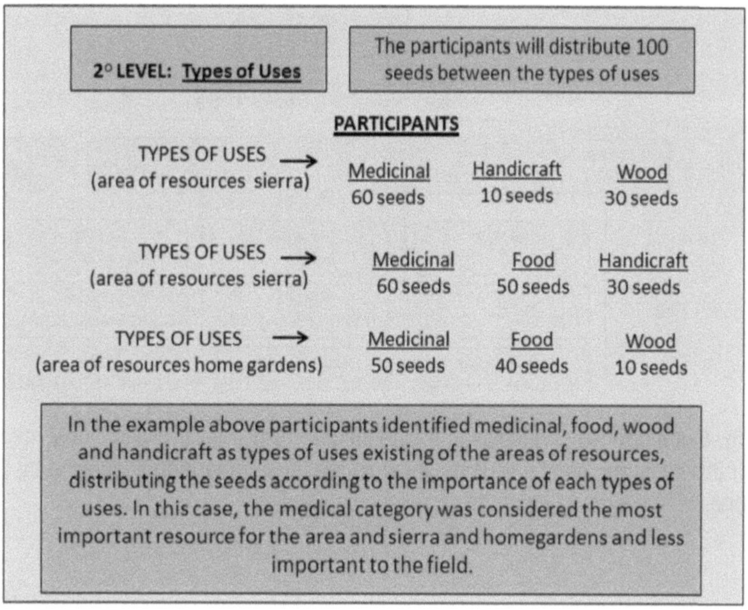

Fig. 4 Importance of medicinal plant collection location suggested by participants

drawing and name of each type of site or species existing in the community and must be placed on the floor in a manner accessible to all of the participants [8, 17]. The score for each card, which is the total number of seeds distributed by the participants, should be added and annotated onto a specific form and should also include all information provided by the participants on the reasons why the scores were given. To evaluate the importance of the species in each category and/or resource zone, one can use the technique of hierarchical weighting, which was described in the work of Sheil et al. [17]. In this technique, the number of seeds assigned to each resource evaluated is proportional to its importance to the participants.

As a fictional example, imagine a rural community in which native species are extracted for medicinal use. The importance of each collection site (zone with resources) (Fig. 4) to the use of medicinal plants can be analyzed (see diagram below) (Fig. 5). However, this analysis does not need to be restricted to a particular category of use, especially if the researcher wants to study other types of uses (medicinal and firewood, for example). Instead, the scoring exercise can be applied simply for each category of use, and this factor can be included in the multiplication, as shown in Fig. 5.

Likewise, at a lower level, we can analyze the relative importance of each species for each use category within a resource zone. In our case, we are using the category of "medicinal use" within the resource zone "serra" (forest) (Fig. 5), but we could have used "species used in other resource areas in the community," as an example, to analyze all of the hierarchical levels (Fig. 6).

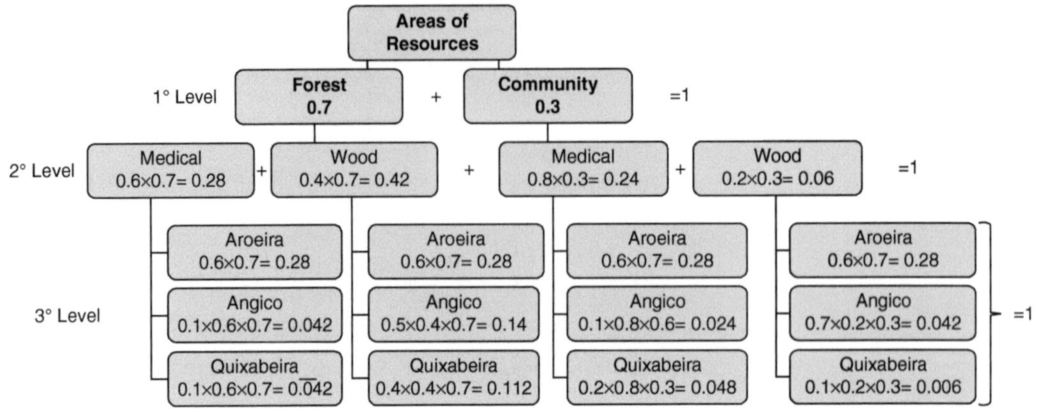

Fig. 5 Importance of species in each category of use (in this case, only medicinal) for each collection site (in this case the forest place appointed as the most important) suggested by participants. For each resource zone, we have the value of the corresponding plants

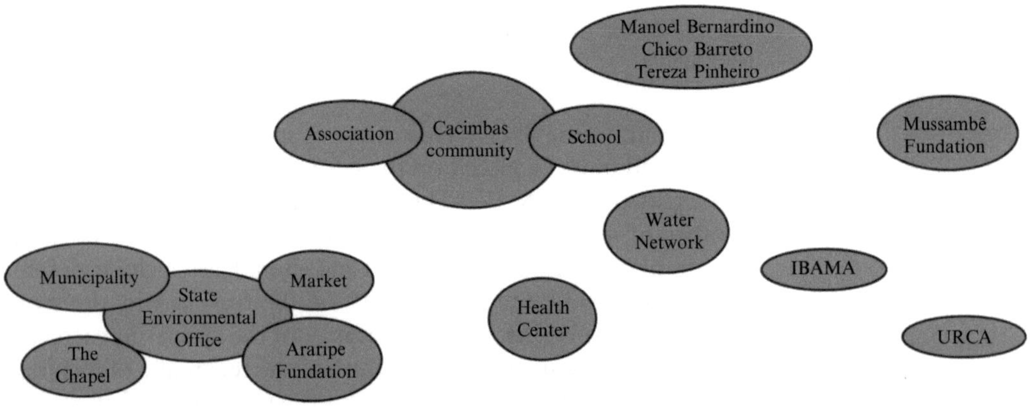

Fig. 6 Venn diagram conducted in the Cacimbas community, in the city of Jardim, Ceara in northeastern Brazil

Thus, we can calculate the relative importance of species, considering the "serra" (forest) as the main area of medicinal resources:

IR = Relative Importance

IR (serra) = 50 seeds = 0.50 (from 100 seeds)

Aroeira: 25 seeds = 0.25; IR = $0.25 \times 0.50 = 0.125$

Quixabeira: 10 seeds = 0.10; IR = $0.10 \times 0.50 = 0.05$

Imburana: 15 seeds = 0.15; IR = $0.15 \times 0.50 = 0.075$

Jurema: 10 seeds = 0.10; IR = $0.10 \times 0.50 = 0.05$

Catingueira: 12 seeds = 0.12; IR = $0.12 \times 0.50 = 0.06$

Jucá: 8 seeds = 0.08; IR = $0.08 \times 0.50 = 0.04$

Braúna: 10 seeds = 0.10; IR = 0.10 × 0.50 = 0.05

Pereiro: 5 seeds = 0.05; IR = 0.05 × 0.50 = 0.025

Mororó: 5 seeds = 0.05; IR = 0.05 × 0.50 = 0.025

$$\sum IR\ of\ species = IR(aroeira) + IR(quixabeira) + IR(imburana) + IR(jurema) + IR(catingueira) + IR(jucá) + IR(braúna) + IR(pereiro) + IR(mororó)$$

IR (for all species) = 0.50, which corresponds to the relative importance of the resource area "serra" for the type of use "medicinal."

In the above example, the medicinal species of greatest importance to the community in the resource area "serra" is the aroeira, and its general importance in all of the resource areas ("serra," fields and home gardens) is calculated by the sum of all its relative importance values at each collection site. However, the most important species for the other resource areas (fields and home gardens) would also need to be surveyed. For example:

$$IR(aroeira) = IR(aroeira\ from\ sierra) + IR(aroeira\ from\ field) + IR(aroeira\ from\ home\ gardens)$$

This analysis can be performed for more than two hierarchical levels, depending on the focus of the study and the realities of each community. One could, for example, analyze the importance of the medicinal category relative to the other categories. In a more complex example, three hierarchical levels can be utilized (Fig. 5).

In the above example, *aroeira* was considered to be more important for medicinal use in the forest compared with its use as firewood (IR of 0.336 and 0.028, respectively). For use as firewood, *angico* was regarded as the most important (IR of 0.14) in the forested area. In the community, *aroeira* was also seen as more important for medicine (IR of 0.168), but for firewood, *angico* was considered to be more important (IR of 0.042). These results are proportionate and relate the relative importance of the three hierarchical levels, taking into account the species used in each use category and in each resource area.

Thus, the use of these methodologies can favor collective spaces that enable discussions and reflections on the use and management of plant resources, demonstrating how and why biodiversity is important to the community [17]. In this way, through a process of knowledge recovery and appreciation, researchers, together with a community, can arrive at possible management solutions that facilitate the maintenance of diversity in the area studied.

2.6 Venn Diagrams

A Venn diagram is a mathematical tool used to illustrate the relationships between different sets [21]. This tool has been used in an adapted form for ethnobiological studies with various objectives, including studies of folk taxonomy, in which the authors used Venn diagrams to illustrate popular taxonomic groupings [22, 23].

However, in this study, the Venn diagram will be described as a tool for collecting data in ethnobiological and ethnoecological participatory research (Fig. 6).

Therefore, ethnobiological and ethnoecological investigations that attempt to access the social structure of a community and make inferences about its relationship with natural resources can be quite successful if they choose to use the Venn diagram in a participatory manner as a tool for data collection. The Venn diagram is a collective technique that aids in identifying the formal and informal institutions that act directly or indirectly in an area, according to the perceptions of the community [1].

This activity can be developed as follows: (1) make a large circle of paper that will represent the community; (2) use smaller circles of different sizes to represent the institutions with which the community relates; (3) ask the participants to choose larger circles for the most important institutions and smaller circles for the least important institutions; (4) the distances between circles when placed on the large circle will represent the relationships between locations of the entities; (5) conduct a debate to discover all of the institutions and their links.

2.7 Participatory Seasonal Calendar

Participatory seasonal calendars are designed with several goals, including understanding the community's perception of the events and seasonal variations in relation to the availability of a resource [21], understanding changes over time relative to the resource's availability [24, 25], as well as changes in weather and climate. The diagrams formed in a seasonal calendar represent graphic illustrations of the interrelationship of knowledge with the studied phenomena, such as climate characteristics, reproductive periods, availability of resources, periodic incidence of pests and diseases, rural development and the availability of non-timber forest products (NTFP), among others [26, 27].

To illustrate the application of the seasonal calendar, an example will be used from the communities surrounding the Araripe National Forest (FLONA), Ceará, Brazil. A participatory workshop with local communities was proposed to assess the perceptions of key informants regarding the different phenological events occurring that were relevant for the edible native species that they preferred. The phenological events were leaf fall, budding, flowering, and fruiting.

For the application of this methodology, a matrix was constructed (Fig. 7) to visualize the different phenological events for the selected species. Each event was given a score by the

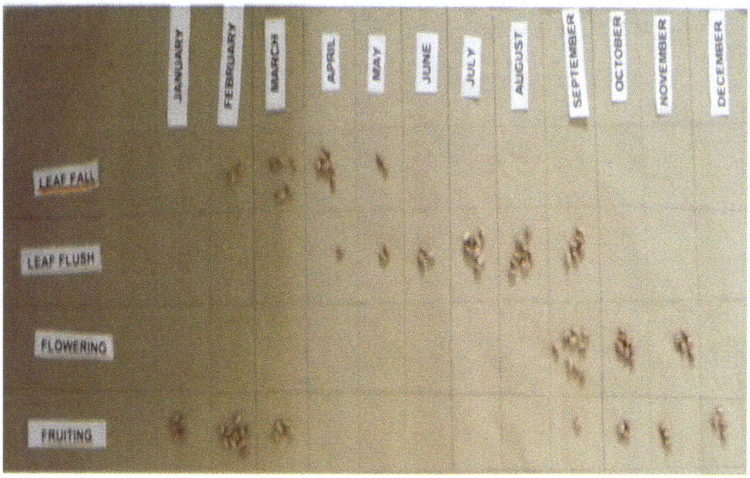

Fig. 7 Seasonal calendar in the Cacimbas community from the city of Jardim, Ceara, northeast Brazil

respondents for each of the 12 months of the year according to the intensity of the evaluated phenophase. To score each phenological event, seeds were used. In this example, the intensity of each event was scored from zero for the absence of any occurrence to ten for the period of greatest intensity.

The period represented by the seasonal calendar does not necessarily need to be 12 months. The time span may vary depending on the events that are analyzed and the interval of time during which the events occur. It is interesting to ask whether there are variations from year to year in the events being studied or if the events remain static. If the events undergo variations, it is important to identify these variations and, if necessary, to obtain information concerning the intensity of the variation in the analyzed events.

Application of the seasonal calendar is also commonly used and is recommended for ethnozoological work. Currently, this methodology has been applied to assess, for example, the breeding seasons of fish species that are important to local communities and the greater availability periods for these species from the knowledge of fishermen [9, 28, 29].

When applying these methodologies, careful attention is needed from the mediator. It is important that the mediator be aware of and learn to control the engagement of all participants so that an informant who communicates easily is not the only one to voice his or her opinion.

2.8 Four-Cell Analysis

The four-cell analysis method was developed and implemented in Nepal to assess the risk of losses in the species diversity of food crops [30]. In this study, the researchers considered the average area of cultivation for each species/variety and the number of

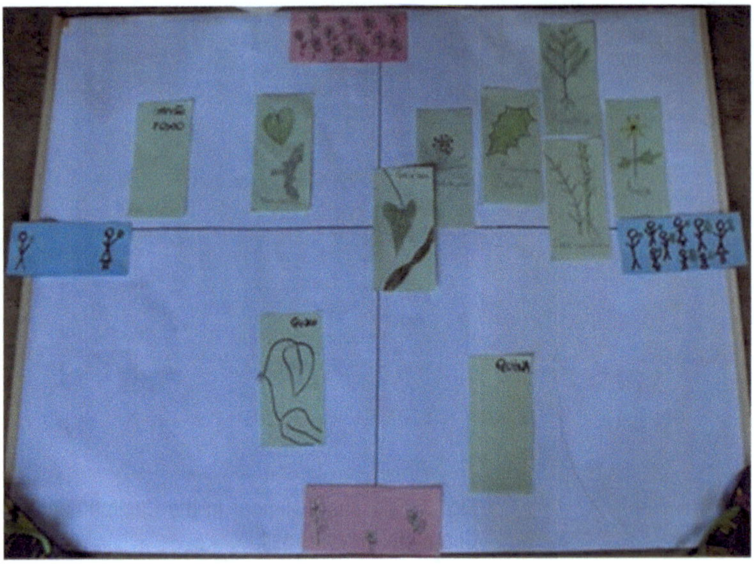

Fig. 8 Four-cell analysis applied in the community of Areais da Ribanceira, Imbituba (Santa Catarina, Brazil), with the goal to assess possible threats to the extraction of native medicinal plant species

families who grew the species and, thus, were able to identify the food resources that were common in the community, as well as those that were rare or unique [30].

In Brazil, four-cell analysis was applied in an ethnobotanical study of local yam varieties (*Dioscorea* spp.) in the municipality of Caapiranga, Amazonas, where the practices for in situ conservation of this species were investigated [31]. This tool can also be adapted to assess the risks of the loss of diversity of plant species that are extracted from natural environments.

To illustrate the use of this tool, in the farming community of Areais da Ribanceira, which is in the municipality of Imbituba (Santa Catarina, Brazil), four-cell analysis was applied to investigate the possible impacts of the extraction of native medicinal species (Fig. 8). In this context, two factors were considered in the analysis: environmental availability and intensity of extraction. Ten species of medicinal plants that were indicated by the community as being the most important were analyzed. During this research, selected species were illustrated on individual cards and labelled with their popular names. Another option would be to use collected specimens.

To apply this method, draw a large cross on the ground, or on paper, and differentiate the four categories. In the study conducted in Imbituba, the categories for the analysis were as follows: (1) high availability and low extraction intensity; (2) high availability and high extraction intensity; (3) low availability and low extraction intensity; and (4) low availability and high extraction intensity.

The meaning of each square is displayed by drawing different numbers of plants (environmental availability) and different numbers of people (extraction intensity) in the squares. It is important to create a consensus among the participants as to what will be considered high or low concerning the levels extraction and the environmental availability. Afterwards, select a species and encourage the farmers to discuss into which quadrant it should be placed, considering whether there is high or low extraction and whether it is highly abundant or not abundant in the environment. The same exercise is carried out for each species.

Once all species are placed within the quadrants, discuss the results with the community, with special attention to species located in category 4 (low availability and high extraction intensity) because these are the species that are possibly threatened.

2.9 Participatory Methods in "Feedback" Activities

In this topic we will demonstrate, briefly, some types of participatory methods that can be used in ethnobiological and ethnoecological research aimed at giving feedback to the studied community. These types include the following: *Election of priorities* consists of a presentation and discussion of all existing community demands, followed by the ranking of these demands in order of priority. The election of priorities must be performed separately for men and for women because priorities can be different according to gender. *Reality and desire* is an efficient activity for illuminating the community's perspective relative to previously established themes, expectations, and processes for carrying these activities out. Its application starts from the generation of a matrix in which the words "reality," "desire," and "what to do" are arranged horizontally, while items related to the community according to the theme of the appraisal are arranged vertically. *Past*, *present*, and *future* participants are divided into groups and are then asked to write, on cards, about a subject in the past, one in the present and about what is expected in the future. This can be an interesting tool to verify the community's perspective relative to research work that has been developed in the community [32].

3 Final Considerations: Problematizing

According to the above discussion, participatory methods have been demonstrated to be powerful tools that supplement the ethnobiological research results collected using conventional methods. However, this scientific research scenario still does not promote the full inclusion of the communities in decision-making; rather, the communities are used as a source of information that provides an efficient space for discussion and reflection. Furthermore, the participatory process of data collection can result in increased awareness and reflection on the part of the community regarding their knowledge and management practices. This perspective contributes to the process of community empowerment, to the extent that communities will appropriate the knowledge and decision-making concerning the management and conservation of their resources.

In this sense, participatory ethnobiological research plays a fundamental role, beginning with the traditional constructed knowledge and taking into consideration all the aspects that involve the ways in which the people in the community perceive the environment, including their social, economic, and historical views. These aspects, if considered, lead to the hypothesis that although people are often responsible for the degradation of the natural environment, they are also responsible for its conservation and may potentially invent alternative uses for the species in their environment. The community members are, or should be, the main stakeholders because they depend on their local environmental resources to survive.

The application of PMs demands a detailed organizational plan for each stage of the diagnostic research (see Chart 3).

Chart 3 Plan for Each Stage of the Diagnostic Research in PMs

Stages

- Establish contact and relationship with the community.
- Discuss the research objectives.
- Explain the research to the community leaders.
- Choose and prepare a mediating group.
- Identify participants.
- Record the information provided by all participants.
- Design a form for recording interesting information.
- Set up a meeting to present the research and facilitate a discussion.

Recommendations

- Research articles about participatory methodologies.
- Study selected tools that will be applied to ensure that they are adequate for the research proposal.
- Have clear research objectives that can be easily achieved. Select facilitators with ample dialoguing capacity and include both males and females.
- Systematize the diagnostics so that others have access to the information.

Materials Needed

- Large paper (A0 or A1) for the creation of maps.
- Large markers.
- Colored cardstock.
- Tape.
- Photos taken in the community, for illustration.
- Flip-chart for facilitating visualization.

Thus, research should seek methods that are carefully prepared and guided by community participation because the problems experienced by communities have a social component and require collective solutions [6]. From this perspective, we emphasize that the construction processes are as important as the data that are ultimately collected. These aspects should be taken into consideration by the researchers, especially when they have objectives, goals, and results to attain. When only the results are considered, the interpretation of the research takes a certain direction, but when one considers that the process of data collection also has relevance, the interpretive path assumes a different meaning that involves the appropriation of knowledge by the community. Thus, participatory methods identify what really matters to the community so that local preferences are made relevant in the process of decision-making, which can be limited in conventional research protocols [5].

An interesting example can be found in the work of Medeiros [32]. In the perceptions of the people studied, areas of sugar cane were the most important cause of deforestation, with a high IR (relative importance) value. However, in this case, the results demonstrated that the two groups had very different perceptions when considering the information from each group. According to the author, this fact indicated that, for the purposes of their work, the IR obtained could not be a valid parameter, noting that people might not realize that their use of the forest had a significant impact. In this context, we emphasize that participatory methods provide a variety of tools that are appropriate for the conditions of a community with respect to its social, economic, and cultural aspects, as well as the environmental conditions that are specific to each place. At this point, we assume that there are specificities in each location that constitute the complexity of its reality and should not be generalized. Finally, we note, as previously mentioned, that the process used to obtain a result is as or more important than the result itself.

If a researcher assumes that his or her research demands community participation, even if on a smaller scale, he or she must also assume that there is no better result but that there is a complex combination of factors that acts on the study environment and that these factors should be considered. That is to say, there are many aspects (qualitative and quantitative) that are expressed through the participatory process and that may be essential for an understanding of what actually occurs in the community.

There are difficulties and variations in the process of gathering information by consensus because participants have different opinions and past realities, even though they live within the same environment. When using different groups, the researcher can facilitate interesting discussions among people with a common history or not, depending on the objectives of the researcher.

The traditional option is to use the standard sociodemographic variables, such as gender, age, and social category; geographic segmentation, such as urban/rural [16]; or random grouping, as in the study by Medeiros [32]. Accordingly, proper planning is part of the objective of participatory work, even without taking into account the advantages of the researcher mastering the mediation process and understanding information when the groups are subdivided. If groups are composed of people who interact and collectively share a past in the same social environment, the exchange of experiences in participatory workshops is facilitated [16].

In addition to creating a detailed plan for the study, the researcher–facilitator must also be flexible and creative to negotiate the changes that arise during the actual application of the tools and processes of collective learning with the community.

Participatory methodologies cannot be used only as tools in the process of data collection with the intention of seeking more detailed results without the effective participation of the community. This is what occurs in many academic studies, especially if they involve research of a scientific nature that is not characterized by extension or active research. The factor that determines the participation levels of people in the study is their degree of involvement with the process of the studies' development, particularly in rural extension projects that require technical assistance. Participation evolves through different stages that can be achieved gradually as the community establishes its autonomy in the process of generating solutions and directing actions [2]. Levels of participation can begin with the complete passivity of those involved, lacking any influence on the decision-making process, and shift to interactive participation, involving the participants in the entire process from the creation of the project to its implementation and evaluation. Dynamically, participation can evolve to the final stage, which includes the autonomy of the community and the development of their capacity for self-management [2, 15]. It is important that participatory methods work in both ways, serving both the researcher, who is generating information for his or her research, and the community, which gains awareness and appropriation of the knowledge of its members.

Participatory methodologies facilitate these processes and promote discussions, following paths towards local awareness in pursuit of the autonomy of the community involved [5].

References

1. Chambers R (1994) The origins and practice of participatory rural appraisal. World Dev 22(7): 953–969
2. Geilfus F (1997) 80 Herramientas para el desarrollo participativo: diagnóstico, planificación, monitoreo, evaluación, 1st edn. IICA, San Salvador, p 218
3. Santos AD (2005) Participatory methodologies: ways to strengthen spaces public environmental, 1st edn. Peirópolis, São Paulo, p 190
4. Gunther M, Vogl CR (2010) Visual participatory methods: an overview. Working group: knowledge systems and innovations division of organic farming department for sustainable

agricultural systems, 1st edn. BOKU—University of Natural Resources and Applied Life Sciences, Vienna
5. Chambers R, Guijt I (1995) DRP, cinco años después ¿Dónde nos encontramos? For Trees People Newsl 26(27):4–13
6. Park P (1992) Qué es la investigación-acción participativa. Perspectivas teóricas y metodológicas. In: Salazar MC (ed) La investigación-acción participativa: inícios y desarrollos. Editorial Popular, Madrid, pp 135–174
7. Colette MM (2001) Moderação. In: Brose M (Org) Participatory methodology: A introduction to 29 instruments. Tomo Editorial, Porto Alegre, pp 17–24
8. Lulekal E, Kelbessa E, Bekele T et al (2008) An ethnobotanical study of medicinal plants in Mana Angetu district, southeastern Ethiopia. J Ethnobiol Ethnomed 4:10
9. Woodward E, Jackson S, Finn M et al (2012) Utilising Indigenous seasonal knowledge to understand aquatic resource use and inform water resource management in northern Australia. Ecol Manag Restor 13(1):58–64
10. Von Der Weid JM (2001) The trajectory approaches participatory development practice of NGOs in Brazil. In: Brose M (Org) Metodologia participativa: uma introdução a 29 instrumentos. Tomo Editorial, Porto Alegre, pp 105–112
11. Conway GR, McCracken JA (1990) Rapid rural appraisal and agroecosystem analysis. In: Altieri MA, Hecht SB (eds) Agroecology and small farm development. CRC Press, Inc., Florida, pp 221–236
12. Petersen P, Romano JO (1999) Participatory approaches local development. AS-PTA-Brasil, Rio de Janeiro
13. Evans K, Jong WD, Cronkleton P et al (2006) Guide to participatory tools for forest communities. Center for International Forestry Research, Bogor, p 37
14. De Boef WS, Thijssen MH (2007) Participatory tools in working with crops, varieties and seeds. A guide for professionals working with participatory approaches in agrobiodiversity management, crop improvement and development of the seed sector. Wageningen International, Wageningen, p 87
15. Verdejo ME (2006) Participatory rural appraisal: practical guide. MDA/Secretaria da Agricultura Familiar, Brasília
16. Gaskell G (2007) Individual and group interviews. In: Bauer MW, Gaskell G (eds) search qualitative text, image and sound: A practical Handbook, 6th edn. Vozes, Rio de Janeiro, pp 64–89
17. Sheil D, Puri RK, Basuki I et al (2004) Explorando la biodiversidad, el medio ambiente y las perspectives de los pobladores en areas boscosas: métodos para la valoración multidisciplinaria del paisage, 2nd edn. CIFOR, Bogor, pp 1–65
18. Sieber SS, Medeiros PM, Albuquerque UP (2011) Local perception of environmental change in a semi-arid area of northeast Brazil: a new approach for the use of participatory methods at the level of family units. J Agric Environ Ethics 24:511–531
19. Boissiere M, Basuki I, Koponen P et al (2006) Biodiversity and local perceptions on the edge of a conservation area, Khe Tranvillage, Vietnam. Center for International Forestry Research (CIFOR), Bogor, p 106
20. Silva TC, Ramos MA, Alvarez IA, Kiill LHP, Albuquerque UP (2011) Representations owners and employes on farms changes and conservation of riparian vegetation the banks of the Rio São Francisco (Northeast Brazil)-Environmental Representations populations the banks of the São Francisco River 11:279–285
21. Piroznia M, Nagaranja V, Deng Y (2007) Gene Venn—a web application for comparing gene list use Venn diagrams. Bioinformation 1(10): 420–422
22. Ferreira EN, Mourão JS, Rocha PD et al (2009) Folk classification of the crabs and swimming crabs (Crustacea—Brachyura) of the Mamanguape river estuary, northeastern Brazil. J Ethnobiol Ethnomed 22(5):1–11
23. Chaudhary P, Wangdi S, Mao A et al (2011) Consistency of local perceptions of climate change in the Kangchenjunga Himalaya landscape. Curr Sci 101(4):504–513
24. Mikkelsen B (2000) Methods for development work and research. A guide for practitioners, 6th edn. Sage, New Delhi
25. Catley A, Osman J, Mawien C et al (2002) Participatory analysis of seasonal incidences of diseases of cattle, disease vectors and rainfall in southern Sudan. Prev Vet Med 53: 275–284
26. Clarke PA (2007) Aboriginal people and their plants. Rosenberg Publishing, Kenthurst
27. Aswani S, Vaccaro I (2008) Lagoon ecology and social strategies: habitat diversity and ethnobiology. Hum Ecol 36:325–341
28. Moreira RF, Funi C, Silva CR (2011) Use Participatory Assessment Methodologies in the Dynamic Fishing in the Central Basin Araguari River, Amapá. Paper presented at the X Congresso de Ecologia do Brasil, São Lourenço, 16–22 Sept 2011

29. Rana RB, Sthapit BR, Garforth C et al (2004) Four-cells analysis as a decision-making tool for conservation of agrobiodiversity on-farm. In: Sthapit BR, Upadhyay MP, Shrestha PK, Jarvis DI (eds) On-farm conservation of agricultural biodiversity in Nepal. International Plant Genetic Resources Institute, Nagarkot, pp 15–25
30. De Castro AP, Fraxe TJP, Pereira HS et al (2012) Ethnobotany of local varieties of yams (Dioscorea spp.) grown in communities in the municipality of Caapiranga, state Amazon. Acta Bot Brasilica 26(3):658–667
31. Gomes LJ, Santana VB (2004) Organization community participatory management of order a conservation: the case of Serra-Itabaiana. Ijuí Editora Unijuí 4(6):31–38
32. Medeiros, Patrícia Muniz, Almeida, Alyson Luiz Santos, Silva, Taline Cristina, Albuquerque, Ulysses Paulino (2011) Pressure indicators of wood resource use in an Atlantic forest area, Northeastern Brazil. Environmental Management (New York), 47:410–424

Chapter 4

Oral History in Ethnobiology and Ethnoecology

Maria Franco Trindade Medeiros, Taline Cristina da Silva, Rosemary da Silva Sousa, and Rafael Ricardo Vasconcelos Silva

Abstract

Oral history is a methodology in which statements about history as it is experienced in its many dimensions are recorded. To access memory of facts, research that employs oral history techniques uses the interviewee's orality and memory. In ethnobiological and ethnoecological studies, in which information acquired from the informant's memory is also used, oral history may be very useful with regard to gaining a detailed understanding of the interrelationship between people and the environment; in addition, it may provide techniques for the systematization and analysis of data. By reading this chapter, the reader will become familiar with the primary conceptual and methodological premises of oral history, will understand how ethnobiological and ethnoecological research may use oral history both to collect and to analyze data, and will also gain access to an analysis of the use of oral history in ethnobiology, ethnoecology, and environmental history research.

Key words Oral source, Technique of historical documentation, Oral history, Ethnobiology, Ethnoecology

1 Introduction

Recognizing the other as a witness to and co-participant in the processes that create the present realities, and living in the group, establishing conditions to apprehend the phenomena in order to pick the best apprehension up of the studied group, that is in fact one of the secrets to obtain a great interaction in the research joining the Oral History with the Ethnobiology and Ethnoecology.

The complexity of this field approach lies in the knowledge of the self-explanatory details of the other including his culture and his environment. Notions of time; logic of the family structure; solutions for feeding, dwelling, clothing, social organization, and health treatment; perspectives on life and death; perceptions of the environment; calendar organization and celebration processes; as well the changes that have occurred over time regarding these

Ulysses Paulino Albuquerque et al. (eds.), *Methods and Techniques in Ethnobiology and Ethnoecology*, Springer Protocols Handbooks, DOI 10.1007/978-1-4614-8636-7_4,
© Springer Science+Business Media New York 2014

numerous aspects; constitute inherent aspects to an understanding of human culture that one is desiring to study.

In the researches that employ Oral History techniques, the orality and memory are the substantial elements to access the facts and reminds. Oral History essentially involves work with memory and, particularly in Ethnobiology and Ethnoecology, these memories are of elements of nature, landscapes (cultural ecotopes), facts, etc. With regard to this research, memory has increasingly become a rich source of information that may provide important data for the search for explanations and solutions for current problems that the local human populations and environments studied are experiencing. Memory is complex and guides us to the persons who have these memories. Memory is constructed and structured according to the social process through the intersections of the memories of each person. So, through collective relationships and experiences it is formed the expression of the collective memory, also called the collective memory. Persons' memories complement one another, in either the same or conflicting directions. The so-called collective in Oral History is the "result of experiences that link some people to others according to assumptions that articulate the construction of identities from their memories, expressed in community terms. The individuals, in this context, have autonomy in the procedure in that their wills uniquely build the combination of personal and biological factors and the influence of the environment in which they live. Thus, the experiences of each individual are authentic and are related to those of others through the construction of a common identity. In Oral History, the 'collective' does not correspond to the sum of parts. What ensures the unity and coherence of the interviews gathered in a single group is the repetition of certain factors that, in the end, characterize the collective memory. Respect for the person as an individual, however, is the basic condition to formulate respect for the individual experience, which justifies the work with interviews, but it is valid for the group. In this regard, Oral History is always social. It is social, above all, because the individual is only explained in the context of community life [1]."

With the effective participation of persons—"living documents"—the researcher has contact with individuals who have experienced events or who have heard others speak of them is possible. There is a possibility of establishing a link among the subjectivities, specifics, and totality of the topic chosen for research.

Considering the aforementioned aspects, in this chapter, the reader will find the primary conceptual and methodological assumptions of Oral History procedures; the reader will also observe how this methodology can be used for data collection in ethnobiological and ethnoecological research, in addition to gaining access to the analysis of the use of Oral History in research that has already been performed in the fields of Ethnobiology, Ethnoecology, and Environmental History.

2 What Is Oral History?

Oral history was created as a technique for historical documentation in 1948 through memory recording performed by historian Allan Nevins of Columbia University (USA) [2].

According to the Statute of the Brazilian Association of Oral History (Estatuto da Associação Brasileira de História Oral), "oral history is understood as research work that uses oral sources in different modalities, regardless of the field of knowledge in which this methodology is used" [3].

The definition presented by Delgado adequately covers the scope of Oral History, as stated: "oral history is a methodological procedure that seeks, through the construction of sources and documents, to record, through induced and stimulated narratives, testimonies, versions, and interpretations of History in its multiple dimensions: factual, temporal, spatial, conflicting, consensual. It is not, therefore, a behavior of experienced history, but rather the record of testimonies of this experienced history" [4].

Oral History appears to be a path to understanding and thinking about contemporary society. Through recorded interviews, Oral History works with the expression of the present moment, which is related to an aspect of interest to the researcher. Oral History articulates the ideas exposed by the persons to record or explain a particular social aspect.

Oral History is a manifestation of oral documentation, which is characterized as material support derived from oral expression. Oral documentation designed using electronic recordings with the objective of recording a particular phenomenon is understood to be an oral source. Oral sources are varied, and oral history integrates this set of documents, with interviews being the most widespread form. The gathering of interviews through the use of Oral History procedures may allow for both the recording and the archiving of these interviews, contributing to the understanding of particular situations related to Ethnobiology and Ethnoecology and thus serving a practical and utilitarian purpose.

3 Methodological Assumptions and Social Utility in Oral History

"Oral history is a set of procedures" [1] and is founded on the following points: (1) the design of a project that guides interviews; (2) the need for live persons, with whom a dialogue is established, and electronics (digital recorders) as an essential medium of executing this task; (3) oral history is based on memory, establishing links with the identity of the interviewed group, and requires that the narration modes of each individual participating in the interview be recognized; and (4) the temporal dimension is the present, and the final product is a document [1].

An Oral History project involves the design and structuring of the procedures to respond to three questions that guide said procedures: Whose? How? Why? The establishment of a group of people to be interviewed is one of the issues justifying Oral History. The project predicts the manner in which recordings will be executed, determining the following: interview location, which is generally determined with respect to the choice of the interviewee; interview duration, which is a function of the availability of the interviewee and of the interviewer and involves an attempt to establish a standard duration among the group of interviewed people; and other environmental factors that allow for adequate privacy and content recording. The objectives of the project, along with the reason that this person's participation in the research has been requested and the eventual purpose of the recordings, are explained to the interviewee before the beginning of the interviews [1].

The plan is also concerned with establishing the following: the transcription and establishment of the text; the prediction of a delay in obtaining the results until reaching a final solution for the text; the recounting of the written product to the interviewees, which implies authorization for the use of the product with a letter of assignment (similar to the Informed Consent Form usually used at the beginning of Ethnobiological and Ethnoecological research [5]) that accompanies the text and that specifies its full or relative use; filing of the material generated in a particular collection, the procedure for which, in addition to clarification that the use of the material is not restricted to a single project, must be communicated to the interviewee; and publication of the results, with an initial focus on communicating the results to the group participating in the interviews.

The Oral History interview is a premeditated process, which is performed through the use of a programmed dialogue between at least two people. In general, the interviewer is the director of the project. Some projects include more than one interviewer, transcriber, or proofreader, which always necessitates the inclusion of an individual to head the project—a director. The interviewer's performance is very important for the attainment of a broader and deeper understanding of the meaning of the recorded interview. The interviewer must be alert to the signs and manifestations conveyed by the interviewee. The interviewer's perceptions when in the recording environment are especially valuable. The interviewee is the person who is heard in the interview. Both the interviewer and the interviewee must be recognized as collaborators at the time of the interview. Before the recordings are set up, proper apparatus functioning must be ensured; the recording location, date, project name, interviewee name, and names of all other people involved in the act of recording, for all interviews, are recommended. The interview is the result of the recording of the meeting between the interviewer and the interviewee. The interview produces at least one documentary material: the recording. The written text is derived from the recorded interview and must be approved by the interviewees. In the Oral History context, questions are asked regarding the use of the document. Will the document be a digital recording? Will the document be a literal transcription in text form? Alternatively, will the document be a reviewed text? A concrete point is that "always, in a recorded meeting, at least one piece of documentary support will emerge going from the oral to the written" [1].

There are three modes of conducting interviews in any Oral History project: (1) Life Oral History, (2) Thematic Oral History, and (3) Oral Tradition. The specification of the criteria for capturing the narratives is a point of distinction between Oral History interviews and conventional interviews [1].

Life Oral History is often referred to as biography, life stories, a biographical report, the biographic method, biographic notes, or even an autobiography [1]. Scientific studies in this line of research emerged at the beginning of the twentieth century, first in the field of Sociology with the seminal work of Thomas and Znaniecki. The academic gathering of life histories serves as a source for the study of the structuring and functioning of societies. Life Oral History focuses on subjectivity, individual versions of events, and narratives originating from memory. This perspective is based on supporting evidence, allowing for the improbable and uncertainties in lieu of exact references. Open interviews are utilized.

Thematic Oral History is considered to be the form that most thoroughly meets academic requirements because Thematic Oral History involves an understanding of Oral History in relation to the production of conventional documentation [1]. In Thematic Oral History, there is a central focus that justifies the interview. The investigated aspect is outlined, which leads to greater objectivity. What distinguishes life oral history is the preparation of a questionnaire, which defines criteria for addressing topics; in other words, the sequence of questions should lead to the clarification of a topic. In this methodological approach, the interviewer should be prepared beforehand and should be well-informed regarding the subject matter. With this deep underpinning with regard to the subject in focus, it is possible for the interviewer to outline questions and prepare a roadmap for situations that he may deem deserving of attention, such as little-explored details of an event. Participants in Thematic Oral History research are fundamental, as testimonial content is key to elucidating the questions referring to the events under investigation. An accurate characterization of the interviewee thus becomes necessary: knowing who the person is; the person's location during the event under investigation; what the person saw, heard, and said; and the person's reaction when faced with different versions of the event. In Thematic Oral History, the hypothesis is tested during the interview, and the interviewee remains conscious of the theme of the questions that the interviewer asks of him. As a narrative of an event obtained from someone who was present, Thematic Oral History considers details obtained from the personal history of the interviewer only when they reveal aspects that help the interviewer to clarify a version of the central topic. This method allows for a confrontation of information, or formed opinions, because it gathers data on a topic, defining dates, facts, names, and situations. The nature of Thematic Oral History is essentially social, promoting discussions, debates, and the comparison of versions and, furthermore, can be combined with other sources. These projects are thematic, aiming for the clarification of versions and triumph over doubts. Therefore, this method can be considered the interviewing mode that most closely satisfies the inherent needs of the majority of ethnobiological and ethnoecological research, which generally stems from rather clear and specific questions and topics.

Oral Tradition involves the continued observation of elements of collective memory and complex, non-synthetic examinations of these same elements, which comprise human groups' beliefs and ways of life. Oral Tradition is a form of Oral History in which the execution of an interview is guided by observation such that the interviewee's narrative is directed toward a practice expressed on paper. Oral Tradition is not restricted to interview, but rather involves living with the group for better recognition of the other. Oral Tradition is similar to ethnography in that it is based on the detailed description of daily life and on self-explained details that are exclusive to the culture of the other, including attitudes, behaviors, and notions of the past and present. Oral Tradition involves a view of the world of communities that reveal mental assumptions with references in the past. The subject is more collective and less centered on the individual. Oral Tradition is qualified by the description of events by someone who experienced them, allowing these people to recall the past and permitting documents to be produced that will form a useful collection for investigations dedicated to the internal and external explanations of that social group. The acquisition of results is time-consuming because it requires a deep investigation of a specific situation and of the mythological set that governs the organization of the worldview regarding the community's origin and future. A set of myths provides support for the construction of documents and the analysis of oral traditions. For the researcher to develop a universalist view on the society under study, some principles should be considered, such as origin myths (the beginning of the world, human beings, or life), references to basic instincts (reproduction and feeding), explanations of stories, indications of personal destiny, and explanations for extraordinary behaviors [1].

Archiving and saving the interviews results in a so-called "history bank" or in "sound file documents," the purpose of which should be to find projects that are aimed toward combining and characterizing individual experiences that form the expression of the collective. In the case of historical database projects, the analysis is gradual, entailing no immediate or exact use. Thus, the history bank requires proposals for future uses that allow for collection categorization planning. In addition to archiving rules, the project should predict situations such as use rights, solutions regarding the eventual establishment of the written text, and storage location. History banking is a procedure used in cases characterized by many narrators, a sense of urgency in the records, a lack of investigative proposals, and a need for immediate use. The narrators of history banks are committed to a future dialogue, telling their stories as if they were assessing the past and recapturing the chronology of the events. Archiving and public availability of the documents are what allow for future reordering of the texts that are used for analyses that recognize possible meanings or analyses that are part of a specific project.

It should be noted that a project may propose the creation of both a collection and a history bank (Pure Oral History) and can foresee intersection of the interviews with other types of sources or documents (Hybrid Oral History). In Pure Oral History, attention should be paid to variation in the narratives, which depends on variants such as gender, age, social class, level of education, and type of interview. In this case, the narratives should maintain a common logic that permeates the constitution of all the narratives, thus emphasizing polyphony, opposition to individual versions, and the social nature of Oral History. At the same time, Hybrid Oral History, which more closely satisfies the needs of ethnobiological and ethnoecological research, is based on the combination of sources, iconographic or textual documents, thus relativizing oral expression. The focus is on thematic objectivity and the specific subject rather than on the logic of oral narrative construction.

4 Data Analysis

The analysis of the interviews results from the finalization of the project, which will first determine the genre of the interview to be performed and the treatment to be used after the text is produced. Once planned, the analysis integrates the final part of project development after creating the documents that emerge from the interviews and after choosing to prepare either a Pure or a Hybrid Oral History. The analytical procedure involves indication of the points of intersection, resonance among the narratives, and findings that diversify the logic of each group under study.

Transcription is an important procedure in Oral History. Transcription involves transforming the oral into the written. To perform this transition, it is necessary to have an adequate understanding of what has been verbally expressed and to retain sentence construction, vocabulary, metaphors, linguistic quirks, and errors. The transition from the oral to the written includes absolute transcription, in which the words are written as they were expressed, including errors, repetitions, and sounds occurring during the interview, and the questions and responses in the interview are indicated. Therefore, exercising critical judgment and restraint is the oralist's responsibility, as in this transition from the oral to the written, interference with the text can occur. "The notion of a transition gains new meaning in the oral history, as it suggests the fatality of the transcriptions as an act of recreation to better communicate the meaning and the intention of what was recorded. [...] Because all written interviews materialize the act of speaking, they should contain, as much as possible, inclusions of interview situations. For example, if the interviewee cries, but does not speak, this should transpire in the text as an integral part of the performance" [1]. After this stage, textualization occurs, in which questions are eliminated and grammatical errors and noise that may have occurred during the interview are excluded. The words are also repaired without semantic weight. A topic sentence or "vital tone" is selected and extracted from the interview as a whole. This method "is a resource used to reclassify the interview according to its essence because it is assumed that every speech has a more important general meaning, [and] it is the task of the person who establishes the text to understand the meaning of this message and to reorder the interview according to this axis" [1]. The topic sentence will determine what can and cannot be excluded from the text. During a third and final stage, the text is presented in its final form and, following the authorization of the collaborator, it is incorporated into the series of interview groups that comprise the project itself. Furthermore, it is important to state that, "in oral history, the recognition of the text, followed by recounting and authorization, determines whether the collaborator identifies with the result or not. This is a great testament to the quality of the final text" [1]. This product may also eventually be integrated into a history bank.

To monitor the progress of the project, control records are designed for use in gaining an understanding of the interview situation, of the project, and of the interviewee, as demonstrated below (see model below—Table 1). The field journal (or field diary) is also an instrument that allows for the recording of observations made when monitoring interviews and observations on the project's evolution. In this journal, information on when contact was established with the collaborators, the route used to reach these individuals, the recording process, difficulties encountered,

Table 1
Example of control records (adapted from [1])

1—Project data Project name: Project director: Institution: Interviewer(s): Type of interview (genre): Interview location: Interview duration:
2—Data from the collaborator Full name: Place and date of birth: Address: Identity document: Location and agency of emission: Level of education: Current profession: Previous professions: Observations:
3—Contact and interview data Indication of contact: Date of contact: Other contact: Form of contact: Date(s) of the interview(s): Location of the interview(s):
4—Data on the progress of the stage and the preparation of the final document 1—First transcription: 2—Textualization: 3—Transcription: 4—Recounting: 5—Letter of assignment: *Monitoring* Name of collaborator(s) 1: Stages concluded: 1 () 2 () 3 () 4 () 5 () Name of collaborator(s) 2: Stages concluded: 1 () 2 () 3 () 4 () 5 () Name of collaborator(s) 3: Stages concluded: 1 () 2 () 3 () 4 () 5 ()
5—Submission of correspondence Date of the cover letter: Date of acknowledgment of the interview: Date of shipping the interview for recounting: Data of the letter of assignment:

impressions, ideas established between the interviews, and future hypotheses is recorded. The journal is an intimate record created by the researcher that allows for a dialogue with the initial project and also serves as a reference source during the conclusive stages of the study.

5 Definition of the Sample for Use of Oral History in Ethnobiological and Ethnoecological Research

In addition to the essentially interdisciplinary nature of Oral History research, the diversity of topics and problems that permit the application of oral history procedures in ethnobiological and ethnoecological research represents important challenges associated with sample definition. In this regard, the basic rule is to attain clarity regarding the problem and the researched reality sufficient for adoption of the sampling methods and techniques that are the most appropriate for each situation.

In this context, Meihy [6] suggests that research utilizing oral history follows procedures and principles for the definition of the sample such as the following: the selection of "colonies" and the formation of "networks" (Table 2). The "colony" is to be composed of all the individuals who have provided testimony or who have played a particular role in the *research problem*. Essentially, we could understand the "colony" as being the research's "sample universe" (N). At the same time, according to the same author, the selection of the "network" represents a subdivision of the colony, or rather, the sample (n) that will determine how the interviews will be articulated. Therefore, the network would be composed of the people identified as potential collaborators in the research.

In many studies, defining a "historical landmark" or, rather, a particular event of relevant interest with regard to the research that serves as a temporal point of reference for interviews, is recommended. In the previous example, provided in Table 1, the historical framework could correspond to the date on which the law creating the reserve officially took effect. In this case, establishing a "historical landmark" could contribute to the definition of the sample by establishing a minimum age for the local collaborators in the research to ensure the selection of individuals who were directly affected by the changes resulting from the law. Therefore, it is confirmed that the use of Oral History is not necessarily limited to the participation of older persons, which depends on the objective of the study, and that different social actors in different age groups and with different life

Table 2
Example of the definitions of "colony" and "network" in ethnoecological research that employ oral history procedures

Research problem	Colony	Network
Identify the changes in the way of life for a local human population after the creation of a natural reserve using a governmental law that restricts the use of resources	Inhabitants of a protected natural area	Inhabitants who make use of the resources (e.g., fishing and firewood collection) in the natural area before the restrictions imposed by law

histories may participate. The use of the "snowball" technique may be employed to select key individuals/collaborators.

Additionally, it is worth stressing that among the possible strategies for identifying and selecting local research collaborators, dialogue with representatives of governmental and non-governmental institutions, as well as local leaders, represents an important contribution and occurs in more varied ethnoecological and ethnobiological research approaches.

Following identification of the collaborators and verification of their fulfillment of the network and colony criteria, these individuals should be invited to participate in the research. They should be observed, and any ethical guidelines and rules delineated in the laws that regulate research involving human beings in the country in which the study will be performed should be respected. Researchers are encouraged to have each collaborator receive and sign a document regarding informed consent to the interview (the aforementioned Informed Consent Form) containing all of the information necessary to perform the study. In situations in which there is no legal obligation to use an Informed Consent Form or to have the interviewee sign other formal documents, researchers are encouraged to record oral consent for use of the information at the end of the interview.

Another important aspect is the selection of collaborators of different genders, ages, and socioeconomic profiles, as long as doing so is possible and necessary with regard to the research question. The use of this procedure may contribute to a broader understanding of the researched topic.

The number of interviews to be performed will vary in accordance with the nature of each study. In general, researchers are encouraged to follow the "*law of decreasing returns*," as proposed by Thompson [2], upon determining that information has begun to repeat itself and denotations have become constant, thus indicating that further interviews are unnecessary.

6 Relationship Between the Researcher and the Researched in Oral History in Ethnobiological and Ethnoecological Studies

As in all research of an ethnobiological and ethnoecological nature, the relationship between the researcher and the researched person must be developed in the best manner possible, based on trust between these two actors, and obtained during the *Rapport* stage [7].

When handling data collection achieved through Oral History methods, concern regarding this good relationship should be redoubled, given that the establishment of a harmonious relationship of trust between those involved in this type of research ensures greater involvement of the researched person and, consequently, greater trust and information richness.

Generally, data collection accomplished through Oral History methods is time-consuming, and it may become tiring for both individuals involved. The following are important suggestions for establishing a relationship of trust and achieving a higher degree of success in collecting information: (1) knowing how to be quiet and listen; (2) seeking to adapt to the psychology of the informant; (3) being patient during the conversation; (4) stimulating informant memories using discrete questions; (5) in the case of Thematic Oral History, not preventing the informant from speaking about other topics; (6) not speaking at the same time as the informant; (7) not being insistent when the informant avoids speaking about an uncomfortable topic; (8) waiting for the length of time necessary for the interviewee's memories to surface; and (9) repeating questions in a different manner to attempt to overcome resistance [8].

7 The Use of Oral History in Ethnobiological and Ethnoecological Research and Related Approaches

Given its theoretical nature and the advantage of obtaining details of events that may be added to other versions or historical events that may not yet have records, Oral History has been used in several studies that involve historical people and moments, social institutions, and natural resources. The majority of studies that use this tool are social science studies using historical or sociological approaches, as they involve the use of written and recorded historical documents with the aim of understanding the relationship between people and the environment.

At the same time, ethnobiological and ethnoecological research has particular characteristics similar to those of oral history because it originates from peoples' memories on the topic that the researchers wish to investigate (plants, animals, ecosystems, etc.). Additionally, although it is one of the data collection forms most criticized by science, the use of memory has been the most effective manner of searching for information that has yet to be recorded in written documents, resulting in very important data that can be triangulated with information from other sources. According to Silva et al. [9], Oral History may be quite variable, particularly when there is a scarcity of written or iconographic documentary sources. In addition, Lane [10] declares that this methodological procedure allows for a more in-depth understanding of peoples' historical experiences with natural resources, aiding in the description and systematization of knowledge and reconstruction of environmental history. This method can also serve as a practical and utilitarian principle in ethnoscientific research.

However, it has been observed that although Oral History is an appropriate tool for historical contextualization of the relationship between people and their environment, it has been used very little or superficially in ethnoecological and ethnobiological studies (Table 3), possibly because these studies are more dedicated to the

Table 3
Description of ethnobiological and ethnoecological research and related research that employ oral history techniques

Reference	Research line	Objective	Sample definition	OH approach	Systematization of the data (transcription and archiving)	Data analysis in OH
[11]	Environmental history	Investigation of the use of oral history as a tool for creating a record of floods	9 Local land owners 11 local managers	Life history	Transcription	Triangulation of the results
[12]	Ethnopalynology	Update of the interpretation of palynological data from a verdict, involving a scientific test of oral history as part of a request for Delgamuukw indigenous land	People of the indigenous tribe who know of the Delgamuukw case	Thematic	Transcription	None
[13]	Environmental history	Reconstruction of the traditional use of nonwood forest products in the pine forest area of the Rhone valley in Switzerland	12 Elderly individuals from both sexes who used forestry, agricultural, and forestry products	Thematic Life history	Transcription	Content analysis
[14]	Ethnomedicine	Analysis of the uses of plant resources in Rama medicine	10 Traditional medicine men (8 midwives and 2 shamans) in the 35–70 age group	Thematic	Compilation of medicinal species and their uses	None
[15]	Ethnoecology	Evaluation of inhabitants' practices regarding their interactions with the forest fragment	People who have a relationship with the fragment	Thematic	Absolute transcription Textualization (topic sentence)	Content analysis
[16]	Historical ethnobotany	Test of the versatility and diversification of hypotheses using the information obtained from Iquito on changes in pharmacopoeia	2 Elderly women and 2 elderly men from Iquito in the Peruvian Amazon	Thematic	None	Data triangulation

cognitive aspects of these relationships [9]. The interview method that is most often observed in these studies is Thematic Oral History. This type of execution is best aligned with academic demands, as the understanding that it produces results in the production of conventional documentation [1]. In addition, the use of thematic Oral History in studies may also be related to the topic that each scientific study proposes to investigate. Furthermore, it is noteworthy that in ethnobiological and ethnoecological research, the interview has been one of the most widely used forms of obtaining data, and in Thematic Oral History, a central focus that justifies the interview is permitted. A definition of the researched aspect is developed, which leads to greater objectivity.

8 Final Considerations: Oral History—Why, Whose, and for What?

Recognition of the importance of historical contextualization as a manner of obtaining a broader understanding of reality has contributed to the increasing value placed on the use of the Oral History technique in ethnobiological and ethnoecological studies. However, this apparent acceptance of Oral History allows for the following questions: Why use Oral History in an ethnoecological or related study? Whose Oral History and for what?

The debate regarding these questions is pertinent and quite relevant in the field of ethnoscience. Without a doubt, slightly deeper analysis of these questions reveals the political bias of Oral History, which even contributes to greater proximity between the university and society as a whole [11]. It is worth noting that from this analysis perspective, oral history has potential as a tool that gives a voice to the social groups studied in ethnoecological and ethnobiological research, which were silenced in some manner during the process of constructing the officially accepted history disseminated in formal historical records.

By adopting these analytical measures, we will understand how scientists and researchers have the opportunity to use Oral History as a broader dialogue strategy with the studied human populations, in addition to the character of oral history as a vehicle for complaints and the formation of the basis of public policies [17]. Such a strategy would perhaps provide opposition to a simplistic fascination with this tool's potential. A primary question with regard to the qualification of Oral History use in ethnoecological and ethnobiological research would therefore relate to its social function.

Finally, it is noteworthy that based on the obtained results, the search for generalizations is not necessarily consistent with the basic attributes of oral history, including the possibility that individuals interpret a single historical event differently. Therefore, it becomes necessary for the analysis of the results to be complemented by other sources of information whenever possible, provided that broader or general inferences are sought with regard to the researched reality.

References

1. Meihy JCSB, Holanda F (2007) História oral: como fazer, como pensar. Contexto, São Paulo
2. Thompson P (1992) A voz do passado: história oral. Paz e Terra, Rio de Janeiro
3. Estatuto da Associação Brasileira de História Oral, de 1994, Artigo 1°, Parágrafo 1°. 1998. Revista de História Oral 1:14
4. Delgado LAN (2006) História oral: memória, tempo, identidades. s. ed., Belo Horizonte
5. Ministério da Saúde (2002) Manual operacional para comitês de ética em pesquisa. Ministério da Saúde/Série CNS Cadernos Técnicos, Brasília
6. Meihy JCSB (2002) Manual de história oral. 4th. ed. Loyola, São Paulo
7. Albuquerque UP, Lucena RFP, Cunha LVFC (eds) (2010) Métodos e Técnicas na Pesquisa Etnobiológica e Etnoecológica. NUPEEA, Recife
8. Marotti PS, Santos JE (2001) Narrativas orais como subsídio para um Programa de Educação Ambiental direcionado a uma Unidade de Conservação. In: Santos JE, Sato M (ed) A contribuição da educação ambiental à esperança de Pandora. Rima, São Carlos
9. Silva RRV, Marangon LC, Alves AGC (2011) Entre a etnoecologia e a silvicultura: O papel de informantes locais e cientistas na pesquisa florestal. Interciencia 36(7):485–492
10. Lane R (1997) Oral histories and scientific knowledge in understanding environmental change: a case study in the Tumut region, NSW. Aust Geogr Stud 35(2):195–205
11. Robertson HA, McGee TK (2003) Applying local knowledge: the contribution of oral history to wetland rehabilitation at Kanyapella Basin, Australia. J Environ Manage 69(1):275–287
12. Mathewes RW (2006) Forensic palynology in Canada: an overview with emphasis on archaeology and anthropology. Forensic Sci Int 163:198–203
13. Gimmi U, Burgi M (2007) Using oral history and forest management plans to reconstruct traditional non-timber forest uses in the Swiss Rhone valley (Valais) since the late nineteenth century. Environ Hist 13(1):211–246
14. Coe FG (2008) Ethnomedicine of the Rama of southeastern Nicaragua. J Ethnobiol 28(1):1–38
15. Silva RRV (2010) Conexões entre o saber local e o científico sobre um fragmento de floresta ciliar na zona da mata norte de Pernambuco. Dissertation, Universidade Federal Rural de Pernambuco, Recife
16. Jernigan K (2012) Plants with histories: the changing ethnobotany of Iquito speakers of the Peruvian Amazon. Econ Bot 66:46–59
17. Meihy JCSB (2000) Desafios da história oral latino-americana: o caso do Brasil. In: Ferreira MM, Fernandes TM, Alberti V (org) História oral: desafios para o século XXI. Editora Fiocruz/Casa de Oswaldo Cruz/CPDOC - Fundação Getulio Vargas, Rio de Janeiro

Chapter 5

Procedures for Documentary Analysis in the Establishment of Ethnobiological Information

Maria Franco Trindade Medeiros

Abstract

Ethnobiological studies depicting the relationships between people and their living environment will be considered in this study from the perspective of the historical moment. This chapter aims to introduce the reader to a research practice in ethnobiology that combines biological assumptions with research in history. Thus, this historical ethnobotany study will greatly emphasize the care with which documentary analysis and processing of collected data are performed.

Key words Primary sources, Secondary sources, Documental fonts, Documentary analysis, Historical ethnobiology

1 Introduction

Ethnobiology research develops from the study of the relationships between people and the environment in which they live [1]. As the man is socio-cultural-historical in his nature, within ethnobiology, historical ethnobotany highlights the fact that the interrelationships between man and plant are processed through events guided by ecological and cultural contexts that change over time due to their dynamic characteristics [2]. Research in ethnobiological science focuses on understanding the dynamics of the past–present relationship that occurs between societies and their environments, within the botanical, zoological, anthropological, ecological, and historical fields. Given the *Krónos* perspective, i.e., the chronology of historical events, the remote or the near past of this relationship is preserved by the documents' existing information, enabling the researcher to understand the ways of thinking of society at the time under study.

Ulysses Paulino Albuquerque et al. (eds.), *Methods and Techniques in Ethnobiology and Ethnoecology*, Springer Protocols Handbooks, DOI 10.1007/978-1-4614-8636-7_5, © Springer Science+Business Media New York 2014

Thus, seeking to introduce the reader to the *procedures for documentary analysis in the establishment of ethnobiological information*, this chapter focuses on disseminating an ethnobiology research practice belonging to historical ethnobotany. Therefore, combining the biological viewpoint with historical analysis, the present study will greatly emphasize the care with which documentary analysis and processing of collected data are performed.

1.1 Aspects of the Research

Habits, customs, and forms of social life that originated in the past and the research aimed to understand this formation, process of events, adjustments, and modifications over time lead to inferences about these factors' influence on today's society [3]. For example, to understand the current idea of using Quina (*Cinchona* spp.—Rubiaceae) or the knowledge of a particular animal species and its use as a food item by Europeans on the Atlantic Coast, a research project is conducted studying the past to discern the different elements traced to various types of use, the characters involved, and milestones of its evolution in society. The purpose, therefore, is the analysis and understanding of the genesis and development of a particular phenomenon. In the words of [4], "the historical method fills the voids left by facts and events, supported in a time, even if artificially reconstructed, which ensures the perception of continuity and the intertwining of the phenomena."

The comparison also leads to significant findings when considering similarities and differences among the information from different societies, contributing to a broader understanding of the human knowledge of a species and verifying the similarities and differences among different types of groups. A comparison can be made among groups in the present or the past or between the present and past.

We will now discuss the procedures involving documentary analysis.

1.2 Theme

Being aware of the various types of information, the researcher who will draw on documentary sources to gather data on the man–environment relationship can address as the central theme of his research the recognition of the origin, distribution, and diversity of cultivated plants in relation to their cultivation over time. This process also includes the approaches aimed at the presentation of myths and rituals that involve certain species according to the past beliefs of certain societies. Thus, the study will proceed through documentary analysis, and under the perspective of the historical moment, different studies can be performed.

1.3 Identification

Given the selected theme, identification of the subject related to the theme will be performed through consultation of catalogs or other search tools available in libraries, archives, or search engines to locate books, academic papers, journals, and documents.

During this stage of the literature search, the researcher should not neglect the aid of the librarian or the archivist because they will help to guide the search. As stated by Eco [5], these information professionals, in general, are delighted to demonstrate their good memory and the richness of their collections. However, it may be that when reading a material said to be of low value, the dedicated researcher will find an idea between the lines that will be a decisive "reading key." Therefore, despite having this specialized help, personal searches should still be conducted because only the researcher himself or herself will know with certainty what is truly relevant to the aims of the project. Once in possession of the material, the first step will be to look at the index or the table of contents and at the abstracts to obtain a real dimension of the importance of the material for the subject under study.

1.4 Location

After identification of the works or the bibliographical survey, the researcher should locate the bibliographic records.

1.5 Compilation

Through transcription of the original or photocopying, the material of interest should be collected. In the context of material compilation, [6] emphasizes that the general conditions for admission of the inquirer (researcher) will aid in the optimization of the search. Once aware of these conditions, the researcher will be able to articulate the material more efficiently by knowing whether all of the information should be transcribed or if reproduction or loan is allowed, which forms of reproduction are the responsibility of the institution, and which ones are permitted by the researcher.

1.6 Cataloging

Transcription of the elements that identify the document should be performed in files. These elements will allow ordering and facilitate access at any stage of the research to reference sources on the subject matter. Through bibliographic files, identification of the work, knowledge of the content, composition of the bibliographical references, and the location where the work is deposited can be obtained (Fig. 1). For each new material, the files will be opened. Thus, at the end of the consultation of the source material, it will be possible to visualize rapidly what has been observed. The references for which the work cannot be accessed for various reasons should also be annotated. Thus, the bibliographic archive will record all books to be searched, and it should be present during every visit to the collections. These files contain critical information that identifies the material and its location and may contain annotations, such as "buy this publication," "find without fail," or even "evidence suggests that it is very important." This type of archive will be useful not only for the current research but also for future studies, as it may serve as a secure list of source material.

Another type of cataloging is the file that will include, in addition to the bibliographical reference, a header, the location of the

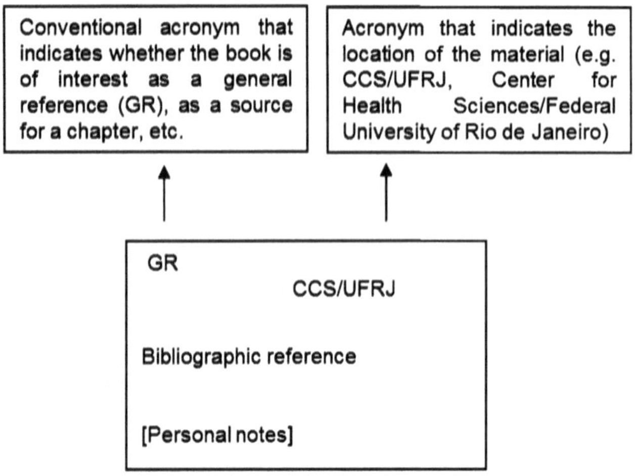

Fig. 1 Explanatory model of the bibliographic file

search, and the body of the text. The header, located at the top of the file, presents the remote generic title (subject). In the following line, the bibliographic reference is placed, and on the right margin, there is an acronym that represents the collection where the consulted material is located. These acronyms should be annotated in a separate listing that will prevent difficulty in identifying them later. Immediately after the second line of the header, the body of the text starts. If there is a need to continue to the body of the text on another file, each file should again contain the specifications aforementioned and be numbered sequentially to the right of the bibliographical reference. The body of the text may contain the following information: general information about the work; a more detailed summary, indicating on the left margin the page on which the author's idea is expressed; a personal critical interpretation of the author's opinion; or a direct quote of excerpts considered relevant and that will be annotated in quotation marks and will be textual, in full, considering the spelling errors, after which the term *sic* is placed in brackets. The page from which the quote was extracted should be indicated in the left margin of the file before each transcript. Suppression of parts of the transcribed text should be indicated by three dots before or after the passage in question, and if the suppression occurs in the middle of the text, the punctuation should be in parentheses.

1.7 Analysis and Interpretation

The analysis and interpretation of documents are divided into four phases: critique of the documentary sources, decomposition and classification of the essential elements, generalization and critical analysis, and interpretation of the data that will confirm or disprove the hypothesis. The critique of the documentary sources is subdivided into external and internal critique, as observed below (Table 1).

Table 1
Discrimination of the critique of the documentary sources (based on [4])

External critique	*Text critique*	To verify whether the text has undergone alterations, forgeries; if it has been written by the author himself or if it was revised by the author; who published it (the author or someone else); indicate what changes occurred from one edition to another
	Authenticity critique	To identify the author, time, place, and circumstances of the composition
	Provenance critique	To find the origin of the documents. To verify the fidelity of translated texts with the originals
Internal critique	*Interpretation critique or hermeneutics*	To understand the meaning that the author prescribed to the text, requiring knowledge of the terms contained therein, the historical and environmental circumstances, philosophical currents, training and thoughts that influenced the author and his or her work
	Critique of the contents' internal value	To make a judgment on the authority of the author and the value of his or her work

1.8 Reading

The document is the initial focus of this type of study. Thus, reading constitutes the means of obtaining basic, peripheral, or specific information. For a broader understanding of the topic under development, it is necessary to read systematically to identify the primary and secondary elements constituting the existing knowledge regarding the selected subject. A key step, given the vast amount of reading material available in the collections that will be consulted, is to select the correct material. The search process will optimize the reading process and will lead to the collection of essential data for the work. During the reading, the researcher should not lose sight of the goal because often other subjects can delight and distract the mind without bringing direct contribution to the work. Certain words should be consulted in specialized dictionaries or specialized glossaries to obtain semantic accuracy.

The analysis, or separation and understanding of the theme into parts, will lead to better understanding of the organization in the document. Combined with the analysis, a critical approach will permit further development of the ideas in the text, thereby correlating them with the central theme of the research.

Reading as a whole involves several stages as follows: prior reading, when the researcher will search for the existence of information concerning the subject through the index or the summary of the work; prereading, which is when the information will be located and the elements of the work evaluated (e.g., introduction, footnotes, bibliography); selective reading, as the name suggests, will be the phase in which the reader will select the information relevant to the purpose of the work; reflective reading will seek to recognize and evaluate the information; the critique will differentiate the main ideas from the secondary ideas, aiming to obtain a general idea of the text, understanding the purposes and intentions of the author subsequently to express one's own opinions and conclusions on the subject; interpretive reading will associate ideas that help to define concepts and to justify or not justify questions; and finally, explanatory reading is when the reader will examine the foundations of truth presented by the author of the work.

Still under procedures, let us examine what is recommended by [7]:

Reading and Data Collection

Before proceeding with more detailed and systematic readings of the document, it is necessary to be clear on the information that is sought after and to collect it in a standardized way to take full advantage of the time devoted to fieldwork as well as the organization and data analysis.

The transcription of data and information must conform to a minimum standard of fields, elements, or categories. It is advisable to create a collection card or spreadsheet, as this will allow the inclusion of information categories into a database, to ask questions about them, and to compare the obtained results [8]. According to [9], transcription of the documents can be carried out line by line or continuously.

In regard to the transcription of textual elements, the preservation of the original orthography of the popular names of organisms is part of an accurate transcription [7].

The Taxonomic Identification

Consider now another aspect related to popular names: the taxonomic identification. [8] states that the contraposition of popular names found in historical documents with current names, present in the same region where the record was performed in the past, can lead to the ancestral terms or the protolanguage that make up the current language. Thus, when working with a material devoid of illustrations, of preserved collection material that can be consulted, and even of indication of the species related to the popular names given, an approximation of these terms with the species present in specialized works from the time when the historical material was prepared should be performed as well as the consultation of contemporary works to have an update of these scientific names while still resorting to botanical experts and zoologists. A similar situation was experienced by Luczaj [10], where his textual material also contained three-dimensional material, composed of the collection of the specimens used as food in post-World War II Poland. In this case, the researcher can compare the botanical material with other material deposited in herbarium collections, in addition to consulting botanical expert to obtain a precise identification. [11], a biography of Manuel Arruda da Câmara, an important Brazilian figure in the eighteenth century, had the collaboration of taxonomists who helped in the identification of plant and animal specimens depicted in the illustrations that make up the scientific works left by Câmara [7].

1.9 Redaction

During the redaction phase of the scientific research, in the process of construction of the scientific discourse, researchers must ensure that their own interpretations do not obstruct the concepts of those who produced the consulted documents. A clear distinction and control of the personal contribution should also be presented in the document that is being prepared, that is, the world view and beliefs of the researcher preparing the document, for the past pre-understanding of the data analyzed [12, 13].

1.10 Indirect Documentation

In this field of work, the data are compiled from a source material, which generates a background on the context that is being studied, also providing clues for the search for other sources of data collection. The collection of information is performed through documentary research (primary sources) and bibliographic research (secondary sources). The data sources include documents that are written or not written, made at the time of or after the fact, i.e., contemporary or retrospective sources (Table 2).

What differentiates documentary from bibliographic research is that the former is restricted to documents and the latter covers all secondary data (bibliography) on a study topic already made public, such as books, magazines, newspapers, newsletters, loose publications, theses, dissertations, monographs, research, cartographic material, and recordings of oral and audiovisual communications (Tables 3 and 4). The bibliographical research intends to place the researcher directly in contact with all the existing material on the subject under focus, be it written, oral, or audiovisual sources. Through secondary sources, the researcher has the ability to manipulate information, analyze the data under a new light, and express new considerations about a known problem as well as on an issue not yet fully resolved.

2 Final Reflection

The challenge of searching for a text or a material and of locating evidence that is patiently aggregated, piece by piece, to form a general idea about what is being researched brings immense pleasure. As a researcher and observer of the clues that the documents leave us, little by little, we understand the ways of thinking of the society under study. It is important to let the documents lead us to

Table 2
Overview of the differences between documentary and bibliographical research (based on [4])

	Documentary research (primary sources)		Bibliographic research (secondary sources)	
	Written	Others	Written	Others
	Compiled at the time by the author	Conducted by the author	Transcripts from contemporary primary sources	Conducted by others
Contemporary	Parliamentary and administrative publications Letters Contracts Statistics	Photos Recordings on magnetic tape Movies Graphics Maps Other illustrations	Research reports based on fieldwork of assistants Historical study using original documents Statistical research based on census data Research using other people's correspondence	Cartographic material Commercial movies Radio Cinema Television
Retrospective	Diaries Autobiographies Reports of visits to institutions Stories about trips	Objects Engravings Paintings Drawings Photos Folklore	Research using diaries or autobiographies	Commercial movies Radio Cinema Television

reflect on the subject we are researching and attempting to understand without closing ourselves to changes in the initial course of our questions because the content of the documentary analyses often reveal other realities and direct us along surprising paths.

The ideas presented here are not the end but the beginning of procedural structuring for future studies based on source material. Each study will find the intricacies that will lead to outstanding analysis of its material. However, in the present study, we focused on what we believe is the basis for a meticulous research of decodification and contextualization of documents.

Finally, we wish to draw the researcher's attention to the issue of handling of evidence. During research, systematic consultation of the material must be carefully performed by the researcher. We should not contribute to the degradation of the sources any more than what is naturally expected due to mechanical action. Given this directive, we will help to preserve the memory, facilitating the access of collections of scientific interest to other people.

Table 3
Sources and types of documents in documentary research (based on [4])

Documentary research (primary sources)				
Sources of documents	**Types of documents**			
Public archives (national, state, municipal)	*Official documents*	*Parliamentary publications*	*Legal documents*	*Administrative publications*
	More reliable data source; express individual acts and acts of political life. The researcher must interpret and compare the material because the manner in which these documents were created is not known Examples: Royal orders, laws, letters, reports, annuals, permits, etc.	Records of the activities of the House and the Senate, being a trusted source, hardly questionable Examples: Minutes, debates, bills, reports	Source of sociological information; demonstrate the regulation of social behavior and how social problems are formed Examples: Records of births, marriages, deaths, wills, etc.	Of lower reliability than the previous documents, represents the external organization (with the public in general) and the philosophy of the administrator (when the document is for internal use), requiring a study of the political moment to better understand the described activities
Private archives (private households*, private institutions**—schools, churches, associations, etc.—public institutions***—health clinics, police stations, etc.)	*Private documents* Content of great importance because they do not describe the facts but the meaning they have for those who lived through them Examples: *Correspondence, memoirs, diaries, autobiographies, etc.; **records, letters, correspondence, minutes, memoranda, programs, press releases; ***data on hospitalizations, illnesses, etc.			

(continued)

Table 3
(continued)

Documentary research (primary sources)				
Sources of documents	**Types of documents**			
Different private and official agencies (Brazilian Institute of Geography and Statistics—IBGE, municipal and state statistics departments, etc.)	*Statistical sources* They are surveyed directly and periodically and may cover the entire population (census) or part of the population (sampling techniques), extending the results to the entire population Examples: Data on population characteristics (age, gender, education); factors influencing population size (fertility, births, deaths, diseases); economic factors (active labor force, unemployment, *per capita* income); living conditions (number of households, number of rooms, infrastructure—electricity, water); means of communication (radio, television, telephones, tape recorders)			
Distributed in the previous sources	*Iconography* Unique source of evidence on various aspects allowing, for example, reconstitution of an environment, observation of trends Every image represented by engravings, prints, drawings, paintings	*Photos* Similar to iconography; however, they portray a more recent past	*Objects* Reveal the level and direction of evolution as well as the significance of the symbolic object in a given social context	*Folklore* Represented by the customs, objects, clothes, songs, and dances, reconstitutes the way of life of past society

Table 4
Types of documents and procedures in the bibliographic research (based on [4])

Bibliographical research (Secondary sources)	
Types of documents	**Procedures**
Written press	Data from newspapers and magazines should be analyzed for their independence (determine the theoretical premises of the countries of origin of this material, whether totalitarian or democratic, and, thus, the modes of regulation of the media); regarding the content and orientation (observe the spaces dedicated to the issues addressed in the publication and how the raised issues are addressed); regarding the dissemination and influence (identify the distribution region and the type of population that is reached); and regarding the interest groups (publications targeted at specific groups, such as professional categories, where the ideas, interests, and actions of the representatives of this social group can be expressed)
Audiovisual media	The content of the audiovisual media, film, television, and radio should be analyzed for their characteristics (trends in content, stylistic characteristics); regarding the creators or the causes of the content (intentions of the content); and regarding the audience or the purpose of the content (attention focus, values, or "cultural patterns" of social groups)
Cartographic material	Widely variable but generally include maps with political and administrative division, hydrographic, ethnographic, ecological, indicators of crops, land use mode, etc.
Publications	Analysis of books, theses, dissertations, monographs, journals, research; go through the identifying, locating, compiling, and cataloging phases

References

1. Clément D (1998) The historical foundations of ethnobiology (1860–1899). J Ethnobiol 18(2):161–187
2. Medeiros MFT (2007) Botica Monachorum: farmácia botânica dos monges—Monsteiro de São Bento do Rio de Janeiro, Brasil. Tese de Doutorado, Museu Nacional/Universidade Federal do Rio de Janeiro
3. Lakatos, EM (1981) Sociologia Geral, 4a. ed. Atlas, São Paulo
4. Lakatos EM, Marconi MA (1991) Fundamentos de metodologia científica, 3a. ed. Atlas, São Paulo
5. Eco H (1977) Como escrever uma tese, 10a. ed. Perspectiva, São Paulo
6. Khoury YA (2004) Guia para os arquivos das santas Casas de Misericórida do Brasil (fundadas entre 1500a 1900). Imprensa Oficial do Estado de São Paulo/PUCSP/CEDIC/FAPESP, São Paulo
7. Medeiros MFT (2009) Etnobotãnica histórica: princípios e procedimentos. NUPEEA, Recife
8. Martin GJ (1995) Ethnobotany. Chapman & Hall, London
9. Werneck ALL, Berwanger AR, Bacellar CAP et al (1993) Technical Hespanha A M (2002) Pre-understandings and crooks. História 42:14–15
10. Luczaj L (2008) Archival data on wild food plants used in Poland in 1948. J Ethnobiol Ethnomed 4:4. doi:10.1186/1746-4269-4-4
11. Mello JAG (1982) Manuel Arruda da Câmara—Trabalhos selecionados. Fundação de Cultura da Cidade do Recife, Recife
12. Rocha LMGM (1999) Museu, informação e comunicação: o processo de construção do discurso museológico e suas estratégias. Dissertação de Mestrado, Universidade Federal do Rio de Janeiro, Rio de Janeiro
13. Standards for Transcription and Editing of Handwritten Documents. Available via http://www.portalan.arquivonacional.gov.br/Media/Transcreve.pdf. Accessed 29 May 2008

Chapter 6

Use of Visual Stimuli in Ethnobiological Research

Patrícia Muniz de Medeiros, Alyson Luiz Santos de Almeida, Reinaldo Farias Paiva de Lucena, Francisco José Bezerra Souto, and Ulysses Paulino Albuquerque

Abstract

In this chapter, we discuss and review the use of visual stimuli in ethnobiology and ethnoecology and present a proposal to standardise the terminology for the use of visual stimuli in these fields. We focus on the use of photographs, drawings, voucher specimens, in situ individuals and recently collected plants and animals.

Key words Projective interviews, Local knowledge, Data collection

1 Introduction

Studies investigating groups of humans can employ visual stimuli to obtain information. Stimuli are widely used, especially in the social and medical sciences [1–5], as an additional strategy to help people remember the specific types of information that are of interest to the researcher or even to locate the interviewee contextually. Stimuli are used in different ways, depending on the area of research and the goals of the researchers. Therefore, techniques and terminology can vary considerably.

These different visual techniques are commonly applied in the social sciences, which uses visual stimuli involving the presentation of cards, photographs, and/or films [1, 5, 6]. One example of a known visual tool is the "Rorschach test", in which ten images (inkblots) are shown to people who are asked to describe what they see [4]. The "Rorschach test" is used in psychology, psychoanalysis, psychiatry [4], and anthropology [1] and seeks information regarding personalities.

Visual stimuli have been used increasingly in ethnobiological and ethnoecological studies with the following goals in mind: to ensure that interviewer and interviewee are talking about the same subject

Ulysses Paulino Albuquerque et al. (eds.), *Methods and Techniques in Ethnobiology and Ethnoecology*,
Springer Protocols Handbooks, DOI 10.1007/978-1-4614-8636-7_6, © Springer Science+Business Media New York 2014

(e.g. a plant or animal species) [7]; to identify and collect specimens in the field for storage in a herbarium or a zoological collection; to collect ethnobiological data; and even to verify the interviewee's recognition of certain elements (plants, animals, or soil types) [8].

However, a quick analysis of this process suggests that there has been an increase in terminology and concepts, which may contribute to a lack of clarity, objectivity, and reproducibility in research studies. Therefore, in this chapter, we outline the use of visual stimuli in ethnobiological studies with an emphasis on studies of an ethnobotanical nature to highlight the methodological limitations and to promote the uniformity of the terminology in the field. We focus on visual stimuli, although other stimuli (e.g. smell, texture, and taste) play a synergistic role, and invariably, the recipients grasp more than one aspect of the stimulus.

In the first section of the chapter, we analyse the visual stimuli used in ethnobotanical studies, and ethnozoological studies are included in the second section.

2 Visual Stimuli in Ethnobotanical Research

2.1 Global Features of the Studies

We identified 45 studies, published between 1985 and 2007, that applied different techniques using visual stimuli. Of these studies, 8.9 % were performed in the 1980s, 22.2 % in the 1990s, and the vast majority (68.9 %) were performed between 2000 and the present. Obviously, the predominance of more recent studies is because of our methodological approach; it is possible that older publications used visual tools, but these studies are not available online and could not be included.

Most of studies (57.8 %) that used visual stimuli were investigating various categories of use. The medical category was conspicuous because 33.3 % of the studies were exclusively devoted to medicinal plants; however, this preference does not signify that visual stimuli are easier to use in studies of medicinal plants, but reflects a trend in the interests of researchers for this use category. In addition to the medical category, only the food category contained studies exclusively devoted to the use of visual stimuli (6.7 %). There was an obvious lack of studies focusing on the uses of wood, including firewood, tool making, or rural and domestic building construction.

One of the studies surveyed did not focus on the categories of use but on information regarding the vernacular names for species, noting whether there was agreement among informants regarding the assignment of plant names [9]. Although this study did not produce data on the uses of the plants, the study was included in the survey because it includes a well-defined research question that could be answered using visual cues beyond merely obtaining the popular names of plants for crossing-referencing them with their scientific names.

Regarding the interviewees used in the studies, 24.4 % of the studies were unclear regarding the qualifications of the people interviewed. To avoid overestimating the number of research studies in which the informants were experts, the studies that failed to indicate the use of expert informants were considered to include non-expert informants. Therefore, 51.1 % of the research studies interviewed general members of the community, whereas 28.9 % interviewed only the local experts. In addition, 20 % of the studies reported data from both expert and non-expert informants. The studies must indicate the type of informant consulted, combined with the efficiency of the technique, to determine whether the efficiency is related to the selection of a certain group of informants (such as experts) or if the stimuli work similarly for both experts and non-experts.

In some cases, there is uncertainty regarding the number of people consulted in the research studies. In total, 15.6 % of the studies did not indicate the number of people interviewed, and in the remaining 84.4 %, the number of informants was estimated (e.g. the word "approximately" was used to describe the number of informants). A number of the studies reported the number of interviewees used for one of the visual techniques but omitted the number of interviewees used when a different technique was applied (13.3 %). We believe that the accurate recording of this type of data is essential because inconsistent information produces risky generalisations that do not correspond to the reality of a given culture and jeopardises comparisons with other studies.

2.2 Types of Visual Stimuli

A number of studies used more than one type of stimulus, and the stimuli used most frequently were in situ plant specimens (64.4 %), voucher specimens (24.4 %), and photographs (22.2 %). However, stimuli including fresh plants and drawings, and the plant-derived products present in the homes (for example, tools, toys, and furniture) were included in smaller numbers. In the analysis, 13.3 % of the studies reported the use of plants for the acquisition of information without elaborating on the part of the plant used or the state of the material. The visual stimuli are discussed below.

2.2.1 In Situ Specimens

The advantage of using in situ specimens [9, 10] as a stimulus is the possibility of showing the informant the entire plant in its biological and ecological context, which facilitates the identification of the plant and the retrieval of information. This stimulus is used most frequently, possibly for this reason. Despite this great advantage, the use of in situ specimens as a visual tool suffers from a number of limitations, including the following:

(a) If there are many informants, the use of in situ specimens will demand a lot of time because the informants must be taken to the field. Indeed, the vast majority of studies that used in situ specimens as stimuli selected a small number of informants (between 1 and 45 informants).

(b) Conversely, if the sample of informants is quite small, it is difficult to generalise based on the data obtained using this technique because the sample is non-representative.

(c) A number of informants may not be available to go to the field with the researchers, especially if the journey is long or if many of the informants are elderly [11], which is common in studies using local experts.

Given the significance of the in situ specimen technique, it is perfectly feasible to use this approach, even if the study anticipates limitations, as it can be used to supplement information without the need to take all the interviewees to the field. It is advisable, however, to specify the total number of informants in the group taken to the field for the in situ interviews.

2.2.2 Voucher Specimens

The use of voucher specimens [10, 12] as a visual stimulus is used when the informants cannot be taken to the field. In addition, this technique requires less time than studies conducted outside the homes. These studies represent significant advantages for this type of stimulus and are likely the reason that the use of voucher specimens is the second most commonly used technique in the research studies surveyed. However, compared to the use of in situ specimens, the voucher specimens exhibit clear disadvantages regarding the identification of the plant material by the interviewees. The voucher specimen removes the plant from its original context, limiting the possibility of identification of a number of its parts (e.g. branches and reproductive structures), and these parts may be compromised by drying, which alters their smell, colour, and texture [11].

2.2.3 Photographs

Photographs [8, 13, 14] have the advantage of representing the plant in its original context more efficiently than dry material; however, photographs restrict the informants to identification using sight alone, which excludes the possibility of smelling, touching, or even tasting the plant. A possible issue (which can be avoided) associated with this stimulus is the difficulty the interviewees experience identifying the plants when the photographs are of poor quality [7], whether this is because of the print quality or the inexperience of the photographer. In other cases, even if the photograph is of good quality, a number of cultures may have difficulties interpreting the images, especially if the people in question have no contact with the photographic resource. Therefore, the study must consider these aspects of using photographs as a visual stimulus because the misidentification of species by the informant may compromise the rest of the research, assigning the properties of one species to another.

2.2.4 Fresh Plants

In this case, the stimulus is also limited to a small number of parts of the plant [10, 15]; however, the original properties of these plant parts, such as colour, smell, and texture, are preserved. Few studies use this resource given the logistical difficulty of repeatedly replacing the material for the study; successive sampling is required to ensure that the samples are always fresh. In addition to time demands, the need for constant samplings may, in the more sensitive species, cause damage to the plant populations.

In a way, using stimuli outside the original context of the plant (voucher specimens, photographs, and fresh plants) is disadvantageous and can give rise to possible difficulties in identification by the informants. Therefore, it is important to ensure that the stimulus used is adequate for obtaining the ethnobotanical data within the social, cultural, and environmental circumstances. This can be verified by testing the efficiency of the resource regarding its identification by the informants, i.e. performing assays to adjust the type or the combination of visual stimuli that are most appropriate for a given situation. For example, a preliminary test can be performed using several stimuli to determine the most successful technique. When possible, it is much more interesting to combine two or more stimuli (for example, voucher specimens and photos), providing more opportunities for identification for the interviewees [10, 16, 17].

The category of use in which the plant is included must also be considered because, depending on the species and its use, the type of visual stimulus selected can have different effects. For example, an informant may recognise only the bark from a given species because this is the part used by the individual, and the informant may not be familiar with the leaves and flowers of that plant.

The parts of the plant (dried or fresh) used as stimuli are always the branches with leaves and occasionally contain the reproductive structures, which is the scientific method for identifying species. However, parts of the wood (in woody plants) are not used. In this regard, when investigating the uses of wood in cases where one cannot take informants to the field to observe specimens in situ, it would be interesting to use wood samples as stimuli. We believe that using portions of the wood may help the informant to recall the characteristics of the wood, such as weight, flexibility, and hardness, among other features. These features can help the interviewee to remember the uses for the wood (for example, the hardness of the wood may help the informant remember that it is used for building rooftops). However, this technique has not been tested, which suggests that for the studies investigating the uses of wood, there is the need for research comparing the use of the wood with the use of the other plant parts (for example, the branches) to determine the stimulus that offers the best results regarding the quantity and quality of the data.

2.3 Intent and Efficiency of the Use of Visual Stimuli

Acquiring information concerning the knowledge and use of plants was the aim of 93.3 % of the studies. The remaining study aims were to acquire data on the vernacular names of plants [9] and to test the degree of recognition of species by the informants [8, 16–20].

A number of the studies sought to restrict the research to a given group of plants. For example, Gaur and Bhatt [21] restricted their research to the popular uses for ferns. In a case such as this study, it is difficult to start an interview on a taxonomic group without a stimulus to direct the informant to the scope of interest of the study. The difficulty arises because the population studied may not be able to classify plants corresponding to the group of ferns. Therefore, using visual techniques, the researcher shows the informant the plants that he or she is investigating without the interviewee necessarily understanding that for research purposes, the plants are part of a (taxonomic) group.

Regarding the efficiency of the techniques, only 26.7 % of the studies discussed the success of the stimulus used. On the one hand, Anderson and Posey [15], in a study of the indigenous Kayapó in northern Brazil, reported the ease the informants demonstrated in recognising the plant sample inside (in situ specimens) and outside (fresh plants) of their natural context. On the other hand, Griffin [10] noted the difficulty the informants experienced in recognising the voucher specimens, given the loss of the natural colouration of the samples. Furthermore, Monteiro et al. [13] were unable to use photographs because in the preliminary phase of the study, less than 5 % of the informants in a semi-arid region of the Brazilian Northeast recognised the species.

The recognition of the plants in photographs also caused problems in other studies. Case et al. [16] realised that the oldest informants in a study performed in Papua New Guinea had difficulties identifying photographs because they were accustomed to using other identifying attributes, including smell and taste, or because they had vision problems. Additionally, Garcia [8], in a study performed in India, found that women of the Paniya and Kuruma tribes were unable to identify plants through photographs, which limited the analysis.

Regarding the studies that used the in situ specimen as the stimulus, Milliken and Albert [22] determined that by observing the plants in the forest, the Yanomami (indigenous people of the Brazilian Amazon) were stimulated and were able to remember the medicinal uses for the plants previously overlooked in interviews. Moreover, in a study conducted in northeastern Thailand, Wester and Yongvanit [9] claimed that diagnostic features that enabled the plants to be identified more precisely were not present in certain plant specimens.

Considering the small number of studies that discuss the efficiency of the visual stimulus method, it is difficult to assess the quality of this approach and its application in ethnobotanical and other ethnobiological and ethnoecological studies. Therefore, we propose that studies involving visual stimuli should provide useful information. For example, the comments by the informants on the process of recognising the plants (or parts thereof) indicate the

Box 1 Comparing different visual stimuli

Thomas et al. [20] *compared the efficiency of photographs and herbarium specimens, investigating the type of stimulus that allowed informants to identify plants more precisely in an indigenous territory in Bolivia.*

What works best in an ex situ *study: photographs or voucher specimens?*

Interviews with five key informants that were encouraged to identify 50 species through photographs and voucher specimens (1-month interval between the application of the two stimuli). The interviewees previously identified the 50 species when they were taken to the field to observe the specimens in situ. Therefore, the identification by photograph or voucher specimen was considered correct when the vernacular name given by the informant was the same name that the same interviewee indicated when observing the plant in situ.

The use of photographs proved to be more efficient than the use of voucher specimens because the interviewees recognised on average 94 % of species through photographs compared to 77 % with voucher specimens, and these differences were significant (Mann–Whitney; $p < 0.01$).

informants' general degree of difficulty in identifying plants. The visual stimuli highlight possible difficulties caused by the informants' limitations, for example, visual problems, the inability to interpret photographs, or unfamiliarity with the plant. The problems arising from the type and quality of the stimulus chosen can be recognised, for example, the loss of colouration in voucher specimens or changes arising from the transformation of a part of the plant into a product, such as a toy or a type of tool (Box 1).

2.4 Terminology for Techniques Using Visual Stimuli

In our survey, we determined that 75.6 % of the studies did not assign names to any of the visual techniques used in the interviews. Designations for the techniques were not common in the research studies, and the techniques used included the following: artefact inventory, checklist, ethnobotanical inventory, ethnobotanical notebook, field herbarium, field informant, field interview, inventory interview, reference sampler, plant interview, variant of inventory interview, and walk-in-the-woods. A number of these terms, and their definitions, are discussed below.

Walk-in-the-woods: This technique consists of bringing the informant into one or more areas of vegetation to collect plants and obtain information [11]. The walk-in-the-woods technique is the same as the field interview [23, 24] and the ethnobotanical inventory [25] techniques. The field interview technique is also described in the literature as hiking for the in situ observation of specimens, although it differs from the walk-in-the-woods technique because it is performed with only one informant from the community [26].

Inventory interview: The inventory interview is similar to the walk-in-the-woods technique because in both cases, the informants must be brought to analyse the plants in situ. However, in the literature, the inventory interview is usually coupled with the plant inventory [26, 27], whereas the walk-in-the-woods is more flexible and can be performed with a simple walk in a field.

Artefact inventory: Using this technique, the researcher shows artefacts to the informants and questions them about the plants used to produce the items [11]. The researcher can take advantage of artefacts in the house the informant lives in (for example, tools, toys, windows, and doors).

The other techniques, including the artefact inventory, are similar because they do not require bringing the informants to the field. However, the techniques sometimes differ in the type of stimulus used. The field herbarium and the reference sampler techniques only use voucher specimens [12, 28], whereas the plant interview can use fresh plants or voucher specimens [11], and the ethnobotanical notebook technique is mentioned in conjunction with voucher specimens and photographs [16, 17]. Finally, the checklist technique includes voucher specimens, photographs, fresh plants, or drawings [11]. The overlap in the terminology is clear in these cases because different names are used for the same types of stimuli.

3 Application of Visual Stimuli to Other Branches of Ethnobiology and Ethnoecology

In other branches of ethnobiological and ethnoecological research, the use of visual stimuli has been helpful. In ethnozoology, for example, it is common to use photographs of animals, showing them to informants to identify and comment on such features as the morphology, reproductive ecology, trophic behaviour, and spatiotemporal dynamics of fish stocks. These visual stimuli have been especially useful in ethnoichthyological studies performed in mangrove areas because the great diversity of species in these ecosystems hinders the acquisition and visualisation of specimens that guide the interviews. Mourão and Nordi [29] and Alarcon et al. [30] used drawings and photographs for the identification and ethnotaxonomic classification of fish in the estuary of the Mamanguape River (Paraíba) and in the Itacaré region (Bahia). In ethnozoological studies of terrestrial vertebrates (e.g. [31]), these stimuli are not only useful but also necessary because the acquisition of dead specimens violates the laws protecting the wildlife.

However, similar to plants, the use of photographs removes the animal from its original context and may hinder the identification and consequently the accuracy of the results. This problem can occur in situations in which informants often distinguish species through features only identifiable by touch (for example, skin texture) or the original colour, which is often lost in photographs,

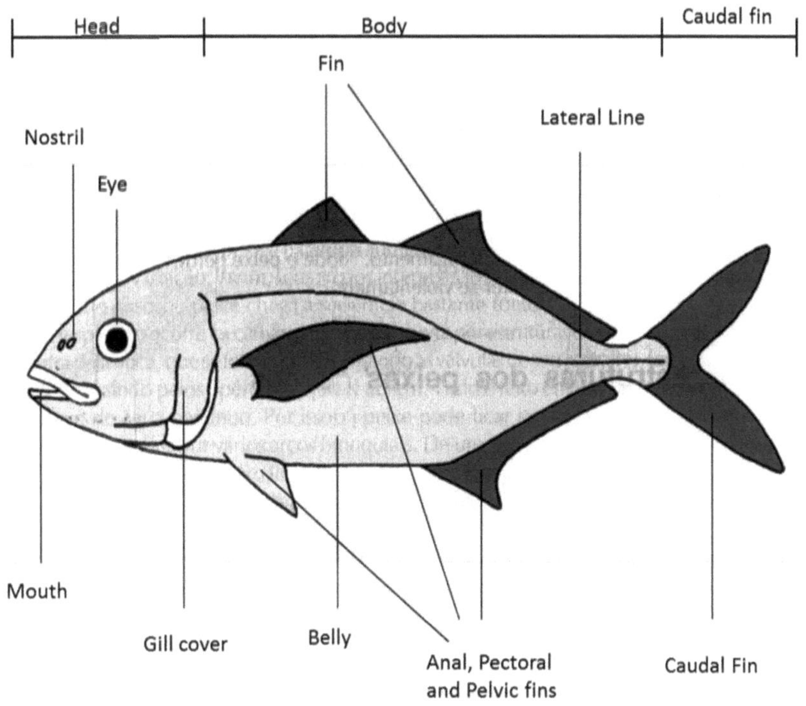

Fig. 1 Body topography of a fish using Acupe-Bahia fishermen's coding. Figure adapted from Szpilman [36] and Souto [33]

either because of the quality/resolution or the use of pictures of specimens from scientific collections.

The use of drawings has the advantage of standardising the stimuli if the study is performed using several animal species or if the acquisition of high-quality photographs of the study-focus animal is precluded. However, in general, illustrations present the same limitations as photographs and may even prevent the identification because of the actual colouration of the animal.

Marques [32], Souto [33], and Maciel and Alves [34] used visual stimuli through drawings in their ethnoecological studies in which one of the approaches was to examine the informants' perception of the morphology of the animals. The use of the terminology from traditional populations to designate the different parts and/or regions of the animal body is termed "body topography" [35]. In these studies, drawings obtained from the literature were shown to informants who identified the different structures and the morphological regions indicated (Fig. 1).

The use of dead specimens as visual stimuli is common in ethnoentomological research. Insects and groups taxonomically close to insects are often sampled and dried and are brought to the community for the informants to provide information on their vernacular names and uses or to group the species according to their local criteria of taxonomy (for example, [37]). The advantage of using dried specimens in ethnoentomology is that the techniques of sampling and drying are simple, and unlike studies using other animals (for example, large mammals), the transport of the material to homes is not complicated (for example, the samples can be transported in entomological boxes).

Ethnozoological and ethnoecological studies also use techniques that take the informants to the field for the identification of animals and for the recording of their uses. Araújo et al. [38] performed guided tours with informants to show and identify birds that were indicative of incoming rainfall. Likewise, Maciel and Alves [34] led their interviewees into mangrove areas to record information on the local knowledge regarding the mangrove root crab, *Goniopsis cruentata* (Latreille, 1803), and its habitats.

In ethnozoological and ethnoecological research, most studies that use visual resources do not name the techniques used. Such studies as Araújo et al. [38] and Maciel and Alves [34] use the term "guided tour" in a manner similar to its use in ethnobotany to describe the technique that takes the informants to the field such that they can provide information on a given group of animals.

4 Final Considerations

In general, there is an increasing need in ethnobotanical research for the classification of the terms for in situ techniques and specimens, including specimens that use parts or representations of plants. In the areas of ethnozoology and ethnoecology, names are seldom assigned to the techniques that use visual stimuli. The names used for each type of technique should be standardised to provide a better understanding for readers. With this necessity in mind, we propose systemising the terminology applied to the techniques. Our suggestion is based solely on the type of stimulus used without considering the number or profiles of the informants. Based on the most popular terminologies, we propose that studies coupling a plant inventory with a concomitant ethnobotanical study use the plant specimen as a stimulus and name the technique an inventory interview [26, 27]. The studies that take informants to the field for the identification of specimens without depending on a plant inventory should use the term walk-in-the-woods [25, 39] or guided tour. These terms should also be used for studies that take informants to the field to acquire information on animals. The research studies that use plant-derived products as stimuli, including tools and house-building materials, could name the technique an artefact inventory. For stimuli outside the original context of the plant or animal (dry or fresh parts of the specimen, photographs, and drawings) we suggest the name checklist-interview. The checklist [11] technique is used to record the knowledge of or use of a given list of plants and animals and can be performed with or without [40] visual stimuli. This terminology has the advantage of encompassing the largest possible number of stimuli, avoiding the pronounced inflation of the classifications. However, when using the term checklist-interview, the studies should indicate whether they use visual stimuli, and if so, the stimulus in question should be named. We believe that standardising the terminology will make it easier to analyse and understand the context of the use of visual stimuli in ethnobiological and ethnoecological research.

References

1. Honigmann JJ (1954) Culture and personality. Harper, New York
2. Zubin J, Eron LD, Schumer F (1965) An experimental approach to projective techniques. Wiley, New York
3. Anderson HH, Anderson GML (1967) Técnicas projetivas do diagnostico psicológico. Mestre Jou, São Paulo
4. Anzieu D (1986) Os métodos projetivos. Campus, Rio de Janeiro
5. Minayo MCS (1993) O desafio do conhecimento científico: pesquisa qualitativa em saúde, 2nd edn. Hucitec–Abrasco, Rio de Janeiro
6. Curtis V, Biran A, Deverell K et al (2003) Hygiene in the home: relating bugs and behaviour. Soc Sci Med 57(4):657–672
7. Martin GJ (2001) Etnobotánica: manual de métodos. Colección Pueblos y Plantas No. 1, Nordan Comunidad, Montevideo
8. Garcia GFC (2006) The mother–child nexus. Knowledge and valuation of wild food plants in Wayanad, Western Ghats, India. J Ethnobiol Ethnomed 2:39
9. Wester L, Yongvanit S (2006) Naming consistency for forest plants in some rural communities of northeast Thailand. Ethnobot Res Appl 4:203–212
10. Griffin D (2001) Contributions to the ethnobotany of the Cup'it Eskimo, Nunivak Island, Alaska. J Ethnobiol 21(2):91–132
11. Alexiades MN (1996) Selected guidelines for ethnobotanical research: a field manual. The New York Botanical Garden, New York
12. Blanckaert I, Vancraeynest K, Swennen RL et al (2007) Non-crop resources and the role of indigenous knowledge in semi-arid production of Mexico. Agric Ecosyst Environ 119(1–2):39–48
13. Monteiro JM, Albuquerque UP, Lins-Neto EMF et al (2006) Use patterns and knowledge of medicinal species among two rural communities in Brazil's semi-arid northeastern region. J Ethnopharmacol 105:173–186
14. Monteiro JM, Almeida CFCBR, Albuquerque UP et al (2006) Use and traditional management of *Anadenanthera colubrina* (Vell.) Brenan in the semi-arid region of northeastern Brazil. J Ethnobiol Ethnomed 2:6
15. Anderson AB, Posey DA (1985) Manejo de cerrado pelos índios Kayapó. Bol Mus Par Emílio Goeldi Ser Bot 2(1):77–98
16. Case RJ, Pauli GF, Soejarto DD (2005) Factors in maintaining indigenous knowledge among ethnic communities of Manus Island. Econ Bot 59(4):356–365
17. Case RJ, Franzblau SG, Wang Y et al (2006) Ethnopharmacological evaluation of the informant consensus model on anti-tuberculosis claims among the Manus. J Ethnopharmacol 106(1):82–89
18. Nguyen LT (2003) Comparison of food plant knowledge between urban Vietnamese living in Vietnam and in Hawai'i. Econ Bot 57(4): 472–480
19. Gavin MC, Anderson GJ (2006) Socioeconomic predictors of forest use values in the Peruvian Amazon: a potential tool for biodiversity conservation. Ecol Econ 60:752–762
20. Thomas E, Vandebroek I, Van Damme P (2007) What works in the field? A comparison of different interviewing methods in ethnobotany with special reference to the use of photographs. Econ Bot 61(4):376–384
21. Gaur RD, Bhatt BP (1994) Folk utilization of some pteridophytes of Deoprayag area in Garhwal Himalaya: India. Econ Bot 48(2):146–151
22. Milliken W, Albert B (1996) The use of medicinal plants by the Yanomami Indians of Brazil. Econ Bot 50(1):10–25
23. Asase A, Oteng-Yeboah AA, Odamtten GT et al (2005) Ethnobotanical study of some Ghanaian anti-malarial plants. J Ethnopharmacol 99(2):273–279
24. Gupta MP, Solís PN, Calderón AI et al (2005) Medical ethnobotany of the Teribes of Bocas del Toro, Panama. J Ethnopharmacol 96(3):389–401
25. Phillips O, Gentry AH (1993) The useful plants in Tambopata. Peru: I. Statistical

hypothesis tests with a new quantitative technique. Econ Bot 47(1):15–32
26. Mutchnick PA, McCarthy BC (1997) An ethnobotanical analysis of the tree species common to the subtropical moist forests of the Peten, Guatemala. Econ Bot 51(2): 158–183
27. Albuquerque UP, Andrade LHC, Silva ACO (2005) Use of plants resources in a seasonal dry forest (northeastern Brazil). Acta Bot Bras 19(1):27–38
28. Villagrán C, Romo M, Castro V (2003) Etnobotánica del sur de los Andes de la Primera región de Chile: un enlace entre las culturas altiplánicas y las de quebradas altas del Loa superior. Chungará 35(1):73–124
29. Mourão JS, Nordi N (2002) Comparações entre taxonomias folk e científica para peixes do estuário do rio Mamanguape, Paraíba—Brasil. Inteciência 27(12):1–6
30. Alarcon DT, Costa RCSD, Schiavetti A (2009) Abordagem etnoecológica da pesca e captura de espécies não-alvo em Itacaré, Bahia (Brasil). Bol Inst Pesca São Paulo 35(4): 675–686
31. Mourão JS, Araujo H, Almeida F (2006) Ethnotaxonomy of mastofauna as practised by hunters of the municipality of Paulista, state of Paraiba, Brazil. J Ethnobiol Ethnomed 2:19
32. Marques JGW (1995) Pescando pescadores: etnoecologia abrangente no baixo São Francisco. NUPAUB/USP, São Paulo
33. Souto FJB (2008) A ciência que veio da lama: etnoecologia em área de manguezal. NUPEEA/Sociedade Brasileira de Etnobiologia e Etnoecologia, Recife
34. Maciel DC, Alves AGC (2009) Conhecimentos e práticas locais relacionados ao aratu Goniopsis cruentata (Latreille, 1803) em Barra de Sirinhaém, litoral sul de Pernambuco, Brasil. Biota Neotrop 9:29–36
35. Silva G (1988) Tudo que Tem na Terra Tem no Mar. Dissertação de Mestrado, PPGAS/MN, Universidade Federal do Rio de Janeiro, Rio de Janeiro
36. Szpilman M (2000) Peixes marinhos do Brasil: guia prático de identificação. M. Szpilman, Rio de Janeiro
37. Posey DA (1981) O conhecimento entomológico Kayapó: etnometodologia e sistema cultural. Anu Antropol 81:109–124
38. Araújo HF, Lucena RFP, Mourão JS (2005) Prenúncio de chuvas pelas aves na percepção de moradores de comunidades rurais no município de Soledade-PB, Brasil. Interciencia 30:764–769
39. Nesheim I, Dhillion SS, Stolen KA (2006) What happens to traditional knowledge and use of natural resources when people migrate? Hum Ecol 34(1):99–131
40. Campos MT, Ehringhaus C (2003) Plant virtues are in the eyes of the beholders: a comparison of known palm uses among indigenous and folk communities of southwestern Amazonia. Econ Bot 57(3):324–344

Chapter 7

Methods in Research of Environmental Perception

Taline Cristina da Silva, Margarita Paloma Cruz, Thiago Antônio de Sousa Araújo, Maria Luiza Schwarz, and Ulysses Paulino Albuquerque

Abstract

In this chapter you will find a brief discussion about the complexity of understanding how people perceive and outsource their vision of the environment. The methods presented here show the diversity of resource that can be used for the construction of investigative tools that can access more wealth and people trust the information and/or social groups.

Key words Ethnobiology, Environmental representation, Data collection

1 Introduction

According to the Portuguese language dictionary Aurélio, perception means the understanding of reality by people, i.e., the perception of colors, sounds, aromas, and flavors, which manifest through chemical and neurological phenomena at the level of the sensory organ and central nervous system and by several psychological mechanisms that tend to adapt this reaction to the perceived objects. However, a number of authors believe that perception is not simply recorded by the senses but is influenced by psychological [1] and cultural factors [2].

Because of these conceptual differences, Gumuchian [3] suggests using the term "representation" to replace the term "perception" because the concept of representation considers the psychological and cultural factors mentioned previously. Cavalcante and Maciel [4] proposed that representation is the externalization of what an individual perceives through physiological pathways, and it is influenced by psychological and cultural factors.

Based on the line of thought advanced by Cavalcante and Maciel [4], in this chapter, the term environmental representation is used in the same sense as environmental perception because

Ulysses Paulino Albuquerque et al. (eds.), *Methods and Techniques in Ethnobiology and Ethnoecology*,
Springer Protocols Handbooks, DOI 10.1007/978-1-4614-8636-7_7, © Springer Science+Business Media New York 2014

representation is how a researcher can access worldviews, sensations, values, and opinions. We agree with Gumuchian [3] that representation is not only what the individual externalizes but includes all the individual meanings built on physiological, cultural, and psychological aspects.

Studies of environmental representations have been used in ethnobiological studies to access the opinions, feelings, attitudes, preferences, and values of people regarding the natural resources and/or environment around them (see Albuquerque and Albuquerque [5] and Silva et al. [6]).

From the access to environmental representations, research objectives can be: to test for changes in the landscape and their possible causes, to devise environmental conservation strategies, to perform environmental diagnosis, to promote the sustainable use of environmental resources, to develop environmental education projects that consider the views of different social actors on the environment, and to consider differences in age and gender.

Nevertheless, in studies of environmental representation, different methodologies must be used because of the great complexity of the studies and adding various types of data facilitates decision-making during data interpretation.

2 Methods for the Research of Environmental Representation

2.1 Prompting Spoken or Written Discourse

When the aim is to study the representation of what a given person thinks or feels in relation to the natural environment or to specific elements of that environment, the use of open questions to prompt spoken or written discourse are key. This type of approach allows the individual to externalize your representations with more freedom compared to structured or semi-structured interviews, which are typically used in ethnobiological studies (see Wezel and Haigis [7], Tatlidil et al. [8], and Soliva and Hunziker [9]). Lefevre [10] considers that restricting an individual to select options among predetermined answers may limit thinking. Notably, the methodology of prompting spoken or written discourse has advantages and disadvantages. An advantage of this method is that the person can formulate discourse without limitations, which is extremely important when studying environmental representations because the person can express greater autonomy to externalize your ideas and representations. In our experience, the main disadvantages of this method include the following: (1) depending on the literacy level of the informant, prompting a written discourse may not be feasible; (2) the use of the prompting method is not recommended if the study is short-term and if there are a large number of participants because more time must be spent with each person compared to interviews with closed-ended questions; and (3) this method produces a large amount of data, which requires more time for analysis by the researcher.

Therefore, a number of ethnobiological studies have used this type of approach, such as the study by Souza et al. [11] testing students' knowledge of the *Achatina fulica* snail using the following prompt: "write what you know about the East African land snail."

Barraza and Ceja-Adame [12] performed an environmental representation study using children and teenagers in Mexico in which the participants had to write essays based on the following prompt: "When I hear about nature I think of...." Later, Silva et al. [6] used a similar method with students at a school near sections of the Atlantic Forest in northeastern Brazil. The students were prompted to write an essay based on the following phrase: "When I hear about the forest I think of...."

Qualitative and quantitative analyses can be conducted from the results of evocative sentences, such as those exemplified above. For example, Lefevre [10] developed a qualitative analysis that reveals what collectives think, termed *Discourse of the Collective Subject Analysis.* This analysis extracts key terms from each answer followed by the sorting of similar central ideas to construct a collective discourse (Box 1).

Another type of qualitative analysis is the content analysis proposed by Bardin [13], which consists of a set of techniques for analyzing communications based on the rigor of the method as a method for detecting the heterogeneity of the object. In content analysis, the researcher should (1) perform a fluctuating reading of the text, (2) extract classification criteria, and (3) break down the observed frequencies (Box 2).

Box 1 Steps to Construct the Discourse of the Collective Subject, Adapted from Silva et al. [8]

Step 1:
Prompt from a phrase, for example: "When I hear about the forest I think of..."

Step 2:
Extract the key terms of the discourse:
"In the woods there are a lot of animals, leaves, birds," *respondent* I
"In the woods I see snakes, butterflies and mosquitoes," *respondent* II
"In the wood there are plants, frogs, armadillos and marmosets," *respondent* III
"It is full of trees," *respondent IV*

Step 3:
Identify the central idea of each key term; in the above case, the key term was the *biodiversity of the forest.*

Step 4:
Construct a discourse connecting similar central ideas

There are many animals, leaves, birds. I see snakes, butterflies and mosquitoes. Plants, frogs, armadillo and marmoset. It is full of trees.

Note
The discourse of the collective subject is usually constructed from several central ideas. The situation above illustrates only a fragment of the full discourse.

Box 2 Steps for Content Analysis, Adapted from Silva et al. [15]

Step 1:
Prompt, "What do you think the woods would tell you if it could speak?"

Step 2:
Answers to the question:
"Please don't cut me down," *respondent I*
"Please don't deforest me," *respondent II*
"Please don't destroy the trees because they are life," *respondent* III

Step 3:
Molar categories (an interview may have more than one molar category)

Interdiction regarding the forest	Frequency (respondents) (%)
Don't cut me down	
Don't deforest me	

Interdiction regarding the trees
Don't destroy the trees because they are life

Note
Percentiles are calculated relative to the total number of answers.

Quantitative analysis can be performed through the quantification of elements of speech or writing in accordance with criteria predetermined by the researcher, such as in the essay regarding the Atlantic Forest, in which the words extracted from the text regarding such elements as the wildlife, plants, and feelings, can be quantified. For example, in the answer "I love the woods," it is possible to detect the feeling concerning the forest through the word "love." However, even if the data can be quantified, the representation studies are qualitative [3, 14, 15].

2.2 Semi-structured Projective Interviews

As mentioned, the use of visual prompts to obtain information from the respondents is performed differently, depending on the research area and goals. Photographic resources can be used in environmental representation studies. Minayo [16] called this type of method a projective semi-structured interview. It is believed that some people have difficulty verbally expressing their opinions. Therefore, the use of images could help the process of externalizing those opinions [17]. Projective semi-structured interviews involve the use of photographs during the interview towards several objectives. For example, using photographs during the interviews, a number of studies accessed the representations about preferences for different types of environment [5, 6, 18]. Gómez-Limón and Fernández [18] performed a study in the central area of Madrid (Spain) using a set of six photographs of different stages of ecological succession caused by the agricultural regime in the region such that different types of users (farmers, tourists, and local communities) would indicate their environmental preferences.

Projective semi-structured interviews can also be used to verify, for example, representations on the change in the abundance of plant resources over time in a given region. Benjamin et al. [19] used photographs of landscapes at different levels of degradation such that during the interviews, the respondents could indicate which of the landscapes were most common in one region of the state of Quebec, Canada.

There are no specific ways to analyze projective semi-structured interviews. To test for a preference for an environment or an element, the salience of the image can be calculated using the *Anthropac* software [20], which considers the order and frequency of choosing the image. Alternatively, if the objective is to test for the preference or incidence of a particular landscape, as performed by Schwarz et al. [21] with the Atlantic Forest landscapes in Joinville (Santa Catarina State, Brazil), the percentage of the preferred or incident image can be calculated using the number of times the image was referenced. A number of precautions are required when using this methodology, such as (1) always use sharp images of the same quality; (2) pay attention to the angle, color, and image framing, i.e., ensure that the image is focused on the subject of study; (3) use replicates, especially if the goal is to test for a preference for any particular environment; this produces more confidence in the analysis of the results because it allows one to check if the grouping is random or the replicates are ranked together; (4) avoid using images of the areas surrounding the study area because the respondents may recognize these areas and bias the results; and (5) it is recommended to label the images with codes (e.g., #, ¥, and *) instead of numbers or letters because the respondent may order the images alphabetically or numerically. Image labels can be useful when plotting data on a spreadsheet.

2.3 Method of Environmental Autobiography

Environmental autobiography is another approach that uses photographs and is applicable in environmental psychology, which also investigates environmental representations. In this methodology, a camera is given to each participant and the individuals are requested to take a certain number of photographs to answer a specific question. Subsequently, the representations of each respondent are verified according to their photographs. Respondents can be prompted to choose the images perceived as most important, to rank the most significant photographs, and to write a paragraph explaining the meaning of each picture [16].

2.4 Prompting for Drawings

A number of ethnobiological studies encourage respondents to express their representations of the environment through drawings [3, 6, 14]. Goldberg et al. [22] reports that through drawing, the individual organizes information, processes his or her

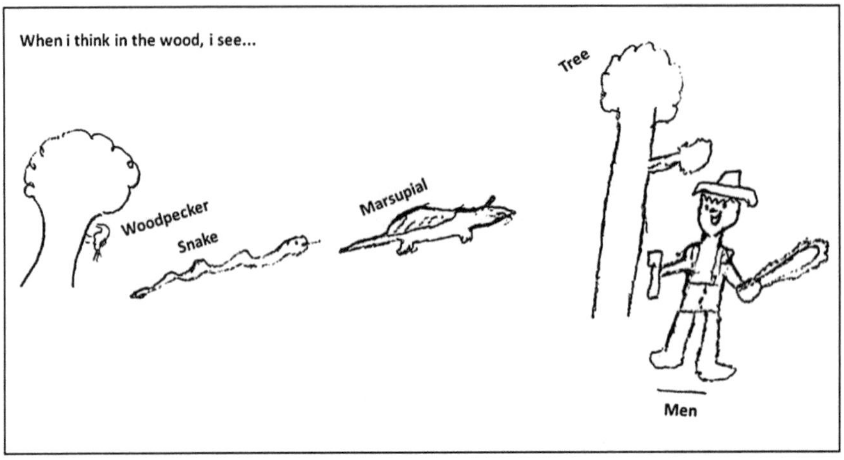

Fig. 1 Biotic elements (woodpecker, tree, snake, opossum, and man) illustrated by a sixth-grade student at the Três Ladeiras Elementary School, Igarassu, State of Pernambuco, Brazil)

lived experiences and thoughts, reveals his or her learning, and may develop a unique style of representation of the world. For example, the drawing method can be applied to investigate representations of the biodiversity of a particular ecosystem. Farias and Alves [14] investigated the representations of a group of students concerning the diversity of birds in a specific region of the Atlantic Forest in northeastern Brazil. In a similar environment, Schwarz et al. [3] and Silva et al. [6] investigated the diversity of the flora and fauna, as represented by children and adolescents [3, 6]. The drawing method can be prompted initially through a sentence or a question (Fig. 1).

To analyze the data resulting from this methodology, each element of the drawing should be identified for the later quantification of the percentage of elements that will be sorted into categories predetermined by the researcher. For example, the elements of the drawings could be sorted into biotic and abiotic components. However, this technique is time-consuming, and the lack of drawing ability of the respondents may limit the data collection.

A number of studies also apply drawing when working with adults. For example, the construction of mental or cognitive maps, which are spatial images that people have of places they know directly or indirectly [23], may contribute to understanding how a place is understood and experienced by a particular social group. For example, researchers attempt to access environmental representations using drawings because it allows them to locate areas of management and uses of natural resources [24]. In addition, drawings encourage the respondent to represent relevant aspects of the perception of the environments related to their preference for the space and the activity performed in the space [25].

2.5 Interviews in Studies of Representation

The interview is a common method used in several areas of ethnobiological research. In these studies, different types of interviews are used that vary from structured to informal according to the degree of control over the data that is required. Semi-structured and structured interviews are the most commonly used interview strategies, as is observed for the study of representations. This method is probably used because of the several advantages it offers, including the ease with which standardizing the responses and analyzing the data can be achieved compared to spoken or written discourse.

For example, Burger et al. [26] interviewed residents of a city near a nuclear plant. In the study, the researchers asked the interviewees their opinions and attitudes regarding the possible threats presented by the presence of the nuclear plant and what they would like to perform in the future.

Regarding the analysis of data from structured or semi-structured questionnaires, in most cases, it is necessary to categorize the responses. This categorization can be performed in two ways, depending on the type of questionnaire. In the structured interview, the categories are established before beginning the interviews; for the semi-structured interview, the categories are created after concluding the interviews because the researcher does not have a lot of control over the data.

After the categorization, the frequency at which each category was referenced is the most simple and common calculation that can be performed. Next, statistical analysis can be performed to test for significant differences between certain groups of people. Comparisons between subgroups can be performed if the data are socioeconomic, such as with gender, age, income, and occupation data.

Although less open interviews offer greater advantages in terms of the ease of quantifying the results, they do not allow the researcher to collect qualitative data, which provide a lot of information but are more difficult to analyze because the data must be categorized at a later stage.

Another option when preparing the interviews is to use the Likert[1] scale, which consists of questions that generate responses, half of which are positive and half are negative. This scale consists of closed responses in a series of scaled statements, such as very good, good, regular, bad, and very bad [27]. This rating scale provides the respondent more possibilities to respond.

2.6 Analysis of Risk Perception

Analysis of risk perception can be a useful tool when searching for solutions to local problems [28] and therefore assists in decision-making, such as the management of natural resources.

[1] Johnson WA, Dixon PN (1984) Response alternatives in likerts scaling. Educ Psychol Meas 44:597.

The word risk appears in several languages. Sometimes it is difficult to translate the word risk without losing the meaning because depending on the location and the social context, variations in the connotation of the word may exist, giving rise to the idea of something undesirable, worrying, or harmful [29, 30]. Therefore, the researcher should be careful to define correctly what is identifying as a risk.

For Sjoberg [29], risk perception is not only a matter of sensorial perceptions but also of attitudes and expectations that can be studied using reasonably well developed methods for measuring attitudes and psychological scales. In the literature, several studies discuss how to access risk perception using primarily questionnaires and scales. We will focus on the method developed by Smith et al. [30], termed the participatory risk mapping (PRM), because it is a practical and fast method that generates easily comparable results.

PRM allows the researcher to discover, classify, and rank risk factors encountered and perceived by populations. The first step for applying PRM is to identify the risk factors from the perspective of each interviewee. Next, the respondent is asked to rank the risk factors that cited in increasing order of severity. Note that the same person can name more than one risk factor. Questions can be posed that are aimed at obtaining more details related to the risk factor, such as the following: in what context can the risk occur, what are the consequences and/or how can the risk be avoided, and what is the behavior of the people who are experiencing or have experienced this risk factor? In some cases, the risk factors need to be categorized, and these questions, which are designed to understand the context more clearly, can facilitate categorization.

Next, two types of calculations can be performed. The first calculation is the ***risk incidence*** (I), which is the proportion of the respondents that identified a given risk source. $I_j = n_r / n_j$, where n_r is the number of times that the risk was cited, and n_j is the total number of respondents. The I values can range from 0 to 1; numbers close to 0 indicate a lower frequency of citations, and 1 indicates that the risk factor was cited by all the respondents (Box 3).

The second calculation is the *severity index* (S), which is based on the number and rank of the risk factors cited by each respondent. The S values range from 1 to 2 and values close to 1 are considered more severe. $S_j = 1 + ((r-1)/(n-1))$, where r is the ranking based on the order indicated by the respondent, and n is the number of risk factors mentioned by the respondent. The average of the S_j values is calculated for the subgroup of people that identified the problem (Box 3).

Box 3 Example of Calculation of the Severity Index (S) and Risk Incidence (I)

Respondents	Ranking (r)	Factors	Factors numbers (n)	S_j
I	1	Unemployment	3	1
	2	Education		1.5
	3	Transport		2
II	1	Transport	2	1
	2	Education		2
III	1	Unemployment	4	1
	2	Education		1.33
	3	Hunger		1.66
	4	Transport		2

Factors	*Severity index (S)*	*Risk incidence (I)*
Unemployment	1	0.66
Education	1.61	1
Transport	1.66	1
Hunger	1.66	0.33

Using both indices, we arrive at a third index based on the combination of the risk severity and incidence, the *total risk* (*RJ*). $R_j = I_j/S_j$. The RJ increases when higher values for I_j, which represent higher incidence, and lower values of SJ, which represent higher severity, are attributed to a given type of risk [31].

Similar to the graphs used in geographic information systems (GIS), the results can be represented graphically to construct a risk map. Using this model, we can determine the space and heterogeneity of the risk incidence and severity.

Note that in Fig. 2, the "unemployment" risk had the highest incidence in addition to a high severity index.

The "transport" and "addiction" risks had a high degree of severity; however, few respondents considered these factors to represent a risk. These risk factors were evoked by the following question: "In your opinion, which (what) is (are) the factor(s) that can adversely affect and impair quality of life for you and your family?" This question was posed to residents between the ages of 25 and 34 years living near sections of the Atlantic Forest of northeastern Brazil (unpublished data).

The risk map model can be employed in ethnobiological studies to assess the risk factors that may affect people in such situations

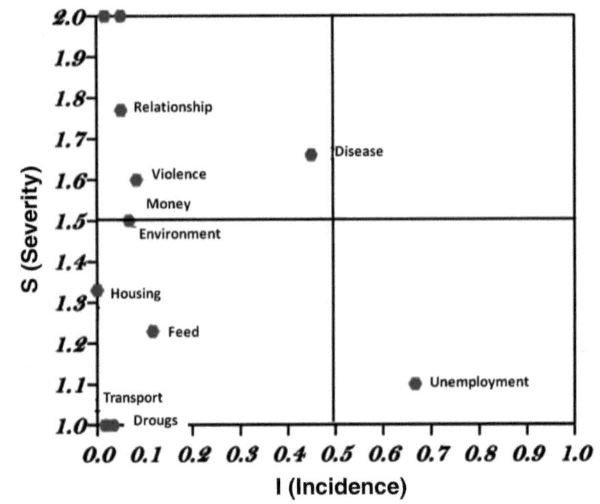

Fig. 2 Risk map model

as deforestation or the construction of a hydroelectric dam in the area. The model can determine which groups or social strata perceive the environmental risks as most severe; therefore, the model identifies the most susceptible groups to participate, for example, in particular environmental conservation activities.

References

1. Bell S (2001) Landscape pattern, perception and visualization in the visual management of forest. Landsc Urban Plan 54:201–211
2. Okamoto J (2002) Percepção ambiental e comportamental. Mackenzie, São Paulo, Brazil
3. Schwarz ML, Sevegnani L, André P (2007) Representações da Mata Atlântica e de sua biodiversidade através dos desenhos infantis. Rev Bras Biocienc 5(1):744–746
4. Cavalcante S, Maciel RH (2008) Métodos de avaliação da percepção ambiental. In: Pinheiro JQ, Günther H (eds) Métodos de pesquisa nos estudos pessoa-ambiente, 1st edn. Casa do Psicólogo, São Paulo, Brazil
5. Albuquerque CA, Albuquerque UP (2005) Local perceptions towards biological conservation in the community of Vila Velha, Pernambuco, Brazil. Interciencia 30:460–465
6. Silva TC, Medeiros PM, Araújo TAS et al (2010) Northeastern Brazilian students' representations of Atlantic Forest fragments. Environ Dev Sustain 2010(12):195–211
7. Wezel A, Haigis J (2000) Farmers' perception of vegetation changes in semi-arid Niger. Land Degrad Dev 1:523–534
8. Tatlidil FF, Boz I, Tatlıdil H (2008) Farmers' perception of sustainable agriculture and its determinants: a case study in Kahramanmaras province of Turkey. Environ Dev Sustain 11:1091–1106
9. Soliva R, Hunziker M (2009) Beyond the visual dimension: using ideal type narratives to analyse people's assessments of landscape scenarios. Land Use Policy 26:284–294
10. Lefevre F (2005) Depoimentos e discursos: uma proposta de análise em pesquisa social. Liber Livro, Brasília
11. Souza RM, Alves AGC, Alves MS (2007) Conhecimento sobre o molusco gigante africano *Achatina fulica* entre estudantes de uma escola pública na Região Metropolitana do Recife. Biotemas 20:81–89
12. Barraza L, Ceja-Adame MP (2005) Los niños de la comunidad: su conocimiento ambiental y su percepción sobre "Naturaleza". México. http://www.ine.gob.mx/ueajei/publicaciones/libros/420/dieciseis.html. Acesso em 17 June 2007
13. Bardin L (1977) Análise de conteúdo. Edição revista atualizada, Brazil
14. Farias GB, Alves AGC (2007) Conhecimento prévio sobre a avifauna por alunos do Ensino Fundamental numa escola pública na Região Metropolitana do Recife: em busca de uma

prática pedagógica culturalmente apropriada. In: Albuquerque UP, Alves AGC, Araújo TAS (eds), 1 edn. Recife, In: (...) Povos e Paisagens Brazil
15. Silva RRV, Marangon LC, Alves AGC (2011) Entre a Etnoecologia e a Silvicultura: O Papel de Informantes Locais e Cientistas na Pesquisa Florestal. Interciencia 36:485–492
16. Minayo MCS (1993) O desafio do conhecimento científico: pesquisa qualitativa em saúde. Hucitec/Abrasco, São Paulo/Rio de Janeiro, Brazil
17. Neiva-Silva L, Koller SH (2002) O uso da fotografia na pesquisa em Psicologia. Estud Psicol 7:237–250
18. Gómez-Limón J, Fernández JVL (1999) Changes in use and landscape preferences on the agricultural-livestock landscapes of the central Iberian Penisula (Madri, Spain). Landsc Urban Plan 44:165–175
19. Benjamin K, Bouchard A, Domon G (2007) Abandoned farmlands as components of rural landscapes: an analysis of perceptions and representations. Landsc Urban Plan 83:228–244
20. Borgatti SP, Natick MA (1996) Analytic technologies. Antrhopac 4.0
21. Schwarz ML, André P, Sevegnani L (2008) Preferências e valores para com as paisagens da Mata Atlântica: Uma comparação segundo a idade e o gênero. Caminhos Geogr 9:114–132
22. Goldberg LG, Yunes MAM, Freitas JV (2005) O desenho infantil na ótica da ecologia do desenvolvimento humano. Psicol Estud 10:97–106
23. Archela RS, Gratão LHB, Trostdorf MAS (2004) O lugar dos mapas mentais na representação do lugar. Geografia 13:127–141
24. Moura FBP, Marques JGW (2007) Conhecimento de pescadores tradicionais sobre a dinâmica e espaço-temporal de recursos naturais na Chapada Diamantina, Bahia. Bio Neotrop 7:119–126
25. Alves SA, Texeira CF, Kowaltowski DCCK et al (2005) Avaliação do ambiente construído através da percepção ambiental: metodologia aplicada à escola PRODECAD—Unicamp. VIII Encontro Nacional e IV Encontro Latino-americano sobre conforto no ambiente construído, 5–7 out 2005, Maceió, NE do Brazil
26. Burger J, Sanchez J, Gibbons JW et al (1999) Attitudes and perceptions about ecological resources and hazards of people living around the Savannah river site. Environ Monit Assess 57:195–211
27. Hill M, Hill A (2005) A construção de um questionário para medir uma variável latente. In: Hill M, Hill M (eds) Investigação por questionário, 2nd edn. Edições Sílabo, Lisboa
28. Sjoberg L (2002) Policy implications of risk perception research: a case of the Emperor's new clothes? Risk Manage 4:11–20
29. Sjoberg L (2000) The methodology of risk perception research? Qual Quant 34:407–418
30. Smith K, Barrett CB, Box PW (2000) Participatory risk mapping for targeting research and assistance: with an example from East African pastoralists. World Dev 28: 1945–1959
31. Quinn CH, Huby M, Kiwasila H et al (2003) Local perceptions of risk to livelihood in semiarid Tanzania. J Environ Manage 68:111–119

Chapter 8

Methods for Data Collection in Medical Ethnobiology

Washington Soares Ferreira Jr., Nélson Leal Alencar, and Ulysses Paulino Albuquerque

Abstract

In this chapter, we discuss methods for data collection in studies of medical ethnobiology. First, we present a brief discussion about traditional medical systems and approaches to study them, including medical ethnobiology. After, we indicate to the researcher some techniques that can be applied to investigations that seek to assess traditional medical systems.

Key words Traditional medical systems, Biological resources, Ethnomedicine

1 Introduction

Every culture has knowledge and practices for the recognition and diagnosis of illness, modes of treatment and health maintenance [1]. This knowledge constitutes a "medical system" that is passed through generations and represents the product of years of observation and experimentation in the use of different resources, such as plants and animals [2]. These systems include medical knowledge and practices related to illnesses (biomedical interpretation of health problems) or illnesses (local interpretation of the health problem by people from a particular group), such as their etiology, the means by which health problems are treated and prevented, and assessments of treatments [1, 3, 4, 6].

Medical anthropology, the science that investigates the nature of illnesses, diagnostic procedures, and healing in the medical systems of different societies, presents a range of approaches to study these systems [2, 5, 6]. The ethnomedical approach, for example, investigates how people recognize and classify ilnesses and organizes procedures for their treatment by considering how people understand the *cause, course, demonstration, and appropriate treatment for each problem* [4, 7, 8]. Although ethnomedicine has a wide scope of investigation, it differs from other approaches of medical

Ulysses Paulino Albuquerque et al. (eds.), *Methods and Techniques in Ethnobiology and Ethnoecology*,
Springer Protocols Handbooks, DOI 10.1007/978-1-4614-8636-7_8, © Springer Science+Business Media New York 2014

anthropology mainly in its study of how people treat illnesses [5]. Considering the range of areas to which the term ethnomedicine has been applied, Nichter [7] presents various approaches in the literature, including the following:

1. Studies involving the identification of illness and their concepts, classifications, and treatment systems in human populations.
2. Comparative studies of illness and their means of treatment in different cultures to determine standard aspects.
3. Studies of the effectiveness of techniques used in treatments and procedures to meet specific cultural needs.
4. Documentation of the relationship between beliefs, social norms, and illness as well as investigation of the role of illness in the maintenance of social order.
5. Studies of positive and negative values assigned by human groups to bodily states, which aid in the understanding of representations of people's bodies and bodily expressions related to states of health and illness.

Research within the field of medical ethnobiology provided major contributions to the study of traditional medical systems through multidisciplinary studies of the knowledge and practices involved in these systems [9–15]. However, medical ethnobiology differs from other research fields, such as ethnomedicine, in its focus on the utilization of natural resources (especially biological resources) by humans for medicinal use [4, 5]. Medical ethnobiology uses ethnomedical tools to investigate the role of biological resources in medical systems. Given the importance of this field, in this chapter we present some of the main methods for data collection in medical ethnobiology.

This chapter is arranged as follows. First, we present techniques used to collect data on cultural understandings of anatomy and physiology of the human body. Then, we present information needed for the construction of explanatory models and ethnomedical formularies [4, 16]. We describe tools to investigate the relationships between the illnesses observed in a culture based on information about their symptoms and treatments as well as methods to understand the choices of treatments, such as plants or animals, by different human groups. Finally, we present a method to assess periodically the illness events and treatments employed by a group of people.

2 Collecting Anatomical and Physiological Data

Disturbances in health or well-being lead to changes in the body that can influence the diagnosis of illnesses. Hence, there is a need to investigate cultural knowledge of the body parts that undergo changes in various illnesses [4, 17]. Researchers must include anatomical and physiological data in these investigations related to local understandings of the anatomy and physiology of the human body.

One of the simplest ways to collect anatomical data specifies the structures of interest by asking an informant "What is the name of this?" A schematic drawing of the human body can be presented, e.g., [18], and photographs can be displayed in which anatomical structures are presented and respondents are asked to name them [4, 19]. Another way to collect this information is to ask the informant to draw human body parts and to indicate the name of each structure represented. For example, Herndon et al. [19] investigated the anatomical nomenclature of the human body among healers in two Trio villages in Suriname in the Amazon. Pictures of the structures of the human body were presented to the respondents during interviews, and they indicated the names of the structures they recognized.

3 Methods for Collecting Data

Data collection on various illness conditions is strongly influenced by the human sciences, especially anthropology. Information is collected primarily through interviews with informants involved with the phenomenon under study. Informants may include patients involved in particular treatments or healers (local experts, healers, shamans, etc.). Parallel with the collection of anatomical and physiological data, information about the different types of illnesses recognized in the community can be collected [4]. For example, a researcher could ask an informant to list the names of all recognized health problems using a free list technique. During this process, names of illnesses may emerge that include a number of other conditions, such as subcategories of illnesses. For example, the illness category *X* cited by an informant may include the subcategories *a*, *b*, and *c*. In these cases, it is suggested that the researcher asks, "What are all types of X?" [4] to identify these subtypes. This question should be asked when a researcher realizes that variants of the same illness are mentioned in an interview with an informant.

One way to collect data and record the meaning of each health problem involves the use of questionnaires to deduce ethnomedical explanatory models that can be constructed from a set of questions about illnesses in the study population [4]. These models provide a detailed breakdown of how people understand health problems [7]. The following illustrates one type of information necessary for developing the model, based on Fabrega [17] and Berlin and Berlin [4] (Box 1).

Once information about the recognized illnesses is collected, it can be enhanced with information about the biological species used to treat each condition collected with an ethnomedical formulary. Berlin and Berlin [4] show the construction of the formulary in two stages. In the first stage, the researcher can obtain information on all of the treatments known by the informant (Box 2):

Box 1 Important information to develop ethnomedical explanatory models, based on Fabrega [16] and Berlin and Berlin [4]

"*Investigating behavioral restrictions of an individual from the illness:*

1. Possible limitations of behavior (e.g., confinement in bed because of illness).
2. Possible restrictions of roles (e.g., inability to perform the role of father or housewife).

Investigating the causes of the illness:

3. One can ask the informant, *Why did you get sick from ___?* or *Why did you get ___?*
4. *Could ___ develop as a complication of another illness?*"

"*Investigating the symptoms and the treatments used:*

5. Indicators and sensations that only the patient perceives: *How do you know you have ____?*
6. Possible treatments for the illness.
7. Treatment strategy: *What types of treatments are attempted first?*

What follows?

Who can treat or cure?

8. *Are there any special diets or restrictions during and/or after treatment?*"

Box 2 First stage of collecting information needed for building ethnomedical formularies. Adapted from Berlin and Berlin [4]

"1. The name of each species used in the treatment.
2. The order of preference in which the informant would use these species when all species are equally available."

The second stage involves each species mentioned in the previous item (Box 3):

The information collected from the explanatory models and ethnomedical formularies can be used to construct a therapeutic formulary. This is a protocol that informants use for the treatment of illnesses, including information about the duration of treatment, the materials or parts of plants or animals used, and the method of preparation and the amounts of these materials administered. Additionally, the formulary reports the time interval required between each administration. For example, a therapeutic formulary for the treatment of a particular illness may be composed of a daily dose (cup, handful, droplets, or even local measures) of a tea prepared from the leaves of plant *x* for 7 days.

3.1 Checklist-Interview Technique

Another way to obtain information about the illnesses known by informants is the checklist-interview technique. In this technique, the researcher can present visual stimuli, such as photographs of

Box 3 Second stage of collecting information needed for building ethnomedical formularies. Adapted from Berlin and Berlin [4]

"1. Local name
2. Scientific name
3. Strength of treatment (expressed on a five-point scale), organoleptic properties (color, odor—five-point scale), and taste (three-point scale)
4. Any distinguishing features (produces foam, caustic soda)
5. Properties (latex presence, resin)
6. Parts used
7. Amount added for each part of the treatment
8. If the ingredient is mandatory or optional in the formula
9. Ecological habitat
10. Scope of growth/occurrence (in hot, temperate, cold regions)
11. Time of availability (seasonality)
12. Preferred time of collection (diurnal, rainy or dry seasons)
13. Special conditions of collection (e.g., rituals)
14. Preliminary preparation of each substance (washing, heating, grinding)
15. Containers for preparation (if special, e.g., clay pots, leaves)
16. Order in which the ingredients are added (e.g., before/after boiled water), shape (chopped, sprinkled, whole), preparation (cold water, boiled, distilled, plunged in cold water), preparation time (minutes baked, time of day)
17. To whom the species is given (age, gender, life stage)
18. Mode of administration (drink, scrub, wash, stew, etc.)
19. Dosage (cup, handful, drops, native measures if relevant)
20. Special diet required (prescriptions/proscriptions)
21. Special behavior required (bed rest, avoid sun, steambath)."

illnesses for which the researcher seeks information, while the interview is developed and contextualized.

The checklist-interview fulfills two main functions: first, it allows the researcher and the informant to talk about the same or illness, which prevents errors in data collection when the researcher mistakenly interprets comments about a illness condition; and second, it can enhance information about each illness presented in the photographs and how to investigate concepts of local illnesses and treatments [20]. This information can lead to the construction of an ethnomedical explanatory model, for example.

3.2 Pile-Sorting Technique

According to Berlin and Berlin [4], after listing the illnesses recognized by the informants, it is important to investigate the relationships between illnesses. One simple technique to understand how informants organize and classify illnesses is the pile-sorting technique [21, 22]. This technique involves writing illnesses on cards, presenting them to informants, and asking the informants to organize the cards into piles of health problems that are similar to one another. After this organization, the researcher can ask the informants to justify the formation of the piles [21]. For informants who cannot read, the researcher can state three conditions (for example, illnesses *a*, *b*, and *c*) and ask the informant which conditions are similar and different [4]. However, for illiterate informants, a large number of conditions can make this process difficult. Therefore, it may be more effective to use pictures or drawings to indicate different conditions.

The data obtained from the pile-sorting technique can be organized into a similarity matrix in a manner similar to a cross-tabulation, with illnesses occupying the rows and columns of the matrix to indicate the conditions that appear together most often [23]. A multivariate analysis can be used, such as non-metric multidimensional scaling (MDS), to facilitate observation of the illnesses that present in clusters [24].

3.3 Successive Free Listing

Standard free listings, in which the informant list items for the domain in question, are generally used to identify items within a specific cultural domain (i.e., the researcher can ask the informant to "list all types of ____") [25]. Depending on the research objectives, the researcher can first ask the informant to list illnesses and to develop a free list of known conditions. Then, the informant can be asked to list the symptoms and treatments for the illnesses indicated, creating additional free lists derived from the first list in a process known as successive or multiple free listing [21].

The data from these multiple free lists can be transferred to a illness (symptoms and treatment) *x* items matrix, in which illnesses occupy the lines and symptoms/treatments occupy the columns. This matrix can contain binary data (0 and 1) according to the type of item (symptom and treatment) for each illness on the corresponding line. The case receives a value of 1 if the item corresponds with the illness and 0 otherwise, as shown in Fig. 1. Once the matrix is constructed, the researcher can compare the different illnesses based on various items (symptoms and/or treatments), and a correspondence analysis can be performed (Fig. 1). This analysis indicates whether the illnesses are closely related based on the characteristics of recognized symptoms and treatment [25]. That is, the more similar the symptoms and treatment of two illnesses, the closer these illnesses will be than other health problems at the end of the analysis. Finally, an analysis of similarity (ANOSIM) can be used to check the consistency of the different groups formed by the correspondence analysis.

A researcher can also conduct separate analyses to understand how illnesses are organized based on symptoms. Similarly, one can

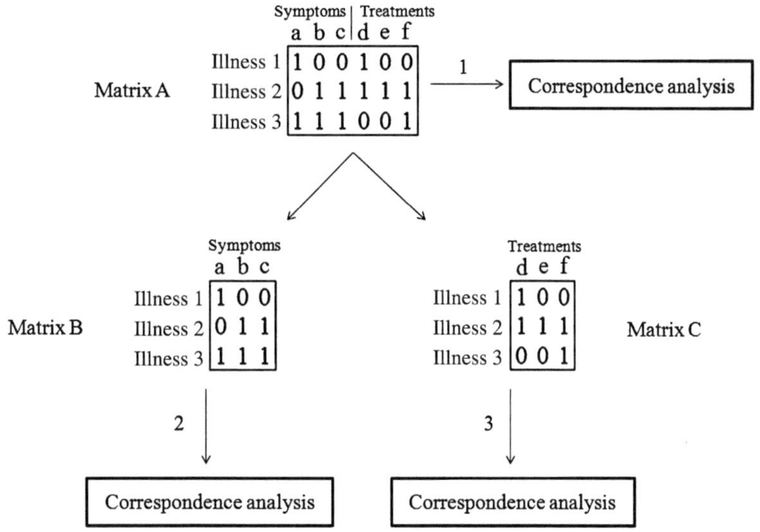

Fig. 1 Example matrices constructed from successive free lists and the correspondence analysis. Matrix A: presence and absence of symptoms presence and absence of symptoms and treatments obtained in free lists for each illness. Matrix B: presence and absence presence and absence of symptoms that describe each condition. Matrix C: presence and absence of treatments presence and absence of treatments (e.g., species) given by informants for each condition. (1) The use of a correspondence analysis to observe the illnesses proximity to its symptoms and treatment characteristics. The researcher can use the correspondence analysis to understand how illnesses are organized based on symptoms (2) and treatments (3), separately. Adapted from Ryan et al. [25]

perform separate analyses of illnesses and treatments to obtain responses about the organization of illnesses (i.e., illnesses nearer to the symptom characteristics or the treatment used) (Fig. 1). As noted, in addition to being simple to administer, successive free listing can be used with both literate and nonliterate informants to provide an analysis that allows for the observation of multiple relationships between the items studied [25].

3.4 Constructing Decision Models Applied to Medical Ethnobiology Studies

The criteria used by people to make decisions can be accessed through decision models [26]. These descriptive models aim to investigate the criteria used by informants to choose one specific treatment (plant or animal) over another to cure illness conditions. In using these models, it is assumed that people have a pattern of response to certain illness situations in daily life. That is, a illness situation (a problem or challenge) can be resolved with a treatment or a combination of treatments that are considered the best solution to the problem or illness. This solution is retained for future use if the same problem occurs. However, the model cannot be regarded as static because changes occur when new solutions are added and others decline with the passage of time [26].

Box 4 The main questions that can guide research using decision models. Adapted from Hill [26]

1. What are the treatments (plants and/or animals) that members of the group consider useful in treating the health problems?
2. What are the criteria that people use to select these treatments?
3. What are the sets of rules to organize criteria (i.e., how are the criteria organized to make a choice)?

The criteria used by individuals to make decisions may be arranged into rules that form a set of criteria for choosing a particular treatment. For example, a rule can be written as follows: "IF a child has blood in the stool AND swollen glands OR is vomiting, THEN ask medical advice" [26]. In this example, three criteria are established to choose an action for medical treatment. Two criteria must be presented together, such as "bloody stools" and "swollen glands," to perform the action "seek medical advice" (Box 4).

First, it must be determined which cultural domain needs to be investigated, such as a illness condition or a specific set of illnesses of interest for the study. Thus, a researcher can obtain information from interviews using the free listing technique, in which the informant is asked to list the treatments used for each health problem. Once the treatments are listed, the researcher can seek to identify the criteria used to choose them. To do so, the treatments can be written, or photos can be shown and the informants choose the treatments used for each illness. After this, the informants can be asked for the reasons used to select each treatment, yielding the choice criteria. Thus, the researcher can construct a model of how decisions are made based on the example below (Box 5).

3.5 Therapeutic Itinerary

The therapeutic itinerary follows principles similar to those described for the decision models. The therapeutic itinerary is a descriptive, qualitative approach in which the researcher seeks to understand the paths taken by a person faced with a health problem that demands attention and the treatment options available locally. Based on the itinerary, the researcher can assess the influence of social, economic, and other factors that may interfere with the formation of illnesses as well as the study subjects' choice of treatment [27, 28]. An effective study requires an in-depth analysis. Therefore, it is practically impossible to follow multiple people in a community.

The researcher reports the accounts of the patients and/or their families to reconstruct the illness episode and the approach used to solve the problem. An interesting aspect of this approach is that the researcher can access people's interpretations of the

Box 5 Example for the construction of decision models. Adapted from Hill [26]

```
Rule 1
IF      criteria (1) (for example, a person has the
        condition X) OR
        criteria (2) (for example, a person has the
        symptom Y) OR
        criteria (3)
THEN    treatment A (for example, plant a)

Rule 2
IF      criteria (4) (for example, a person has the
        condition Z)
THEN    treatment B

Rule 3
IF      criteria (5) (in this case, the treatment B
        did not cure the condition Z)
THEN    treatment C

Rule 4
IF      criteria (6) (in this case, the treatments
        used in rules 2 and 3 did not cure the condi-
        tion Z) AND
        criteria (7) (in this case, the disease has
        lasted more than a week)
THEN    treatment D (for example, take the person to a
        doctor)
```

Adapted from Hill [26].

problem and the actions taken. It should be emphasized that the therapeutic itinerary is developed from in-depth interviews, so the researcher must work with a small set of interviews (10–20 interviews) and must be careful not to extrapolate to an entire community by assuming that these routes are standardized within a culture or are identical for the same types of culturally recognized illnesses (nosological categories). These studies may take different approaches; for more information, see, for example, Kleinman [16].

3.6 Therapeutic Calendar

By studying traditional medical systems, some researchers record information provided by the members of a culture and rely upon the informants' memory. The memory of a phenomenon can have various influences that can certainly influence the reports obtained. When, due to logistical issues, a researcher cannot use participant observation to access directly a phenomenon, a therapeutic calendar can provide an interesting alternative. The therapeutic calendar consists of a monthly calendar on a sheet of paper on which the researcher records the events of the illness and treatment of a family and/or person through periodic visits at intervals defined by the research objectives (Box 6).

Box 6

The following information must be collected to build a therapeutic calendar:

1. Health problem.
2. Symptoms recognized.
3. How was the problem handled?
   3.1. What were the remedies used?
   3.2. How was each remedy prepared?
      3.2.1. What was the dosage of each medication?
4. Did the patient improve?
   4.1. If so, what were they?
5. What was the duration of treatment?

The researcher can adopt three approaches:

1. Periodically visit the family to record a recent episode. Limitation: time availability.
2. Ask the patient or family member to record the episodes. Limitation: need for literate informants.
3. Integrate into the research team a community health agent to record illness events in families. Limitation: membership of these professionals.

4 Final Considerations

There is a variety of methods for studying medical ethnobiology, and some of the main methods were addressed in this chapter. However, the researcher can choose which methods to use or to adapt depending on the objectives of the research. Many of these methods have been widely used in anthropological studies, but few have been explored in ethnobiological studies. These studies must incorporate the methods described above, as well as others, to understand how natural resources are integrated into different forms of treatment within different concepts and perceptions of health and illness in human groups.

References

1. Banerjee BG, Jalota R (1988) Folk illness and ethnomedicine. Northern Book Center, New Delhi
2. Bhasin V (2007) Medical anthropology: a review. Stud Ethno-Med 1(1):1–20
3. Kleinman A (1978) Concepts and a model for the comparison of medical systems as cultural systems. Soc Sci Med 12:85–93
4. Berlin E, Berlin B (2005) Some field methods in medical ethnobiology. Field Methods 17(3):235–268
5. Waldstein A, Adams C (2006) The interface between medical anthropology and medical ethnobiology. J R Anthropol Inst N. S.:S95–S118

6. Singer M, Baer H (2007) Introducing medical anthropology: a discipline in action. Altamira, Exeter, UK
7. Nichter M (1992) Anthropological approaches to the study of ethnomedicine. Gordon and Breach Science, Amsterdam
8. Fabrega H (1975) The need for an ethnomedical science. Science 189:969–975
9. Albuquerque UP (2010) Implications of ethnobotanical studies on bioprospecting strategies of new drugs in semi-arid regions. Open Complement Med J 2:21–23
10. Alencar NL, Medeiros PM, Medeiros MFT (2010) Medicinal plants prescribed in the hospital of the São Bento Monastery between 1823 and 1824 in Olinda—Northeastern Brazil. Open Complement Med J 2:74–79
11. Costa-Neto EM, Motta PC (2010) Animal species traded as ethnomedicinal resources in the Federal District, Central West region of Brazil. Open Complement Med J 2:24–30
12. Dounias E (2010) An ounce of prevention is worth a pound of cure. Fang homegarden organization as means to prevent from health risks. Open Complement Med J 2:31–41
13. Haverroth M, Negreiros PRM, Barros LCP (2010) Ethnobiology and health among the Kulina people from the Upper Envira River, State of Acre, Brazil. Open Complement Med J 2:42–57
14. Trillo C, Toledo BA, Galetto L, Colantonio S (2010) Persistence of the use of medicinal plants in rural communities of the Western Arid Chaco (Cordoba, Argentina). Open Complement Med J 2:80–89
15. Zamudio F, Kujawska M, Hilgert NI (2010) Honey as medicinal and food resource. Comparison between polish and multiethnic settlements of the Atlantic Forest, Misiones, Argentina. Open Complement Med J 2: 58–73
16. Kleinman A (1980) Patients and healers in the context of culture. Regentes, California
17. Fabrega H (1971) Medical anthropology. Bienn Rev Anthropol 7:167–229
18. Netter FH (2007) Atlas de anatomía humana. Elsevier, Barcelona
19. Herndon CN, Uiterloo M, Uremaru A, Plotkin MJ, Emanuels-Smith G, Jitan J (2009) Disease concepts and treatment by tribal healers of an Amazonia forest culture. J Ethnobiol Ethnomed 5:27
20. Balick MJ, Cox PA (1996) Plants, people, and culture. The science of ethnobotany. Scientific American Library, New York
21. Winch PJ, Makemba AM, Kamazima SR, Lurie M, Lwihula GK, Premji Z, Minjas JN, Shiff CJ (1996) Local terminology for febrile illnesses in Bagamoyo District, Tanzania and its impact on the design of a community-based malaria control programme. Soc Sci Med 42(7):1057–1067
22. Roos G (1998) Pile sorting: “kids like candy”. In: Munck VC, Sobo EJ (eds) Using methods in the field: a practical introduction and casebook. Altamira, Walnut Creek, CA
23. Hair JF, Tatham RL, Anderson RE, Black W (1998) Análise multivariada de dados, 5th edn. Artmed, São Paulo
24. Calvet-Mir L, Reyes-García V, Tanner S (2008) Is there a divide between local medicinal knowledge and Western medicine? A case study among native Amazonians in Bolivia. J Ethnobiol Ethnomed 4:18
25. Ryan GW, Nolan JM, Yoder PS (2000) Successive free listing: using multiple free lists to generate explanatory models. Field Methods 12:83
26. Hill CE (1998) Decision modeling: its use in medical anthropology. In: Munck VC, Sobo EJ (eds) Using methods in the field: a practical introduction and casebook. Altamira, Walnut Creek, CA
27. Izugbara CO, Etukudoh IW, Brown AS (2005) Transethnic itineraries for ethnomedical therapies in Nigeria: Igbo women seeking Ibibio cures. Health Place 11:1–14
28. Bove RM, Vala-Haynes E, Valeggia CR (2012) Women’s health in urban Mali: social predictors and health itineraries. Soc Sci Med 75:1392–1399

Chapter 9

Introduction to Botanical Taxonomy

Alessandro Rapini

Abstract

Botanical taxonomy delimits groups of plants and describes and names taxa based on these groups to identify other members of the same taxa. The circumscription of taxa is directed by the principles of classification, and the name assigned is governed by a code of nomenclature. However, changes in the principles of classification and information accumulated from different sources affect taxon circumscriptions and, consequently, the meaning of scientific names. This process is continuous but, by governing the application of names, nomenclature has enabled the construction of a sizable body of plant knowledge. Taxonomic works store botanical information, and scientific names permit the access to and linkage of this information synergistically, thus enhancing the knowledge regarding plants and disseminating it in space and time.

Key words Classification, Description, Identification, Naturalists, Nomenclature, Phylogenetics, Plant systematics, Scientific names

> "Nature produces individuals, and nothing more. She produces them in such countless numbers that we are compelled to sort them into kinds in order that we may be able to carry them in our minds. This sorting is classification—taxonomy." [1]

Ethnobiology investigates the folk knowledge of human interactions with organisms [2]. Traditional societies are intimately dependent upon local environment where they have found material to supply most of their demands—food, shelter, clothing, implements, utensils, medicaments, instruments, etc.—for a long time [3]. Therefore, ethnobiology can provide critical knowledge in assessments of biodiversity and sustainable conservation and is currently considered a priority area in biological sciences because traditional knowledge and biological diversity are both being rapidly lost [4].

Although folk classifications are structurally arranged following general principles [5], a plant may have different local names and a vernacular name may be used as reference to different plants.

Ulysses Paulino Albuquerque et al. (eds.), *Methods and Techniques in Ethnobiology and Ethnoecology*, Springer Protocols Handbooks, DOI 10.1007/978-1-4614-8636-7_9, © Springer Science+Business Media New York 2014

Consequently, common names cannot substitute scientific names, and appropriate procedures for recording plant species should be followed to reach best practice and guarantee high standards of scientific works in ethnobotany: first, a voucher specimen identified by the local informants should be collected and placed in a herbarium to be used as reference to this record; second, the herbarium specimen should be identified by a competent botanist; and third, the correct scientific name of the plant and its author(s) should be mentioned in the article as well as the voucher specimen and the botanist responsible for its scientific identification [6]. Therefore, the accuracy, reliability and quality of plant information to accomplish the best practice in Ethnobotany rely essentially on the taxonomy, which is a basic requirement to any biological science.

Taxonomy is the discipline responsible for ordering the diversity of life: it provides a synthetic method to classify and designate organisms, allowing efficient communication. The discipline of taxonomy arises from the combination between the principles used to arrange biodiversity and the procedures established to name its components. As an inductive method by which we organise the universe, classifications are anthropocentric constructions devised with particular interests and assumptions: they group objects and create classes (intention) based on their properties in such a way that other objects (extension) of that same kind will also belong to those classes. In taxonomy, the organisms are the objects, and taxa are the classes [7] or natural kinds of a sort [8]. Taxa are concepts or statements defined by common properties or the relationships of their members: they are created but do not have physical existence; they can be modified but do not evolve; they can be abandoned but do not become extinct [9].

The function of taxonomists is to delimit, describe and name taxa in such a way that organisms are identified by their conformity to these taxa. The primary role of taxonomy, therefore, is to circumscribe taxa and create a hierarchical system of classes. These classifications can be constructed with different aims. Some are only operational and are designed to be simple and stable, whereas others favour the empirical content, synthesising the knowledge of groups or investing in predictions. Lastly, some classifications are theoretical and represent entities in action or the result of natural processes. Considering that plants are highly diverse and widespread and also that they have long been used by humans in different parts of the world and with different purposes, it is not surprising that different systems have been proposed to classify this group. In this chapter, botanical taxonomy is introduced considering three integrated components of the systematisation of plant diversity: (1) the principles of classification, (2) botanical nomenclature and (3) description and identification.

1 Systems and Principles of Plant Classification

Although Aristotle is widely known as the first naturalist, Theophrastus (371–287 B.C.), his pupil, is recognised as the father of Botany (important events for the history of plant taxonomy since then can be found in [10, 11]). Theophrastus described approximately 500 plants. He was the first to name plants and contemplate how they should be ordered by using characters, such as their habit (herbs, sub-shrub, shrubs and trees) and their utility for humans (wild and cultivated plants), as criteria for their grouping. After the books in Alexandria had been burned, Theophrastus' contribution became largely ignored for more than 1,500 years, until his *Enquiry to Plants* was translated back into Latin by Teodoro of Gaza in the fifteenth century from an Arab manuscript discovered in the Vatican library.

Theophrastus was followed by Dioscorides, a Greek physician who produced a field guide of useful plants, particularly for medicine, and by Pliny, a Roman lawyer who compiled most of the natural history information that had accumulated up to his lifetime. Dioscorides' *Materia Medica* (approx. 600 plants) and Pliny's *Natural History* (approx. 800 plants) were published almost at the same time (60–80 A.D.) and became the main botanical references during the Middle Ages (fifth to fifteenth centuries A.D.). In Medieval times, Arabian contributions to plant knowledge were restricted to innovations in agriculture and medicine, in addition to reproductions of earlier manuscripts.

Natural history was reborn during the Renaissance. The German herbalists Brunfels, Bock, Cordus and Fuchs produced their own botanical-medical books, and the herbals spread to Holland, France and England. With the great exploratory travels during this period, many curious new plants arrived in Europe from the Orient and the New World, and in the sixteenth century, universities began to devote attention to the study of plants. Plantsmen were also interested in describing plants and not merely their uses, as most herbalists did. In Italy, Luca Ghini set up the first European botanical garden in Pisa, and his student Andrea Cesalpino prepared a herbarium (collection of dried plant specimens) in Bologna, now housed in the Natural History Museum in Florence.

Cesalpino renewed Theophrastus' search for order and advocated for a method to organise plants according to their affinities. He also used plant growth, a character recurrently employed in several systems, from Theophrastus (approx. 300 A.D.) to Hutchinson (the 1970s), but associated it with the properties of fruits and seeds. Cesalpino's method for plant classification was further improved by John Ray. Ray had included more than 18,500 "species" of plants in his *Historia Plantarum* (1686–1704), therefore a method for the arrangement of these plants was needed. Ray employed plant growth as a primary criterion, associating it with the number of cotyledons, a character influential in subsequent systems. According to Ray's *Methodus Plantarum Emendata* (1703), the method of classification should be suggested by the plants and not imposed to them: groups should be ordered according to their affinities, and they should be clearly recognised and stable enough to avoid confusion. Ray was establishing a new scientific discipline for plantsmen: Taxonomy (a term only introduced by Augustin Pyramus de Candolle more than a century later), thus establishing a course for Linnaeus' contributions.

In *Species Plantarum*, published in 1753, the Swedish botanist Carl Linnaeus named 5,890 species of plants and 1,097 genera from different parts of the world. Linnaeus was the first to apply Latin binomials consistently and to use a hierarchical system of classification considering five categories—species, genera, orders, classes and kingdoms; most previous treatments used polynomial tags describing diagnostic features and scarcely recognised categories above genus. His concept of species was derived from John Ray, whereas his concept of genera was based on the nearly 700 genera considered by the Frenchman Joseph Pitton de Tournefort in the 1700s. The simple standardisation promoted by Linnaeus for naming taxa has been incorporated in botany since and is still in use today. His sexual system was one of the first to emphasise floral features in classification, recognising 24 classes based on the number and arrangement of stamens and subdividing these classes into orders according to the number of pistils. Nevertheless, Linnaeus was not strict with his own artificial method and eventually included species with different numbers of stamens in the same genus, species that otherwise would be treated as different classes [12]. Following Ray, Linnaeus also thought that characters should be designated by taxa, and not the contrary, and believed that natural groups existed, though it was impossible for these groups to be revealed at that time. An advancement in this direction would have to wait for *Genera Plantarum*, which was published by the Frenchman Antoine Laurent de Jussieu in 1789.

Describing genera and orders (currently, termed families) and grouping them into classes (now, orders), de Jussieu's work on 100 families is the basis of our current system. He recognised groups with more than 1 and less than 100 members, a reasonable range to favour memorisation. Similar to the earlier Frenchman Michael Adanson, de Jussieu also proposed a synthetic system based on general similarities, defining groups based on a combination of different characters; however, unlike his fellow countryman, he considered the features within groups to be invariable. De Jussieu's system was followed by other natural systems. The Swiss professor of Botany Augustin Pyramus de Candolle, for instance, initiated the Herculean effort to describe all vascular plants in his

Prodromus Systematis Naturalis Regni Vegetabilis. This work was started in 1824 and was continued by his son and grandson, Alphonse and Casimir de Candolle, respectively, not to have the dicotyledons completed until 1873. By this time, the number of species was approaching ten times the number that Linnaeus had considered. Natural systems were also constructed by, for instance, the Englishmen Bentham and Hooker, who prepared the 3-volume *Genera Plantarum* (1862–1883), and later, by the Germans Engler and Prantl in their 23-volume *Die Natürlichen Pflanzenfamilien* (1887–1915).

Although artificial systems are based on a single or few characters selected a priori, such as in the sexual system of Linnaeus or in herbals in which plants are often arranged according to their uses and effects, natural systems are polythetic: they are based on many characters, grouping plants with a large number of correlated attributes. Accordingly, a natural classification is expected to be more useful and informative than an artificial one and to offer a higher predictive value [13]. However, different from what has been largely disseminated since the mid-twentieth century [10, 14–16], these pre-evolutionary systems were not essentialist. They used exemplars (or types) as references to associate other members according to their overall resemblance. Therefore, groups were formed around these models and not defined because of essential features found for Platonic types [17]. In fact, evolutionary principles had little effect on taxonomic methods; what changed was the way classifications were explained [12].

Darwin was aware that taxonomic categories were constructions defined by convenience and that classification was a logical process that synthesised much information in few words [18]. However, he did not agree with the natural systems of his time because they did not consider genealogical relationships in their classifications. For him, degrees of similarity should be used only to reveal descendant from a common ancestral stock upon which taxa should be based [19]. Although Darwin was already appealing for a type of phylogenetic classification, more than a century would be required for it to become consolidated in botany.

In 1915, the American Charles Bessey, a student of Asa Gray, proposed an influential system, in his *Phylogenetic Taxonomy of Flowering Plants*. Despite the title, his system was evolutionist. Following Ray and de Jussieu, Bessey also employed the number of cotyledons as an important character, and most subsequent evolutionary systems to the end of twentieth century, such as those proposed by Armen Takhtajan, Arthur Cronquist, Robert Thorne and Rolf Dahlgren, continued to divide flowering plants into dicots and monocots. These evolutionary authors employed an enormous amount of information, from phytochemistry and anatomy to ordinary morphology and ecology, to infer evolutionary relationships. Taxa were then constructed based on groups, as revealed by a combination of common evolution and degree of similarities and differences.

Given the complex mosaic of features, evolutionary authors devote particular attention to different parts of the plant. Weights are usually given to some characters, and classifications depend greatly on the author's judgment. Consequently, evolutionary systems also address a high degree of subjectivity and are usually supported based on the authority and experience of their proponents rather than on their utility or objectivity. As a reaction, a numerical taxonomy [20] emerged. Because plant features can be observed but ancestral relationships can only be inferred, some authors advocated for classifications objectively based on the amount of similarity; they were calling for repeatability and more objectivity in systematics.

Although the German entomologist Willi Hennig published the book *Phylogenetic Systematics* in the mid-1950s, phylogenetics only became popular in botany decades later. Introduced by Bremer and Wantorp [21], the phylogenetic approach soon changed the way plants were classified. In a phylogenetic system, classifications must consider only monophyletic taxa (also treated as "holophyletic" by some). Monophyletic groups (or clades, then cladistics) are those that include a common ancestor and all its descendants; consequently, these groups are recognised by synapomorphies, i.e. features shared because of their common ancestry. Therefore, a phylogenetic classification requires, before anything else, hypotheses of ancestral relationships, which are usually depicted in a phylogenetic tree: a branched diagram composed of internal nodes, representing hypothetical ancestors, and terminals, representing organisms. Similar to an artificial system, phylogenetic classifications are also based on a single property of their members, also established a priori: common exclusive ancestry, even though many characters are usually considered to reconstruct this relationship.

Both evolutionary and phylogenetic systems recognise only groups derived from a common ancestor, thus preventing the recognition of polyphyletic groups, those recognised by homoplasies (convergences or reversions), i.e. features that appear more than once independently. Nevertheless, in addition to clades, evolutionary systems also recognise grades (paraphyletic groups), which are conveniently decoupled from the immediately nested clades because of their degree of dissimilarity. Grades are recognised by symplesiomorphies (or the absence of synapomorphies), i.e. features that were present in their common ancestor but derived in some of its descendants (Fig. 1). Grade members are not necessarily closely related to each other and may share a more recent ancestor with members of another group. By accepting only monophyletic groups, phylogenetic systems reduce the subjectivity of recognising grades based on the amount of their similarity. Nonetheless, many alternative clade classifications are also possible based on a

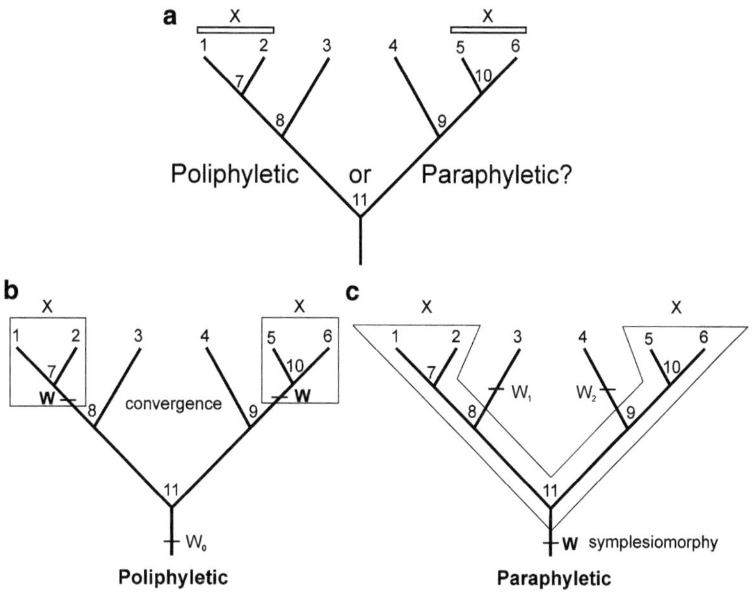

Fig. 1 Phylogenetic tree, with the terminals (*1–6*) and internal nodes (*7–11*) numbered. Taxon X is recognised by the feature W and includes terminals *1*, *2*, *5* and *6* (**a**). Through this information, it is possible to conclude that X is not monophyletic. However, to decide the phylogenetic nature of a taxon, it is necessary to ascertain whether the common ancestor of its members (*11*, in this example) is included, which can be accessed only by convention or by understanding the evolution of taxon intension (W, in this example). If feature W appeared twice independently (convergence), then Taxon X is polyphyletic (**b**). However, if feature W appeared only once, Taxon X is recognised by a feature present in the ancestor but not in all its descents (symplesiomorphy); that is, X is recognised by the absence of W_1 and W_2 (**c**)

phylogenetic tree. Thus, secondary principles of classification must be considered to assist taxonomic decisions; among them, Backlund and Bremer [22] listed the general stability and phylogenetic content and confidence (support) and diagnosability of clades.

Although different assumptions, analyses and data sources can be used to reconstruct the phylogeny of a group, nothing can guarantee that the true phylogeny will be completely revealed. Adding data and improving analytical procedures are obvious strategies to achieve results that are close to the correct ones, and progress in our knowledge of plant relationships directly affects the taxonomy of these groups. Up to the 1980s, most phylogenetic studies with plants included few terminals and were mainly based on morphological data using parsimony. According to the parsimony criterion, when presented with alternative hypotheses, the simplest should be preferred; phylogenetically speaking, this means trees with less homoplasy. However, parsimony is a philosophical criterion and not a biological one: nature is not necessarily simple, and parsimony, in several situations, can be inconsistent [23]. Regardless, morphology is an extremely complex source of data, and different authors often interpret it differently. Furthermore, morphological comparisons are not possible for or are questionable between distant groups, and features are often affected by the environment, offering adulterated evidence of relationships. Indeed, trees based on morphology alone are usually poorly resolved, and the few clades are barely supported.

In the mid-1990s, evidence based on molecular data was already surpassing morphology, and this soon became the main source of data in phylogenetic studies, either in combination or not with morphological data. This shift was possible mainly due to molecular sequencing and computational progress and was responsible for important advances in plant systematics. Because DNA is the source of heritability, molecular data are the most direct evidence of ancestral relationships available. Nucleotide sequences are stored in electronic databases, such as GenBank (http://www.ncbi.nlm.nih.gov/genbank/), currently with more than 135 million DNA sequences, and can be easily accessed by the scientific community. The task of recognising the four character states of a nucleotide sequence is straightforward, though sequences are not available for every taxon, and their comparisons are not always simple or unambiguous. With the predominance of molecular data, parsimony analyses are also losing space to model-based analyses. Empirical and theoretical advances on molecular evolution favoured the implementation of more reliable models of nucleotide substitution. Currently, most phylogenetic studies in plants use DNA sequences, eventually combined with morphology or, in a genomic approach, considering whole-plastid DNA, and employ maximum likelihood and/or Bayesian inference analyses, often associated with the results from parsimony analyses.

The impact of phylogenetic systematics in botanical taxonomy was enormous. Comprehensive classification systems, such as those proposed for flowering plants [24] and ferns [25], are rooted in phylogenetic results and emerged from consensual collaborative works. Vascular plants were divided into Lycophytina and Euphyllophytina (including seed plants and monilophytes), and the pteridophytes are no longer recognised as a taxonomic group [26]. For monilophytes, the Psilotaceae and Equisetaceae are known to be closer to the eusporangiate ferns Ophioglossaceae and Marattiaceae, respectively [27]. For angiosperms, *Amborella trichopoda* is sister to the rest of angiosperms, forming a grade with Nymphaeales [28], including the graminoid Hydatellaceae [29]. In evolutionary systems (e.g. [30]) *Amborella* (in Laurales) and Nymphaeales belonged to the subclass Magnoliidae of Magnoliopsida (dicotyledons), whereas the Hydatellaceae were treated as Liliopsida (monocotyledons). Traditional groups, such as the dicotyledons, have been abandoned, and informal groups, such as the eudicots, have been incorporated in classification. The changes at lower levels are also remarkable: some families were disintegrated, including Scrophulariaceae, the members of which were divided into five families [31], whereas others, such as Apocynaceae and Asclepiadaceae, were amalgamated [32]. Most phylogenetic results are absorbed relatively quickly by botanists and are incorporated in textbooks for undergraduates (e.g. [33, 34]).

2 Principles of Nomenclature

Our principles of classification direct how plants should be organised. However, taxonomy would be handicapped if there was not a system to govern the application of scientific names independently of these principles. Botanical nomenclature is constantly being refined to better serve taxonomy. Its function is to offer clarity, universality and stability for plant names. New propositions are discussed every 6 years in a nomenclatural section, 1 week before the International Botanical Congress. Although changes have been gradually incorporated into the nomenclature code [35], the Melbourne code is an example of how such changes can be revolutionary. First of all, rather than *International Code of Botanical Nomenclature*, the new code is titled *Code of Nomenclature for Algae, Fungi, and Plants.* More importantly, since 2012, an electronic publication is effective, as long as it is published in a journal with an ISSN or book with an ISBN and in PDF format, and the diagnoses of new names, in addition to Latin, can also be in English. Previously, a publication would have to be printed and distributed in at least two public libraries to be effective, and a Latin diagnosis or description was mandatory for the valid publication of new plant names [36]. These changes began to work before the publication of the Melbourne code and will certainly increase the rate at which new taxa are described, favouring floristic studies, particularly in megadiverse and still poorly known regions.

The code of nomenclature for plants is composed of articles, recommendations, examples and notes and is independent of the codes for animals, bacteria, viruses and cultivated plants. The names are Latin or Latinised and must be validly published to be recognised; they follow the binomial, hierarchical system, established since Linnaeus, 260 years ago. Taxa are named according to their taxonomic rank. Species names are binomial and composed of the specific epithet following the name of the genus; names above species level are uninomial and begin with a capital letter. Names at the genus level and below are written in italics; those below species also consist of an infraspecific epithet and must have their rank explicitly indicated: subspecies (subsp.), variety (var.) or form (f.). Cultivars can be written in modern language using single quotation marks, e.g. 'Grape Cooler'. Hybrids are denoted by the multiplication sign "×" or the prefix "Notho-" and can be written as a formula (or as its own combination), such as *Oenothera biennis* × *O. villosa* (or *O.* × *drawertii*). Above the genus level, categories are usually flagged by a suffix denoting their rank. Seven taxonomic categories are mandatory—species, genus, family (-aceae), order (-ales), class (-opsida), phylum or division (-phyta) and kingdom (-bionta)—but others can be recognised at a level between them, usually using the prefix "sub" or "super," such as subclass (-idae) and superorder (-anae).

Nomenclature does not establish ranking criteria, and taxa are assigned to categories by convention. Categories differ only by their degree of inclusion. When taxa share organisms, those at higher categories are more inclusive unless there is a redundancy between taxa. Circumscriptions of such taxa are broader, and the information is more generalised; consequently, their intensions are more comprehensive than those at lower ranks. Taxa at the same rank are exclusive: they do not share organisms. It is always important to consider that ranks are designed by convention and convenience: they are comparable only by designation and not by any biological or natural meaning [37, 38]. The use of categories has been questioned as a source of instability, and some scientists are accepting only species and clades, as with the Phylocode [39, 40] or "cladonomy" [41]. In the current Linnean nomenclature, however, categories continue to be a requirement for the application of names, as they have informational content and are mnemonic devices that have been successfully and universally applied for a long time [42].

The general principle of nomenclature is that every taxon must have only one correct name; these names must be validly published and legitimated. The discernment between alternative names is taxonomically guided by the type of method and assisted by the nomenclatural principle of publication priority. Since 1958, authors must designate a holotype to validly publish a new name. The holotype is the specimen or illustration designated or used by the author of the species (or a taxon below species) to which the name is attached. Duplicates of a holotype are isotypes. However, types were not always mandatory and, in ancient literature, names were published without an explicit reference to any type specimen. In some of these cases, the holotype can be inferred; but in others, a lectotype must be designated from the original material (syntype) as nomenclatural type. A lectotype must also be designated among the isotypes when the holotype is missing. However, when no original material is available, a neotype must then be designated for the name. Lastly, an epitype can be designated to assist in the precise application of the name when its type is insufficient. Certainly, the type method is central for nomenclature procedure. However, it is important to emphasise that types are designated by convention and are attached to taxon names only; the taxa themselves do not have any type.

When more than one type is considered in the same taxon, there will be competing names for that taxon. In these cases, the principle of priority must be invoked to discern which is the correct name: the name that was published first has priority and is the correct name of that taxon (equivalent to a valid name in zoology); the names that are published later are heterotypic synonyms. This is an ancient principle of nomenclature that was already formalised in the nineteenth century. However, it can be abandoned in favour of nomenclatural stability. To avoid disadvantageous changes, a widely used name, particularly one with major economic importance, can be conserved, regardless of its priority of publication, whereas others can be rejected. The principle of priority is applicable only at the level of family and below, and names have no priority outside the rank in which they were published.

In botany, tautonomy—species names whose generic and specific names are the same—is invalid. To be valid, names must also be accepted as correct by their authors in the original publication. A correct name of a species or infraspecific taxon is the combination of the legitimate final epithet in that rank and the correct name of the genus or species. The authors of the basionym (original combination of a species) of a later combination appear between parentheses, as in zoology; in botany, however, the author names are followed by the authors of the correct combination. A name can be illegitimate either because it is superfluous, i.e. its type was previously applied to another name, or because it is a posterior homonym, i.e. it was already used for another taxon at that rank. Because a scientific name must be assigned to only one taxon, a posterior homonym requires a substitute new name.

3 Taxonomic Works: From Descriptions to Identification

Classification is a continuous process. Taxonomists may disagree about taxonomic concepts and can always find new evidence to change their own taxonomic opinions. Disagreements may have different causes, and evidence may originate from different sources. For instance, although a taxonomist can consider two morphological variants distinct enough to be recognised as different species, another taxonomist may recognise them as parts of the morphological variation of a single species; in the latter case, the two species will be considered synonyms. Given that a taxonomy based on phylogenetics must reflect ancestral relationships, the discovery of unexpected ancestral relationships may require taxonomic changes, and these changes may affect the circumscription of taxa, potentially changing the name of a species. For instance, a new family can be created or an old one resurrected to comprise individuals of a clade without requiring any nomenclatural change at the species level. By contrast, a closer relationship of a species with the species of another genus will require its transference and, consequently, a new combination. Changes may also be required in favour of clarity and universality, for instance, when a name is found to be superfluous or a posterior homonym.

Currently, taxonomic changes most frequently result from new phylogenetic information. This is a sign of scientific progress, in contrast to taxonomic stability, which may only reflect ignorance and not necessarily correctness [43]. Taxonomists are responsible for critically evaluating new information and incorporating this information into classifications. As this process often implies taxonomic rearrangements and nomenclatural changes, it is important also to update identifications continuously.

Identification consists of assigning scientific names to organisms of yet-unknown classification. There are several strategies for plant identification [44]. A common procedure is to assign this task to taxonomists, the specialists of the group, as a specialist can best apply his/her experience to provide a rapid and accurate identification. However, taxonomists are not always available; thus, other strategies must be attempted. In such cases, plant identification may also be achieved by consulting specific monographs, technical works that are complete and scientifically accurate. Monographs usually combine complete nomenclatural information, detailed descriptions and high-quality illustrations, comments on taxonomy, phylogenetics, nomenclature, ecology and geographic distribution (usually with maps of occurrence); analytical keys for taxa identification and an initial section introducing the morphology of the group are also included in a thorough monograph. In these treatments, species are arranged in alphabetic order, favouring direct individual queries, or according to their relationships, favouring comparisons of closely related taxa.

When the complete taxonomic and nomenclatural information is included, the correct name of the species is followed by the authors and original reference of the combination. Starting with the basionym, the other homotypic synonyms, with authors and reference, are included in sequence, after the data for the correct name, according to the publication date. Similarly, heterotypic synonyms are included below the correct name header, each starting on a different line, usually listed by the publication date of their basionym. The types of heterotypic names appear at the end of each header, with an indication of whether they were examined by the authors of the work.

Plant descriptions should be standardised and compact but clear and precise. For instance, verbs, articles and conjunctions can be eliminated, but the type of measures must always be specified, even when this seems obvious. The method for description is frequently the same as that used in ancient times, which Winsor [17] named "method of exemplar". An initial description is prepared based on an example of the taxon, and the variation is added by including attributes of other exemplars of that taxon. Nevertheless, other methods can be employed, such as the combination of DELTA (DEscription Language for TAxonomy; [45]) format and Lucid software (http://www.lucidcentral.com) to construct sharp and parallel descriptions in natural language and consistent keys for identification directly from original databases. Descriptions in taxonomic revisions are complete because these monographs are expected to be more definitive and universal references. In contrast, the intention of a floristic work is to discriminate species from a certain region; their descriptions comprise only the variation found in the study area and should be shorter, long enough only to provide the recognition of those species. The importance of floras in botanical literature is recognised for a long time and they were the first and one of the main sources of plant knowledge for beginners. In the last decades, however, checklists, electronic keys and illustrated guides, rather than classical descriptive floras, have reached a wider audience [46].

Illustrations have been popular in botany since the Juliana book. The work was given to her by the townspeople of Honorata, Constantinople, in 512 A.D. Designed for the illustrations, the book included 383 plants from Dioscorides' work [11]. Since then, illustrations have become more common, helping users to interpret descriptions and identify plants. Pictures can also be very attractive and helpful, particularly for fresh plant identification and are largely used in field guides. Line drawings are recommended to illustrate diagnostic details; they must bear accurate scales, allowing feedback consultation with descriptions, and legends should be detailed, indicating the specimens used for drawings.

As a scientific work, a taxonomic study must include the material examined by the authors, with the following collection information: locality, date, phenological state, collector and number, and herbaria of duplicates. However, some monographs are based on hundreds of specimens. In these cases, it is recommended to indicate a selected material representing the variation and distribution of the species, and include, at the end of the work, a complete list of exsiccates, only with the collection reference (first collector and number) and its identification (the species number, depicted from its order in the treatment).

Since Lamarck, the characters used in identification keys are decoupled from those used for classification, allowing artificial, pragmatic diagnostic keys designed for the easy identification of taxa in natural (and phylogenetic) classifications. A key consists of pairs of contrasting, preferentially mutually exclusive statements, which are sequentially followed by the user as they best fit the attributes of the specimen to be identified until a taxon is reached. The inclusion of more attributes per statement is preferred, as long as these attributes are simple and easy to observe. Keys are traditionally dichotomous and can be numbered or indented. In numbered keys, statements of a pair are adjacent, and the next statements are indicated by numbers. In an indented key, statements are organised sequentially, departing gradually from the left, without the necessity of numbers to indicate further statements. Some keys include references to representative illustrations of the attributes; others can be richly illustrated and presented as a diagram.

Different from printed dichotomous key, electronic keys (e-keys) have multiple access, allowing interactive identifications. Users can start at any character and follow different sequences, delaying the use of missing or difficult characters, or they can use a combination of attributes simultaneously (polyclave key). Depending on the material available, morphological information may not converge on a single taxon through the exclusive use of a key. Therefore, additional aspects, such as ecology and distribution, must be considered when comparing the possibilities to decide which taxon best fits the specimen. E-keys permit constant updates and can be accessed through the Internet or distributed in CD format and can also be incorporated in small devices, becoming useful in fieldwork.

Identifications must always be confirmed. A confirmation can be performed initially by comparing the specimen with the description, illustration and comments of the species and later by comparing the material with the specimens in herbaria, preferentially using those identified by a taxonomist with experience in the group. Exsiccates of different species, including types, can also be accessed at synoptic virtual herbaria. If a monograph or a flora is not available, a checklist can help the user to know whether the taxon has been registered in the region; however, its absence in a list cannot exclude the possibility of a new occurrence.

DNA barcodes are another strategy for identification. They consist of short, standardised DNA sequences that are conservative within species but variable enough to discriminate between species. This technology is particularly interesting because it permits identification without conferring with a specialist and is based only on DNA fragments. For animals, the mitochondrial COI gene is used as a universal DNA barcode; however, this region in plants evolves too slowly, and other regions are being tested. To date, the best strategy found is a combination of two plastid regions (*matK* and *rbcL*; [47]). Although this method was unable to identify more than 25 % of the species sampled, when the strategy is used at the regional level, the success of identification is close to 100 %. Considering that the region with most sequences available for plants, the *rbcL* gene, was sequenced for less than 15,000 species and the diversity of plants is approximately 380,000 species, we may understand why DNA barcodes for use in botany is still in its infancy [48].

Taxonomic works are the documentation of our biodiversity. First, publications function like birth certificate of new taxa. They also register new information as data are being collected and join them together in monographs, making the knowledge of plants available to others and enabling its perpetuation between generations. In spite of that, taxonomic work is losing its place in the scientific community. This is because such efforts are time-consuming and publications are used mostly for consulting, scarcely generating citations. However, taxonomy fits well within the Internet, where space is not a limitation, such that data can be continuously accumulated, illustrations can become available in colour without additional costs, and information can be updated and linked to a great assortment of related matters. Websites of plant systematics can be phylogenetically ordered, such as the Angiosperm Phylogeny Website [49], where orders of flowering plants are accessed using an alphabetical index or using a phylogenetic tree. Most orders of the tree are linked to another tree, but of families. By clicking on a family name, users can access updated information of that taxon, such as its diversity, internal arrangements and relationships, characters, and maps.

Currently, there are many websites that can assist plant taxonomy. TROPICOS (http://www.tropicos.org/), for instance, provides nomenclatural and taxonomic databases. With approximately 1.2 million scientific names, TROPICOS includes authorship, references and type information, in addition to correct names, synonyms and homonyms. For several names, users are only one click away from the original work, some of which was published hundreds of years ago. Images of types, exsiccates, illustrations and pictures of fresh plants are also available. The website informs about the chromosome number registered for the species and more than four million specimen records. Links to other websites allow the netizen to go directly to different digital collections, such as the virtual herbaria of the New York Botanical Garden (http://sciweb.nybg.org/science2/vii2.asp), the *Australian Plant Index* (http://www.anbg.gov.au/apni/), and the *African Plant Database* (http://www.ville-ge.ch/musinfo/bd/cjb/africa/recherche.php) in Geneva. From TROPICOS, users can also visit the *JStore Plant Science* (http://jstorplants.org/), which combines images of exsiccates and the classical literature associated with the species name, and to *The Plant List* website (http://www.theplantlist.org/). Many other doors open immediately from *The Plant List*. At this site, the user can often confirm the status of the name, whether it is accepted or not, and continue to other gateways, such as the *Biodiversity Heritage Library* (BHL; http://www.biodiversitylibrary.org/) and the *National Center for Biotechnology Information* (NBCI; http://www.ncbi.nlm.nih.gov/), with different possibilities of information associated with that taxon, from literature to genetic sequences.

4 Final Remarks

Scientific names unite an intricate network of information. They are like seeds travelling in space and time and work like keys for the entire body of literature of the taxa they represent. Taxonomic names guarantee the accumulation of knowledge and its transference between generations, from naturalists to the current phylogeneticists, and have the power to disseminate this knowledge quickly to any country, regardless of the language. However, plant names should have meaning, and this is provided by taxonomy.

References

1. Bessey CE (1908) The taxonomic aspect of the species question. Am Nat 42:218–224
2. Berlin B (1992) Ethnobiological classification: principles of categorization of plants and animals in traditional societies. Cambridge University Press, Cambridge
3. Gilmore MR (1932) Importance of ethnobotanical investigation. Am Anthropol 34:320–327
4. Newmaster SG, Subramanyam R, Ivanoff RF, Balasubramaniam NC (2006) Mechanisms of ethnobiological classification. Ethnobotany 18:4–26
5. Berlin B, Breedlove DE, Raven PH (1973) General principles of classification and nomenclature in folk biology. Am Anthropol 75:214–242
6. Nesbitt M, McBurney RPH, Broin M, Beentje HJ (2010) Linking biodiversity, food and nutrition: the importance of plant identification and nomenclature. J Food Compost Anal 23:486–498
7. Rapini A (2004) Classes or individuals? The paradox of systematics revisited. Stud Hist Philos Biol Biomed Sci 35:675–695
8. Ruse M (1987) Biological species: natural kinds, individuals, or what? Br J Philos Sci 38:225–242
9. Loevtrup S (1987) On species and other taxa. Cladistics 3:157–177
10. Mayr E (1982) The growth of biological thought: diversity, evolution, and inheritance. Belknap Press of Harvard University Press, Cambridge
11. Pavord A (2005) The naming of names. Bloomsbury Publishing, London
12. Stevens PF (1984) Metaphors and typological in the development of botanical systematics 1690–1960, or the art of putting new wine in old bottle. Taxon 33:169–211
13. Gilmour JSL (1937) A taxonomic problem. Nature 13:1040–1042
14. Hull DL (1988) Science as a process. An evolutionary account of the social and conceptual development of science. The University of Chicago Press, Chicago
15. Ereshefsky M (1997) The evolution of the Linnaean hierarchy. Biol Philos 12:493–519
16. Mishler BD (2000) Deep phylogenetic relationships among "plants" and their implication for classification. Taxon 49:661–683
17. Winsor MP (2003) Non-essentialist methods in pre-Darwinism taxonomy. Biol Philos 18:387–400
18. Darwin C (1987) The correspondence of Charles Darwin, vol 2: 1837-1843. In: Burkhardt F, Smith S (eds). Cambridge University Press, Cambridge
19. Padian K (1999) Charles Darwin's view of classification in theory and practice. Syst Biol 48:352–364
20. Sokal RR, Sneath PH (1963) Principles of numerical taxonomy. Freeman, San Francisco
21. Bremer K, Wantorp H-E (1978) Phylogenetic systematics in botany. Taxon 27:317–329
22. Backlund A, Bremer K (1998) To be or not to be-principles of classification and monotypic plant families. Taxon 47:399–400
23. Felsestein J (1978) Case in which parsimony and compatibility will be positively misleading. Syst Zool 27:401–410
24. APG III (2009) An update of the angiosperm phylogeny group classification for the orders and families of flowering plants. Bot J Linn Soc 161:105–121
25. Smith AR, Pryer KM, Schuettpelz E, Korall P, Schneider H, Wolf PG (2006) A classification for extant ferns. Taxon 55:705–731
26. Kenrick P, Crane P (1997) The origin and early diversification of land plants. Smithonian Institute Press, Washington
27. Pryer KM, Schuettpelz E, Wolf PG, Scheider H, Smith AR, Cranfill R (2004) Phylogeny and

evolution of ferns (monilophytes) with a focus on the early leptosporangiate divergences. Am J Bot 91:1582–1598

28. Qiu Y-L, Lee J, Bernasconi-Quadroni F, Soltis DE, Soltis PS, Zanis M, Zimmer EA, Chen Z, Savolainen V, Chase MW (1999) The earliest angiosperms: evidence from mitochondrial, plastid and nuclear genomes. Nature 402: 404–407
29. Saarela JM, Rai HS, Doyle JA, Endress PK, Mathews S, Marchant AD, Briggs BG, Graham SW (2007) Hydatellaceae identified as a new branch near the base of the angiosperm phylogenetic tree. Nature 446:312–315
30. Cronquist A (1981) An integrated system of classification of flowering plants. Columbia University Press, New York
31. Olmstead RG, dePamphilis CW, Wolfe AD, Young ND, Elisens WJ, Reeves PA (2001) Disintegration of the Scrophulariaceae. Am J Bot 88:348–361
32. Endress ME, Bruyns PV (2000) A revised classification of the Apocynaceae s.l. Bot Rev 66:1–56
33. Simpson MG (2006) Plant systematics. Elsevier Academic, London
34. Judd WS, Campbell CS, Kellogg EA, Steves PF, Donoughe MJ (2007) Plant systematic: a phylogenetic approach, 3rd edn. Sinauer Associates, Sunderland
35. McNeill J (2000) Naming the groups: developing a stable and efficient nomenclature. Taxon 49:705–720
36. Knapp S, McNeill J, Turland NJ (2011) Changes to publication requirements made at the XVIII International Botanical Congress in Melbourne—what does e-publication mean for you? Bot J Linn Soc 167:133–136
37. Stevens PF (1997) How to interpret botanical classification—suggestions from history. Bioscience 47:243–250
38. Stevens PF (1998) What kind of classification should the practising taxonomist use to be saved? In: Dransfield J, Coode MJE, Simpson DA (eds) Plant diversity in Malesia III. Royal Botanic Gardens, Kew, pp 295–319
39. Queiroz K (1997) The Linnaean hierarchy and the evolutionization of taxonomy, with emphasis on the problem of nomenclature. Aliso 15:125–144
40. Cantino A, Queiroz K (2010) International code of phylogenetic nomenclature, version 4c. http://www.ohio.edu/phylocode/index.html. Accessed 7 Nov 2012
41. Brummitt RK (1997) Taxonomy versus cladonomy, a fundamental controversy in biological systematics. Taxon 46:723–734
42. Lidén M, Oxelman B, Backlund A, Andersson L, Bremer B, Eriksson R, Moberg R, Nordal I, Persson K, Thulin M, Zimmer B (1997) Charlie is our darling. Taxon 46:735–738
43. Dominguez E, Wheeler QD (2005) Taxonomic stability is ignorance. Cladistics 13:367–372
44. Lawrence A, Hawthorne W (2006) Plant identification: creating user-friendly field guides for biodiversity managements. People and plant conservation series. Earthscan, London
45. Dallwitz MJ (1993 onwards) Applications and documentation of the DELTA System. http://delta-intkey.com Accessed 7 Nov 2012
46. Frodin DG (2001) Standard floras of the world, 2nd edn. Cambridge University Press, Cambridge
47. CBOL Plant Working Group (2009) A DNA barcode for land plants. Proc Natl Acad Sci U S A 106:12794–12797
48. Ausubel A (2009) A botanical macroscope. Proc Natl Acad Sci U S A 106:12569–12570
49. Stevens PF (2001 onwards) Angiosperm phylogeny website. http://www.mobot.org/MOBOT/research/APweb/. Accessed 7 Nov 2012

Chapter 10

Analysis of Vegetation in Ethnobotanical Studies

Elcida de Lima Araújo and Elba Maria Nogueira Ferraz

Abstract

In this chapter, you will find information about (a) the problem of vegetation sampling regarding methods, criteria, and plant measurements; (b) the phytosociological parameters commonly used to describe communities; (c) the similarity and distance indices employed and the means of their estimation; and (d) recommendations on the use of different indices in ethnobotanical studies.

Key words Vegetation sampling, Plant measurements, Diversity of use of the vegetation

1 Introduction

Recent research in the field of conservation biology has highlighted the importance of studies addressing the cultural relationships between man and nature for defining (socially participative) strategies that will contribute to the conservation of the biological diversity of forests and to the maintenance of ecosystem functions [1–4]. In fact, the significance of such studies has increased in recent decades, mostly because various ethnobiological studies have adopted and/or adapted quali-quantitative ecological methods to measure the value of natural resources for human communities and to evaluate how use and/or use pressure have been modifying the cultural form of the relationship between man and nature.

In addition, ethnobiological studies have sought to test hypotheses regarding resource usage and to correlate them with ecological theories to define current scenarios and project future scenarios regarding the use and conservation of natural resources. These studies question the sustainability of the usage practices that exist in different cultures and suggest alternatives for conservationist uses based on the knowledge of local populations. Thus, the assessment of resource availability is being increasingly associated with quantitative measures, such as use value and local importance, among others. For examples, see [5–11].

Ulysses Paulino Albuquerque et al. (eds.), *Methods and Techniques in Ethnobiology and Ethnoecology*,
Springer Protocols Handbooks, DOI 10.1007/978-1-4614-8636-7_10, © Springer Science+Business Media New York 2014

Regarding vegetation, which is the main subject of this chapter, ethnobotanical research has been employing vegetation sampling methods that characterize the floristics and the structure of plant populations to correlate the quali-quantitative ecological characteristics of vegetation fragments with human perceptions, values, habits, and modes of resource use. It is also common for researchers to begin studies by estimating the potential of useful resources based on local knowledge.

Thus, we present the main problems faced by ethnobotanists when performing ecological sampling of vegetation to obtain a phytosociological characterization of an ecological community. We also explore the parameters and/or indices used to evaluate the community's organization and the diversity of its use. The methods and forms of analysis applied in vegetation studies are diverse and are presented briefly. We will avoid extensive citations throughout the text; however, all of the information and analysis techniques presented here have a theoretical framework, and examples of studies are listed in the bibliography provided for this chapter.

2 Vegetation Sampling and Plant Measures

2.1 Selection of the Study Area and Determination of the Sampling Methodology

The criteria used to select a study area and to determine the sampling methodology to be applied are various and depend on the question addressed by each study. Such variability may be of a political-administrative nature when the research question involves more than one municipality, state or country, or an environmental nature, such as related to the topography, climate, and vegetation. Maps, aerial photography, satellite imagery, and interviews with community members are useful tools for study area selection.

The selection of the vegetation sampling method must consider what kind of statistical analysis will be performed on the dataset. The most common sampling methods are random, systematic, and preferential sampling types (*see* [12]).

When sampling is random, the sampling units are distributed by chance, i.e., the location of the point where each sampling unit will be placed is determined randomly (using a table of random numbers, for example). This kind of sampling permits the use of various statistical analyses and comparison of results within and among the sampled areas.

In systematic sampling, a starting point is selected, and all sampling units are set systematically from this point. The distances between sampling units are generally fixed, and the units may be set along straight lines, perpendicular to each other (fishbone), or in various other orientations, with fixed distances between them.

Preferential sampling is subjective, requires previous knowledge of the vegetation to be studied, and prevents the evaluation of confidence intervals for the obtained data. However, there are

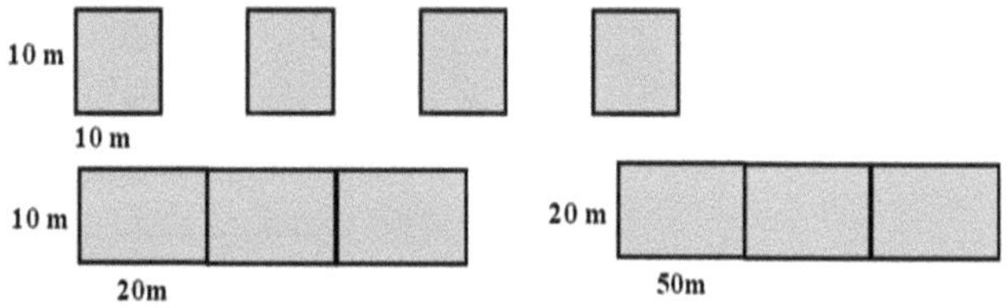

Fig. 1 Schematic representation of the shapes and sizes of sampling plots

studies that, after preferential selection of a stretch of vegetation, distribute the sampling units in a random or systematic manner. In such cases, the obtained data may be subjected to statistical treatment. Ethnobotanists have generally been opting for the latter two procedures, often relying on the help of informants to select appropriate stretches of vegetation for a study.

2.2 Selection of the Sample Size

The question of the representative sample size is in vegetation studies. For example, there was a great deal of variation in sample size between ecological studies conducted in Brazil up to the 1980s [13], which led to problems in comparing the results of different studies. To solve this problem, in recent decades, the scientific community has attempted to standardize sample sizes, with a minimal sample size of 1 ha commonly being used to study woody plants [14]. When the sampled area is smaller than 1 ha, the obtained data are estimated for 1 ha to enable comparisons.

In ethnobotanical studies, the sampled area has ranged from 0.1 to 1.4 ha. The objectives and particularities of each study determine the choice of the size of the study area in ethnobotanical research. However, it is recommended that the sample size be standardized to 1 ha in studies of woody plants to facilitate future comparisons. In the case of studies that opt for smaller sample sizes, the use of a rarefaction method prior to comparisons is recommended; this rarefaction method will be discussed in Sect. 2.7.

2.3 Choice of the Type, Size, and Shape of the Sampling Unit

The most commonly used types of sampling units are plots, or quadrats, and point-quarters (*see* [15, 16]), with the former type being more commonly used by the scientific community, as exemplified by various studies in Brazil [13].

The shape of the plot may be square, rectangular, or circular. The first two types are most commonly used. The numbers and shapes of sampling units vary with their size. For example, different numbers and shapes of plots maybe used to sample 1 ha of vegetation, e.g., 100 10×10 m plots, 50 10×20 m plots, or 10 20×50 m plots may be employed (Fig. 1). These plots may or may not be contiguous. In general, a larger number of smaller plots are preferred because, depending on the spacing between them, there is a greater probability of sampling more species in the community.

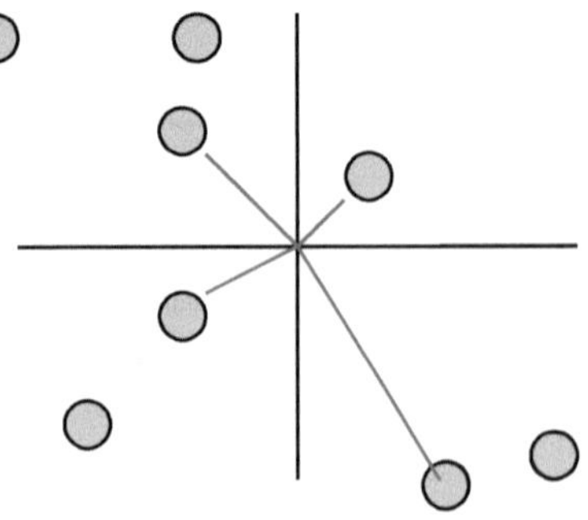

Fig. 2 Schematic representation of sampling by the point-quarter method

The life span of the material used to construct the plots may be long (permanent plots) or short (semipermanent plots). Long-term studies usually employ permanent plots. The use of a string or cord to aid in demarcating the sides of the plots is also common. When plants contact the external edge of the cord, they are, by convention, included in the sample, and usually only two of the cord's sides are considered to avoid overestimation. The use of statistical tests to evaluate the differences in results between areas is also common. In this case, plots represent pseudoreplicates, and the assumptions of the applied statistical tests must be carefully observed.

Sampling by plots is also often adopted for the study of herbaceous vegetation, in spite of the existence of other sampling techniques for this biological group, such as the point method. Under this method, stakes, usually composed of metal, are placed in the soil at fixed distances. All plants that touch any part of the stake are sampled. However, when the objective is to monitor herbaceous plants, sampling by plots is more advisable because the sampling area remains delimited in the field.

Additionally, some studies have adopted the point-quarter method for vegetation sampling [13, 16]. This method differs from sampling by plots because it does not involve a defined area. This method employs a set of quadrants formed by intersecting two pieces of wood so that a cross is formed. The distances between the center of the cross and the plants that are closest to its vertices are measured. These plants will be the individuals that are sampled in the community (Fig. 2).

To avoid having the same plant sampled in consecutive points, it is recommended that the distance between the sampling points be twice the average distance between plants in the community [17]. The number of sampling points may vary between studies depending on the researcher's objective. However, it is recommended that a large number of points should be used, as this increases the probability of acquiring better knowledge of the local diversity and structure of the vegetation. The number of individuals in a point-quarter sample is related to the number of points used, as each point-quarter permits the sampling of four plants.

2.4 Plant Measurements

After defining the sample size and sampling methods, it is necessary to define the biological group(s) of vegetation and the criterion(a) to be employed for including an individual in the sample. For example, the researcher may decide that he/she does not want to sample epiphytes and lianas or that he/she wants to sample all of the possible diversity in the community. The decision regarding what to include clearly depends on the study's objective.

Studies on woody plants usually establish inclusion criteria that define which individuals will be included in the sample. There is currently a trend toward standardizing the criteria for a given vegetation type and/or biological group to facilitate future comparisons. For example, there is a tendency to standardize the inclusion size as corresponding to a 5 cm stem diameter at breast height (DBH) for plants in Atlantic forest communities [14, 18], or a 3 cm diameter at soil level (DSL) for plant communities in the caatinga [19]. A criterion based on height is also used in some cases, which is usually over 1 m for woody plants.

After establishing the sampling units and defining the inclusion criteria, the popular names of plants are noted, and botanical material including flowers is collected for detailed taxonomic identification. The botanical material is herborized [20] and, after identification, is incorporated into an Herbarium's collection.

Manual pruning shears and a tall trimmer are used for collecting plants. Newspaper, paperboard, corrugated metal sheets, a press, and a kiln are used for herborization. During sampling, it is always advisable to fill out a collection form containing the following information: a description of the site where the material was collected (e.g., vegetation type, coordinates); a description of the collected plants (e.g., size, frequency in the area); features that are lost in the herborization process (e.g., color of the corolla, smell); and characteristics regarding plant usage (e.g., medicinal, wood). Each researcher generates a sequential collection number in his/her collection notebook, and this number is added to the plant's collection form. The collector's number accompanies the material when it is incorporated into the herbarium. For publication, it is usually requested that the researcher indicates the name of the herbarium in which the collected material was deposited.

After the collection of botanical material, tree height, stem circumference and, sometimes, canopy area are measured. All individuals meeting the pre-established inclusion criteria for height and diameter are measured.

The instruments commonly used for measuring individuals are the following: (1) measuring tape, to measure the circumference/perimeter; (2) calipers, to measure the diameter (herbaceous individuals or seedlings); (3) larger calipers, to measure the diameter of woody individuals; (4) a graduated rod, to measure height; (5) a clinometer, to measure angles and height; (6) a hypsometer, to measure height; (7) field forms or tables, to record the collected

Table 1
Hypothetical example of a field spreadsheet for recording field data

Plots no. ______				
Number of individual	**Plant name**	**Height (m)**	**Diameter (cm)**	**Observations**
1	*Poincianella pyramidalis*	5.5	17.0	Shows evidence of cut
2	Aroeira	7.0	35.0	
...	...	...	...	

Fig. 3 Area view of caatinga vegetation in the dry season, showing the occurrence of individuals with tillers arising from stem base

data (see example at Table 1); and (8) identification tags, corresponding to the number that the researcher writes on the plaque used to mark the plant. This last procedure prevents the researcher from measuring the same plant twice, enables monitoring studies and allows the researcher to find the plant to collect reproductive material if it is not flowering at the time of sampling.

Field forms or spreadsheets are prepared in the form of tables with several columns. After the plot number, the first column contains the number of the individual in the sample, and the following columns contain the popular and/or scientific name of the plant, its height, the perimeter/circumference, or diameter of its stem and observations that the researcher finds appropriate (Table 1).

Because plants are modular organisms, the criterion used to define an individual plant may vary. Usually, every plant that is not connected to another plant at the soil level is considered an individual. However, it is possible for some plants to present multiple stems above ground without an individualized stem base (Fig. 3).

In such cases, the circumference of every shoot is measured. Their basal areas are then calculated and totaled to obtain a single diameter measure.

The measurement data are used to calculate phytossociological parameters, including density, basal area, dominance, frequency, and the importance value index and cover value index. These parameters permit the community to be described at two levels: architectural, when the parameters are expressed for the community as a whole, without individualizing any species; and structural, when the parameters are expressed for every species in the community [17].

Height and diameter distribution classes are used to describe plant sizes. These classes allow testing of whether the vertical space in the community is occupied in a stratified fashion and what species occur in each stratum. They also enable the perception of whether there is an absence or excess of individuals in a given class. The formulas employed to calculate phytosociological parameters are widely known [14, 17, 21, 22], and these parameters are calculated with computer software such as FITOPAC [23] or Mata Nativa, or directly in Excel, using the formulas shown in Fig. 4.

In general, density measures permit the evaluation, in a total or relative manner, of the number of individuals in the community. The determination of total or species-by-species dominance also permits a partial evaluation of the occupancy of horizontal space above the soil, while calculation of the absolute frequency allows evaluation of the number of sampling units occupied by particular taxa, permitting inferences about the distribution pattern of the population. The importance value and cover value indices are synthetic indices that permit the ordering of species consideration of their densities, relative dominances, and frequencies with a single value. All of these parameters, together with height and diameter distributions, allow inferring how the species are using environmental resources, and are useful for the ethnobotanist who wishes to evaluate the sustainability of resource exploitation.

Herbaceous vegetation may also be described based on the same phytosociological parameters used in studies of woody plants [24]. However, in semiarid environments, many of the herbaceous species are ephemeral, with their life cycles being delimited by the duration of the rainy season, as occurs, for example, in caatinga vegetation. This factor implies that herbaceous plants must be sampled in the rainy season. However, if a researcher intends to evaluate whether there are differences in the relationship of man with native herbaceous species during the dry season, sampling must be performed in both the rainy and dry seasons.

In addition to the examination of phytossociological parameters, Feitoza et al. [25] proposed a new index specifically for studying herbaceous vegetation in seasonal environments, referred to as the mixed ecological importance value index (MEV). This new index allows the persistence time to be indicated for herbaceous plants

1) $DT = N(U/A)$

2) $Da_i = N_i(U/A)$

3) $DR_i = 100(N_i/N)$

4) $DoT = \sum_{i-1}^{N} G(U/A)$

5) $G = P^2/4\pi$ ou $0{,}785.D^2$

6) $DoA_i = G_i(U/A)$

7) $DoR_i = 100(DoA_i/DoT)$

8) $FAT = \sum_{i-1}^{S} FA_i$

9) $FA_i = 100(NUA/NUT)$

10) $FR_i = 100(FA_i/FAT)$

11) $IVI = DRi + DoRi + Fri$

12) $IVC = DRi + DoRi$

Where:

DT = Total density, (ind.ha^{-1})

N = Total number of individuals

U = Area (10.000m^2)

A = Sampling area (m^2)

DA_i = Absolute density of taxon *i*, (ind.ha^{-1})

N_i = Number of individuals of taxon *i*

DR_i = Relative density of taxon *i*, (%)

DoT = Total dominance, (m^2.ha^{-1})

G = Basal area (m^2)

D = Diameter (cm)

π = pi ou 3,14159

DoA_i = Absolute dominance of taxon *i*, (m^2.ha^{-1})

G_i = Basal area do taxon *i* (m^2)

DoR_i = Relative dominance relativa of taxon, (%)

FAT = Total frequency absolute in sample, (%)

FA_i = Frequency absolute of taxon *i*, (%)

NUA = Number of the sampling units with the táxon *i*

NUT = Total number of the sampling units

S = Number of taxon

FR_i = Frequency relative of taxon *i*, (%)

IVI = Importance value index, (%)

IVC = Importance cover index, (%)

Fig. 4 Formulas and units of the parameters usually used in phytosociological studies

within the vegetation during the year (physiognomic parameter) and is calculated using the formula:

$$VEM = PFA + VI_M$$

where:

VEM = Mixed ecological importance value index

PFA (apparent physiognomic persistence) = Number of months in which the species has leaves or is visible in the habitat divided by the number of months in the year

VI_M = the species' importance value (IV) divided by its population size

During the dry period, many herbaceous species represent an important source of food, water, or shelter for various animal species. In addition, herbaceous plants provide protection to the soil and aid in the retention of seeds on its surface. Therefore, the VEM index is quite interesting, as it reveals the species that are temporarily visible in the community and permits the identification of herbaceous plants that may play an important role in the maintenance and functioning of seasonally dry habitats.

Ethnobotanical studies employ phytosociological sampling to relate information about vegetation usage (made available by urban or rural communities) and the actual quantity and frequency of the resource used [26–28]. These relationships enable inferences to be made about plant richness and distributions, patterns of resource use, the degree of management of the studied fragment, and the priority species for conservation plans.

Furthermore, the sufficiency of floristic sampling is a criterion adopted to ensure that the community is well represented in a sample. Thus, in addition to the species included in the sample, it is common to record the species found close to the sampling units. Subsequently, a rule of three is used to verify the proportion of the total community that is represented by the sampled species.

Another way to evaluate the sufficiency of floristic sampling is to plot a collector's curve. This is achieved by plotting the number of species that were not previously recorded in the sample on the y axis, whereas the x axis represents the number of the sampling unit. The idea is to verify at which point in the sampling a tendency toward stabilization of the species appearance curve can be observed. There is also a running average curve of the number of species in the sample [17, 29]. This curve is drawn from the average accumulated number of species per area. A 5 % variation is calculated from the last average, which is 2.5 % above the last point and 2.5 % below it. If at least 10 % of the sampling units are contained in this 5 % interval, the floristic sampling is considered to be sufficient.

2.5 Floristic Similarity Indices

Indices are often calculated to compare the floristic similarity among communities. Among these indices, the Jaccard and Sorensen binary coefficients (presence/absence) are widely used (*see* [30–32]) and are calculated using the following formula:

$$S_j = a / (a + b + c)$$

$$S_s = 2a / (2a + b + c)$$

Table 2
Number of individuals per species in two hypothetical samples

	Species			
Communities	A	B	C	D
1	5	5	5	5
2	1	2	1	20

where:

S_j = Jacard's similarity coefficient

S_s = Sorensen's similarity coefficient

a = Number of species with simultaneous occurrence in sample 1 and sample 2

b = Number of species that occur only in sample 2

c = Number of species that occur only in sample 1

Even though these two indices are similar, the Sorensen index is recommended when researchers want to place more value on the simultaneous occurrence of a species in the areas being compared. These indices may also be used in ethnobotanical research [11] to compare the similarity of species used by different local human communities, whether quali-quantitative methods have been employed for vegetation sampling or not. For this purpose, the constants a, b, and c are considered to represent the species used by the local human community.

2.6 Species Diversity and Forms of Measurement

The determination of species diversity is a subject of wide interest in ecological and ethnobotanical studies because it not only permits an understanding of the local biodiversity but also enables the detection of differences between and within biomes and the identification of the resources that are important for the social and economic development of a region. One of the first questions asked in the diagnosis of an area is the number of species contained therein. This question may be answered using a species list, but such a list does not permit determination of the number of individuals (abundance) of each species present. Indices addressing species abundance were first used in 1953, and they measure heterogeneity as a synonym of diversity. For a better understanding, let us imagine, as an example, two areas with the same number of species, but with different numbers of individuals, as represented in Table 2.

A richness index based solely on the species list would show that community 1 is equal to community 2; however, community 2 is more uniform. Therefore, the two communities are different, as the total number of individuals in the area is much more evenly distributed among species in community 1.

The literature describes several indices that answer questions about the richness and abundance of populations in the community, leading us to question which factor(s) determine community organization and functioning.

The importance of different factors and/or ecological theories associated with the subject of diversity will not be discussed here. We will, however, indicate the types of indices used in ecological studies, some of the interpretations attributed to them and how they can be used in ethnobotanical studies. For example, imagine that in Table 2, instead of species A, B, C, and D, we used the following categories: medicinal, food, ornamental, and wood, with the numbers representing how many people in the community use the vegetation in this manner. In this case, showing only the usage list would lead to the conclusion that community 1 uses the vegetation resources in the same way as community 2. However, if we consider the frequency with which resources are used, it can be observed that there are differences between the communities: according to the calculated index, the extraction of wood would be a factor with a larger impact for community 2.

In general, species diversity measures may be grouped into two categories: (1) species richness indices based on the number of species in a sample and (2) heterogeneity measures that consider abundance models (based on the distribution of species abundances) and diversity indices (based on the relative abundances of species) (*see* [31]). The latter type of measures aims to express richness and evenness in a single value. Species abundance models are one way to evaluate species diversity and employ statistical sampling theory to investigate how communities are structured. In the literature, the most common types of these models are log series, log-normal distribution, and broken stick models. Among the various heterogeneity measures, we only describe diversity indices. Information on species abundance models is available in the bibliography.

2.7 Species Richness Indices

Objective richness may indicate the number of species in the community. However, this parameter is associated with the problem that the total community size is often unknown. In practice, studies are conducted by sampling, and it is therefore possible that only a portion of the species richness is being sampled. This situation may also occur in ethnobotanical studies.

Richness indices usually require knowledge of the number of individuals or the sample size. The richness index indicated in Whittaker [33] may be calculated for any taxonomic level (species, genus, family) using the formula $d = S/\log N$ or $S/\log A$, where the log calculation may be performed using base 10, e or 2, with the natural logarithm (e) being the most common; S is the total number of taxa; A is the size of the sampled area; N is the number of individuals in the community; and d is the species richness in the sample. The units in which this index is expressed depend on the base of the logarithm used: decit if log 10; nats if log *e*; and bits if log 2 (*see* [30, 31]).

The index indicated by Whittaker is very similar to the Margalef and Menhinick indices. *Margalef's index* is calculated using the formula $D_{Mg} = (S-1)/\ln N$, where one species is subtracted from the total sample. *Menhinick's index* is calculated with the formula $D_{Mn} = S/\sqrt{N}$, where the square root of N is used. However, it should be noted that Margalef's index is calculated using base e (*see* [30]).

The richness indices described above may also be adopted in some ethnobotanical studies. Imagine, for example, that some species in a native vegetation area are widely used and that we wish to express the richness of the useful native species in the vegetation. To calculate richness, only the numbers of species that are indicated as useful in interviews (S_u) would be included in the analysis. Sampling of the vegetation would also be necessary to identify the species that occur there and the number of individuals that each species contributes to the sample. Thus, S_u would be divided by the log of the number of individuals of useful species in the sample. The resulting richness value would refer specifically to the useful species, and you would also be able to verify how well this value represents the total species richness in the sampled plant community.

Sometimes, the objective of a study is to visualize differences between the sampled communities with regard to the richness of useful species. However, the sample sizes in the different communities are not always equal (in terms of vegetation sampling or the people who were interviewed), reducing the reliability of comparisons. One way to minimize this problem is to use the *rarefaction method* (see formula in [30, 31]). This is a statistical method that estimates the expected number of species when a random sample of individuals is taken from a collection. This method assumes that all individuals in the community are distributed randomly. Therefore, the rarefaction curve overestimates the number of species in the community if the populations occur in clusters.

The main question addressed by this method is how many species can be found in a sample of *n* individuals. Thus, if we had a community with 4 species and 42 individuals distributed in the following manner: 21 individuals belonging to species A, 16 to species B, 3 to species C, and 2 to species D, what would be the expected number of species if we had only 30 individuals in the sample?

In ethnobotanical studies, the question would be as follows: if 21 people in a community cited plant *x* as medicinal; 16 cited plant *y* as ornamental; 3 cited plant *w* as wood; and 2 cited plant *z* as food, what would be the expected use diversity indicated if the interviews had been performed with only 30 people?

2.8 Diversity Indices

2.8.1 Simpson's Index

The first nonparametric diversity measure was proposed by Simpson. Based on probability theory, Simpson postulated that diversity is inversely related to the probability of two individuals of the same species being selected at random, given that they are part of the same community. For an infinite population, this probability was defined by the formula $D = P_i^2$, where D = Simpson's index, and P_i = the proportion of individuals of species i in the community. This probability, by itself, is not considered a diversity measure. The conversion of this probability into a diversity measure is performed using the complementary formula below (*see* [30, 31]):

$$1 - D = 1 - \sum (P_i)^2$$

where:

$1 - D$ = Simpson's index of diversity

1 = Probability that two individuals picked at random belong to the different species

D = Probability that two individuals picked at random belong to the same species

P_i = Proportion of species i in the community

Some authors have contended that a confusion exists in the literature as to what Simpson's index measures and have proposed a formula that is reciprocal to Simpson's original index, also referred to as the N_2 reciprocal by Hill. The N_2 reciprocal is defined by the formula $1/D = 1/\sum P_i^2$, where $1/D$ is the reciprocal of the original Simpson's index (*see* [30–32]).

Simpson's index $(1 - D)$ ranges from 0 to 1, and the reciprocal of Simpson's original formula ranges from 1 to the total number of species in the community. Simpson's index is quite sensitive to changes in the number of abundant species in the sample. Thus, in plant communities in which few species account for a large number of individuals, this index's value will be close to 0.

In ethnobotanical studies, the use of an adapted Simpson's index permits perception of the diversity of local knowledge about specific resources in the environment. For this purpose, it would be necessary to modify the meaning of P_i, which could then be understood as the proportion of interviewed people that use the plants in the category i.

2.8.2 Shannon–Wiener's Function

Shannon–Wiener's diversity measure (H') is widely employed. Shannon–Wiener's function was derived independently by Shannon and Wiener and is sometimes erroneously referred to as Shannon–Weaver's function. It is based on information theory, as suggested by Margalef, which aims at measuring the degree of order/disorder within any system. The main question addressed by this function is prediction of the expected species in a community sample. This uncertainty is measured using the following formula (*see* [30, 31, 34]):

$$H' = \sum_{i=1}^{s} (p_i)(\log_2 p_i)$$

where:

H' = Shannon–Wiener's diversity index (bits/individual)

S = Number of species

P_i = Relative proportion of abundance of the specie i

This diversity index may be calculated using base 2, e, or 10, and it can also be converted from one base to another. The units in which the results are expressed depend on the adopted base: bits per individual (base 2); nats per individual (base e); or decits per individual (base 10). Shannon–Wiener's diversity index is sensitive to the number of rare species in the community and increases with increasing numbers of species.

There is another way to express Shannon–Wiener's index, which was recommended by Hill, using the formula: $N_i = e^{H'}$ (*see* [30])

where:

e = 2.71828

H' = Shannon–Wiener's function (calculated with base e log)

N_i = Number of species equally common that produce the same value of diversity (H')

The recommendation made by Hill [35] was put forth with the objective of using as the unit of the results only the word species, which is better understood by ecologists and ethnobotanists than logarithmic bases. However, the use of H' is more common than N_i.

In ethnobotany, the H' index is used [36, 37] to express the diversity of plant species. It can also be employed to express the diversity of use of the vegetation of a given area. Three questions are frequently asked. (1) Does the diversity of the plants used correspond to the species diversity in the area? (2) Are the plant species evenly used by the human population? (3) Are there differences in the diversity of use with regard to the structure of the interviewed human population (e.g., regarding age or sex)? For this type of analysis, some modifications in the understanding of the terms in the formula must be made.

If the researcher's question addresses the general diversity of vegetation use versus species diversity in an area, he/she must conduct interviews to identify which species are used and a quantitative floristic survey of the area's vegetation, according to the floristic sampling procedures discussed above, to identify the species that occur in the area and their abundances. Shannon–Wiener's index must be applied to the total dataset resulting from the survey, allowing calculation of the species diversity in the area, and to the data only on the species indicated as useful (S_u). Through these calculations, it is possible to evaluate how much the diversity of used species corresponds to the total diversity in the area.

If the question only involves the diversity of uses attributed by the population, the plant uses indicated in the interviews must be categorized, and the number of people in the community that use the plants in different categories must be determined. In this case, vegetation sampling is not required [37]. In this index, p_i would represent the number of people who cited species i in the interviews. In other words, the abundance of species i is taken as being equal to the number of informants who cited it in the interviews. Thus, the result of the calculation will permit comparison of the diversity of the used plants taking into consideration their relative abundance in the citations. If people's knowledge about the utility of plants is poor, then the diversity based on use may be a subset of the real species diversity in the region.

The application of Shannon–Wiener's index based only on interviews does not always permit assessment of the species actually used by the human population because the informants generally refer to the plants by their popular names, and they may sometimes use the same name for ten different species, which may result in an underestimation of the useful plants in a region. The inverse may also occur if different people use different popular names for the same taxonomic species, leading to an overestimation of the useful species in an area. Therefore, it is necessary to perform sampling of botanical material so that the calculation is carried out using a list of taxonomically identified species, rather than ethnospecies.

It is worth noting that the use of this index based solely on interviews does not allow evaluation of the stocks of resources in nature. Sometimes, the species cited in an interview may be grown randomly in backyards and gardens, without occurring in the native vegetation. Another issue is that the abundance of a species' citations does not necessarily correspond to the abundance of the plant population in nature. Thus, it is possible that in 50 interviews, a given species will be cited 50 times, but only five individuals of this species may be found in 1 ha of the sampling area. Therefore, when the index is calculated only based on interviews, all inferences made from the index results require wider discussion about the resource's conservation and sustainable use.

Recently, a new index for the evaluation of the diversity of people's knowledge about biological resources has become available in the literature (*see* [38]). The application of this index reveals that the adaptation of Shannon–Wiener's index to determine the diversity of useful species in a region does not always permit evaluation of the real knowledge an informant has about the useful plants in an area.

2.9 Evenness

The measure of evenness is also known as uniformity or equitability (*see* [30–32, 34]). Evenness shows how the abundance of species is distributed in a community and may be applied to both Simpson's and Shannon–Wiener's indices. This parameter may be calculated using the following formula (*see* [30]):

$$E = D / D_{max}$$

where:

D = Observed index of species diversity

D_{max} = Maximum possible index of diversity, given the number of species (S) and the total number of individuals (N) in sample

For Simpson's evenness measure, the maximum diversity is obtained when all species have the same abundance ($p = 1/S$). Thus, Simpson's evenness (E) is given by the following formula: $E = \check{D}_{max}$ and $\check{D}_{max} = 1/S$

where:

$\check{D}_{max}$ = Maximum value possible of the Simpson's index

S = Number of species in sample

The evenness measure of Shannon–Wiener's function is calculated using the following formula: $J' = H'/H'_{Max}$

where:

J' = Evenness measure of Shannon–Wiener's function (varia de 0 a 1)

H' = Shannon–Wiener's diversity index

$$H'_{Max} = Maximum\ value\ of\ H' = \log S$$

The application of evenness measures in ethnobotanical studies would clearly show whether the cited species are known equally well by the community as well as which resources are under greater use pressure.

These diversity indices adjusted for ethnobotanical studies become important tools because they permit comparing areas and making inferences about cultural, social, and economic aspects of a region and about the conservation status of the biome in question.

2.10 Techniques to Evaluate the Richness and Diversity Indices

It is possible to verify whether the richness and diversity estimates obtained for a community are close to the community's real diversity. This evaluation is important for all research that addresses sampling, whether the sampling is of vegetation or of human populations by means of interviews.

One of the commonly used approximations for this purpose is the so-called jackknife. This tool involves a nonparametric technique, which therefore does not present demands with regard to the sample's statistical distribution. This methodology permits calculation of the mean and variance of the diversity of species or of the reported uses. It may be applied to all of the indices discussed above (*see* [30, 31]).

This technique consists of a consecutive series of richness estimates with a decreasing sample number for each estimate. In practice, the community's diversity is initially calculated considering all samples. Consecutive estimates are then performed, with one sample removed for each new estimate. The estimated values are called pseudovalues, and their mean and variance are subsequently calculated. These means and variances are expected to be closer to the community's real diversity.

There is another technique related to the jackknife method known as bootstrapping. The bootstrapping procedure is similar to the jackknife method. Estimation by the jackknife method is recommended when the community is represented by a small number of samples, whereas bootstraping is recommended when the communities are represented by a larger number of samples. Computer programs to perform the bootstrapping procedure and calculate the other indices described in this chapter are available. The formulas for the jackknife and bootstrap procedures are available in Krebs [30].

3 Final Considerations

Phytosociological and ethnobotanical studies employ statistical tests and multivariate analyses to test for differences among the measured parameters, correlate variables, and test hypotheses. These tests include the *t*-test, *Kruskal-Wallis* test, multiple linear regression, cluster analysis, and principal components analysis (PCA).

The choice of test depends on the characteristics of the sampled biological data and on the question that the researcher intends to answer. Some biological data exhibit a distribution that meets the statistical assumption of normality (i.e., the mean is equal to both the median and the mode) or exhibit variance homogeneity, permitting the application of statistical tests classified as parametric. However, large portions of biological data (possibly the majority) violate these assumptions. In this case, the researcher attempts to normalize the data by means of transformation (e.g., using logarithmic or square root transformation). When the transformed data still violate the assumption of normality or do not exhibit variance homogeneity, the researchers use tests classified as nonparametric.

The classification and assumptions of each test and the way they are calculated can be found in basic statistical books [34, 39–41]. All of these statistical tests may be performed with specific software, such as SYSTAT [42], PC-Ord [43], or SAS [44]. However, interpretation of the results must be performed by the researcher, considering the hypothesis and/or question established in the study.

References

1. Rudd MA (2011) Scientists' opinions on the global status and management of biological diversity. Conserv Biol 25(6):1165–1175
2. Hardin R (2011) Competing cultures of conservation. Conserv Biol 25(6):1098–1102
3. Painter RLE, Duran A, Miro E (2011) Indigenous alliances for conservation in Bolivia. Conserv Biol 25(6):1084–1086
4. Pietrzyk-Kaszynska A, Cent J, Grodzinska-Jurczak M, Szymanska M (2012) Factors influencing perception of protected areas—the case of natura 2000 in Polish Carpathian communities. J Nat Conserv 20:284–292
5. Phillips O, Gentry AH (1993) The useful plants of Tambopata, Peru: I. Statistical hypothesis tests with a new quantitative technique. Econ Bot 47:15–32
6. Phillips O, Gentry AH (1993) The useful plants of Tambopata, Peru: II. Additional hypothesis testing in quantitative ethnobotany. Econ Bot 47:33–43
7. Galeano G (2000) Forest use at the Pacific Coast of Chocó, Colômbia: a quantitative approach. Econ Bot 54(3):358–376
8. Wezel A, Bender S (2003) Plant species diversity of homegardens of Cuba and its significance for household food supply. Agroforest Syst 57:39–49
9. Albuquerque UP, Araújo TAS, Ramos MA, Nascimento VT, Lucena RFP, Monteiro JM, Alencar NL, Araújo EL (2009) How ethnobotany can aid biodiversity conservation: reflections on investigations in the semi-arid region of NE Brazil. Biodivers Conserv 18:127–150
10. Fernandes-Ferreira H, Mendonça SV, Albano C, Ferreira FS, Alves R (2012) Hunting, use and conservation of birds in Northeast Brazil. Biodivers Conserv 21:221–244
11. Ghorbani A, Langenberger G, Sauerborn J (2012) A comparison of the wild food plant use knowledge of ethnic minorities in Naban River Watershed National Nature Reserve, Yunnan, SW China. J Ethnobiol Ethnomed 8:17. http://www.ethnobiomed.com/content/8/1/17
12. Matteucci SD, Colma A (1982) Metodologia para el estudio de la vegetacion. Secretaria General de la Organización de los Estados Americanos, Washington, p 167
13. Araújo EL, Albuquerque UP, Castro CC (2007) Dynamics of Brazilian caatinga—a review concerning the plants, environment and people. Funct Ecosyst Commun 1:15–29
14. Araújo EL, Ferraz EMN (2010) Análise da vegetação: amostragem, índices de diversidade e utilidades na etnobotânica. In: Albuquerque UP, Lucena RF (eds) Métodos e Técnicas na pesquisa etnobiológica e etnoecológica. Editora Comunigraf/NUPEEA, Recife, pp 161–198
15. Cottan G, Curtis JT (1956) The use of distance measures phytossociological sampling. Ecology 37:451–460
16. Felfili JM, Eisenlohr PV, Melo MMRF, Andrade LA, Meira Neto JAA (2011) Fitossociologia no Brasil: métodos e estudos de caso. UFV, Viçosa, MG, p 556
17. Martins FR (1991) A estrutura de uma floresta mesófila. Editora da UNICAMP, Campinas, p 246
18. Lopes CGR, Ferraz EMN, Araújo EL (2008) Physiognomic-structural characterization of dry- and humid-forest fragments (Atlantic coastal forest) in Pernambuco State, NE Brazil. Plant Ecol 198:1–18
19. Andrade LA, Fabrincante JR, Araújo EL (2011) Estudo de Fitossociologia em Vegetação de caatinga. In: Felfili JM, Eisenlohr PV, Melo MMRF, Andrade LA, Meira Neto JAA (eds) Fitossociologia no Brasil: métodos e estudo de casos, vol 1. Editora UFV, Viçosa, pp 339–371
20. Mori AS, Silva LA, Lisboa G (1989) Manual de manejo de herbário fanerogâmico. Centro de Pesquisa do Cacau, Ilhés, p 104
21. Curtis JT, McIntosh RP (1950) The interrelations of certain analytic and synthetic phytosociological characters. Ecology 31:434–455
22. Mcintosh RP (1967) An index of diversity and the relation of certain concepts to diversity. Ecology 48:392–404
23. Sherpherd GJ (1995) Fitopac1. Manual do usuário. Departamento de Botânica, UNICAMP, Campinas
24. Reis MAS, Araújo EL, Ferraz EMN, Moura AN (2006) Inter-annual variations in the population structure of an herbaceous of caatinga vegetation in Pernambuco, Brazil. Rev Bras Bot 3:497–508
25. Feitoza MOM, Araujo EL, Sampaio EVSB, Kill LHP (2008) Fitossociologia e danos foliares ocorrentes na comunidade herbácea de uma área de caatinga em Petrolina, PE. In: Albuquerque UP, Moura NA, Araújo EL (eds) Biodiversidade, potencial econômico e processos eco-fisiologicos em ecossistemas nordestinos, vol 1. Comunigra/Nupea, Recife, pp 11–38
26. Rico-Gray V, Franco-Garcia J, Chemas A, Puch A, Sima P (1990) Species composition, similarity, and structure of Mayan homegardens in tixpeual and Tixcacaltuyub, Yucatan, Mexico. Econ Bot 44(4):470–487

27. Ferraz EMN, Silva SI, Araújo E, Melo AL (2002) Espécies de interesse econômico na Mata Atlântica de Pernambuco: distribuição e relação entre formas de uso e abundância das populações. In: SECTMA, Silva JM, Tabarelli M (eds) Diagnóstico da biodiversidade do Estado de Pernambuco. Editora Massagana, Recife, pp 689–696
28. Florentino ATN, Albuquerque UP, Araújo EL (2007) Contribuições de quintais florestais na conservação de plantas da caatinga, município de Caruaru, PE, Brasil. Acta Bot Bras 21:37–46
29. Mueller-Dumbois D, Ellenberg H (1974) Aims and methods of vegetation ecology. Wiley, New York, p 574
30. Krebs C (1989) Ecological methodology. Harper & Row, New York, p 654
31. Magurran A (1989) Diversidad ecológica y su medición. Ediciones Vedra, Barcelona, p 200
32. Pinto-Coelho RM (2000) Fundamentos da ecologia. Editora Artmed, Porto Alegre, p 252
33. Whittaker R (1975) Communities and ecosystems. Macmillan, New York, p 385
34. Kent M, Coker P (1994) Vegetation description and analysis: a practical approach. Wiley, New York, p 663
35. Hill MO (1973) Diversity and evenness: a unifying notation and consequences. Ecology 54:427–432
36. Lamont SR, Eshbaugh WH, Greenberg AM (1989) Species composition, diversity, and use of homegardens among three Amazonian villages. Econ Bot 53(3):312–326
37. Begossi A (1996) Use of ecological methods in ethnobotany: diversity indices. Econ Bot 50(3): 280–289
38. Araújo TAS, Santos ALA, Melo JG, Medeiros MFT, Ramos MA, Silva RRV, Almeida CFCBR, Albuquerque UP (2012) A new technique for testing distribution of knowledge and to estimate sampling sufficiency in ethnobiology studies. J Ethnobiol Ethnomed 8:11
39. Sokal RR, Rohlf FJ (1979) Biometria Principios y métodos estadísticos en la investigación biológica. H Blume Ediciones, Madrid, p 831
40. Zar JH (1996) Bioestatistical analysis. Prentice Hall, Englewood Cliffs, NJ, p 660
41. Valentin JL (2000) Ecologia numérica uma introdução à análise multivatiada de dados ecológicos. Editora Interciência, Rio de Janeiro, p 117
42. Wilkinson L (1992) SYSTAT: the system for statistics. SYSTAT, Evaston
43. McCune B, Mefford MJ (1999) PC-ORD. Multivariate analysis of ecological data. MJM Software, Gleneden
44. SAS Statistical Analysis System (1996) SAS system for linear models. SAS Institute, Cary

Chapter 11

Techniques for Collecting and Processing Plant Material and Their Application in Ethnobotany Research

Lucilene Lima dos Santos, Fábio José Vieira, Luciana Gomes de Sousa Nascimento, Ana Carolina Oliveira da Silva, Leidiana Lima dos Santos, and Gardene Maria de Sousa

Abstract

Until they are incorporated into an herbarium, the sampling of plant species and every process involved in their collection are key steps in many areas of botany. When sampling processes are performed properly, they provide information well beyond a plant's scientific name; these processes also emphasize the local importance and conservation status of the species and its significance in different environments. Showing plants in situ, fresh parts or even plant parts in small voucher specimens are becoming increasingly common tools in ethnobotany because the goal is to obtain information and rescue the use of particular plants by traditional communities. Therefore, visualization of the plant sample is critical for ensuring that both the researcher and the community member are referring to the same plant. In this chapter, we will discuss the common techniques of sampling, preserving, and processing plant material as essential tools for basic field procedures and their relevance and applications in methods that assist in ethnobotany research.

Key words Plant collection, Herborization, Herbarium, Complex pressing, Ethnobotanical studies

1 Introduction

The correct identification of plants is essential to ethnobotany research because it is a vital link between scientific information and popular knowledge [1]. The collection, the preparation of voucher specimens, plant identification, and storage of the specimens in an appropriate place (in this case, an herbarium) are important, because allows other researchers using the same material [2]. In botany-related studies, the stages that include the collection, processing, and incorporation of plant material are key because the final product of these procedures is the voucher specimen. The specimen consists of plant parts removed from a given species, including pressed and dried vegetative and reproductive structures (flowers and/or fruits), that have been properly labeled with information

Ulysses Paulino Albuquerque et al. (eds.), *Methods and Techniques in Ethnobiology and Ethnoecology*,
Springer Protocols Handbooks, DOI 10.1007/978-1-4614-8636-7_11, © Springer Science+Business Media New York 2014

about the sampling site and some morphological characteristics. This plant material is deposited in an herbarium and will provide basic information for students and researchers from different fields, such as plant taxonomy, phytogeography, plant anatomy, ecology, plant physiology, and ethnobotany.

Moreover, the sampling of plants for the preparation of a voucher specimen is an essential and mandatory part of any study because it ensures that the plant was actually collected and that misidentification of the species will be avoided.

The information on the tag attached to the voucher specimen aids in plant identification and provides data about the plant's geographic distribution, its conservation status, its potential uses, and its local relevance. Sampling, herborization, and incorporation into an herbarium are essential for understanding the local flora in many environments and serve as a genetic heritage. Some plants that are not available in their natural habitat are only represented by voucher specimens, and their analyses are important for developing conservation strategies for certain species.

2 Methodology for Sampling and Herborization

2.1 Sampling Kit

The equipment and tools required for collecting plant material may vary according to the characteristics of each group of plants. Therefore, before each field trip, the preparation and selection of all of the equipment that will be used should be carried out. Knowledge about techniques and the proper handling of these tools is also important because they will provide convenience, ease of transportation, and safety to the collector while ensuring better use of the samples and that minimal damage to the plants from which the samples were extracted occurs. The essential objects for this phase of ethnobotany research are described below, independent of the plant group to be studied.

2.1.1 Field Notebook

These books are used to record field observations, including a plant's characteristics and descriptions of the environment in which it occurs (Fig. 1). For example, plant characteristics, such as color and aroma, are usually recorded (such information may be lost during the drying process), and its height and diameter at breast height (DBH) (e.g., 1.30 cm from the ground) are also recorded. Characteristics of the environment may include the geographic coordinates of the sampling site, physiognomic aspects of the vegetation, soil characteristics, and the distribution pattern of the species in the area.

The field notebook should be small to ensure that it is easy to carry and handle in the field.

For ethnobotanical studies, a more detailed description of the plant, including any common names, its uses and the parts used

Scientific name			
Family:	Common name:		
Collector:	N°:	Date:	
Determiner and date:	Material collected:		
Altitude:	Latitude (S):	Longitude (W):	Country:
Estate:	Municipality:	District:	
Place:	Vegetation:		
Height:	DAP/DNS:	Soil:	Habit:
Sample collection for studies in laboratory			
( ) Phytochemistry ( ) Farmacology ( ) Wood quality ( ) Biomass ( ) Others ______			
Type of sample (organ):	Quantity of wood:	No:	
Observations: Flowers (color and odor): Local uses:			

Fig. 1 Model of a field notebook for collecting plant material as used in ethnobotanical studies

(roots, bark, and/or bast), is required. The parts that are used can be added to the voucher specimen next to the sample.

All notes should be performed at the moment of sampling; therefore, no information is forgotten.

2.1.2 Common Graphite Pencil, Wax Pencil, or Permanent Marker

During fieldwork, we are always exposed to different weather conditions; thus, the common pencil is recommended for writing field notes because ballpoint pens, although very practical, dissolve easily when in contact with water and ethanol, which can lead to the loss of important information.

2.1.3 Pruning Shears, Pole Pruner, and Machete

These tools aid in the collection/cutting of plant material from small species or trees or even for the removal of an entire plant from the ground. The machete becomes an especially important tool when collecting bromeliad species. The pole pruner is essential for the collection of arboreal plants, allowing the highest branches to be reached.

2.1.4 Field Presses

These tools are wooden structures that assist in the pressing and transportation of plant material that has already been collected.

2.1.5 Newspaper or Blotting Paper

Plant samples are placed on the paper. On the blotting paper, it can be interesting to note information concerning the plant, such as its name, the identity of the collectors, the date, and the sampling site. This measure should be taken because plants are often mixed; if the data are also on the blotting paper, it can aid the fieldwork. We recommend the use of newspapers only if blotting paper is not available because newspapers can cause the spread of plant diseases.

2.1.6 String

String should be made of resistant material because it is used to tie the pressed samples. The ideal string should also be resistant to oven temperatures (above 70 °C).

2.1.7 Plastic Bags

Bags are used when fieldwork requires a large amount of samples. We recommend plastic bags measuring 40 × 25 cm (60 L). Plastic bags are notably practical and facilitate movement during fieldwork, allowing us to collect a large amount of plant material. It is worth noting that samples should be pressed on the day of their collection.

2.1.8 Adhesive Tape

Tape is used to join groups of branches collected from a single individual, i.e., from a single sample number when it is not possible to press the plant at the moment of collection. Adhesive tapes are also used for sealing plastic bags containing fruits or bark.

2.1.9 Plastic Containers

When plants are sampled with flowers, storage of some flowers in plastic containers with 70 % ethanol for subsequent identification is recommended. The same procedure can also be used for collecting fruits for subsequent identification or for inclusion in a seeds and fruits collection. Plastic containers can also be useful for collecting exudates (e.g., latexes or resins) for subsequent phytochemical analyses.

2.1.10 Measuring Tape

Measuring tape is used to obtain different measurements such as the length of plant parts, trunk circumference, which can be the diameter at ground level (DGL) or DBH, and the distance between individuals.

2.1.11 GPS

This equipment is important because it provides the geographic coordinates of the sampling site or of a particular plant.

2.1.12 Envelopes

Envelopes are used to store small plants, loose leaves, and seeds.

2.2 Sampling of Plant Material: General Aspects

During fieldwork, depending on the objective of the study, the researcher should use an experimental design that limits him/her to collecting plants of a specific physiognomy (herbs, trees, vine/creepers, or shrubs) or study all plant strata, as in general floristic surveys.

Generally, collecting plant samples with fruits and/or flowers is recommended because these reproductive structures assist in the correct identification of the specimen. If collecting fruits and/or flowers is not possible, it is important to collect the vegetative structures with the aim of later finding the plant with flowers and fruits to confirm its identification.

Collecting plants with mechanical damage or damage caused by insects or fungi should be avoided; these plants should only be collected if judged to be representative of the species or if no other plant was found. Collecting extra fruits and flowers is also useful to aid the identification of the plant and to facilitate the work of taxonomists, who can remove or dissect fragments of these extra samples.

Collecting 4–6 samples of each individual species is ideal. One sample, known as the unicate, is deposited in the herbarium of the institution at which the study was carried out. The remaining samples, called duplicates, are used for exchange with researchers and experts at other institutions. It is also important to deposit samples in other herbaria as a backup in the event of loss or damage at the herbarium where the unicate was deposited.

Sampling techniques for special groups, which have morphologies that limit certain aspects of sampling, and the importance of collecting plant samples for ethnobotany studies will be described next.

2.3 Pressing

The essential steps for herborization [3–6] are described below.

1. Press samples immediately after their collection or at least on the same day. Never leave plants to be pressed on the following day because certain plants dehydrate easily, which can make pressing impossible.
2. Prepare samples with an average size of 34 × 25 cm. In cases where this procedure results in the loss of essential parts (e.g., very large leaves) or if the sample represents the entire plant, it should be folded into an N or V shape.
3. Branches with several leaves should be thinned.
4. If it is necessary to discard some leaves of the sample, the leaves should be cut such that the vestige of the petiole is evident.
5. Leaves should be pressed alternately in the sample, i.e., some leaves should exhibit the abaxial face and others the adaxial face.
6. Long inflorescences should have their stem lightly creased before being folded.
7. Large fruits that cannot be pressed should be detached from the sample and placed in nylon mesh bags tagged with the collector's name and the corresponding collection number.
8. Place each specimen collected on a sheet of blotting paper, and never include two samples on the same sheet.

Fig. 2 Pressing and drying of plant material: (**a**) Model of the presser assembly; (**b**) plant drying in a field oven

9. The collector's name followed by the collection number should be written in the margin of the sheet (make notes with pencil if the material is preserved in ethanol).

2.4 Drying

The following procedure is mainly derived from two classic works [4, 5].

If a field oven is available (Fig. 2b), the previously pressed material should be arranged in the presser (Fig. 2a) at the end of each sampling day and later placed inside the oven in the following sequence:

1. One of the presser grids
2. A cardboard sheet
3. A corrugated aluminum sheet, if available
4. Another cardboard sheet
5. Blotting paper with one sample

Steps 2–5 should be repeated for each of the samples. This structure is called the "sandwich." After assembling "sandwiches," the presser should be closed using the other wooden grid, tied using the strings and placed in the oven.

Each batch of pressed material should have an average height of 40–50 cm. Because of variations in drying time according to the consistency of the sample, the material should be examined every 3 h. The blotting paper should be replaced if it becomes too wet, always taking care to record the data contained on the old paper. As the sample dries and shrinks, the strings of the pressed batch should be readjusted such that the sample does not wrinkle.

The plant material is considered to be dry when it is rigid, does not bend when it is suspended and is not wet when touched. At this point, the material is removed from the oven. Upon returning from the field, all material should be examined to determine if it should be returned to the oven in the laboratory for additional time.

When a field oven is not available, plant material is dried in the sun. To ensure that the material is preserved, it is sprayed with commercial-grade ethanol at the time of its collection

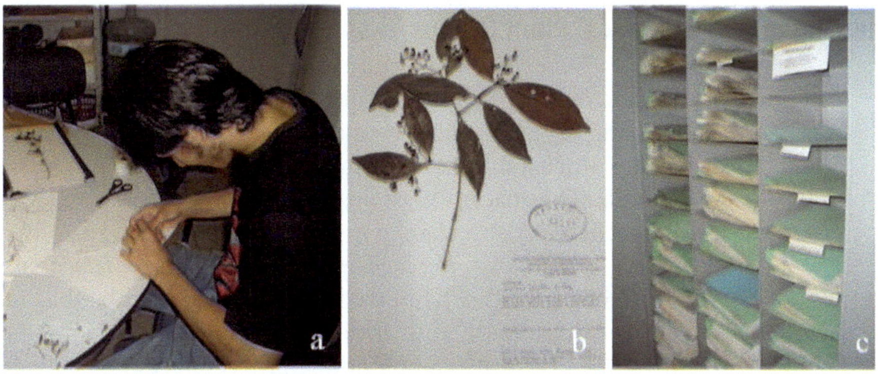

Fig. 3 (**a**) The process of sewing the voucher specimen; (**b**) specimen sewn and identified with tags and an herbarium number; (**c**) Incorporation of plant material into an herbarium

(this information should be recorded on the field tag because samples treated with commercial grade ethanol are not recommended for molecular studies). On the following day, the pressers are removed, and the edges of the package containing samples are covered with two cardboard sheets. Next, the package is strongly tied, placed inside a plastic bag, and sealed with adhesive tape. After returning from the field, the samples should be placed in the oven for immediate drying.

After they are dry, the samples should be placed in a −20 °C freezer for 3 days to avoid different types of infestations. After this period, samples should be returned to the oven to remove any traces of moisture. Later, samples should be sorted by their collection campaign in numerical order of their collection and placed in plastic bags with naphthalene or camphor until being deposited in the herbarium.

2.5 Preparation of Voucher Specimens

As previously mentioned, samples with flowers and/or fruits and preferably more than four samples of each individual should be collected whenever possible with the exception of herbs; in most cases, the entire individual is collected and herborized under a single collection number [7]. With the samples in hand, a specimen is selected to be the unicate, which will be sewn to the cardboard (Fig. 3a); the remaining samples, the duplicates, will be sent to other reference herbaria. It is worth noting that the sample should be completely attached to the cardboard because a poorly sewn sample can result in damage to the specimen due to frequent handling in the herbarium.

2.6 Registration and Incorporation into an Herbarium

After the plants are sewn, the next step is the attachment of the identification tags to the lower right corner of the cardboard (Fig. 3b). The herbarium stamp will be in the lower left corner with the corresponding registration number that the sample receives from the herbarium into which it is incorporated (Fig. 3c).

3 Collection in Special Cases: Plant Groups That Require Complex Pressing

During fieldwork, adverse situations can arise in the collection and pressing of particular groups of plants. Initially, such situations create uncertainty about how to collect and process a given sample. The macambira (*Bromelia laciniosa* Mart. ex Schult. f.—Bromeliaceae), Brazilian twin-stemmed palm (coco catolé) (*Syagrus cearensis* Noblick—Arecaceae), coroa-de-frade (*Melocactus bahiensis* (Britton & Rose) Luetzelb.—Cactaceae), embaúbas (*Cecropia* spp.—Urticaceae), and anthuriums (*Anthurium* spp.—Araceae) are some practical examples difficult to collect plants. These plants are locally important for several communities in notably different environments and require distinct collection and herborization methods. Below, we describe procedures to facilitate the collection and processing of some of these groups of plants.

3.1 Bromeliaceae

The Bromeliaceae are a well-known group of plants that are used by several communities in different Brazilian ecosystems. In Brazil, they are known as "macambira" and/or "bromelia." Bromeliads exhibit morphological peculiarities, which result in a degree of complexity during their collection and pressing.

The entire leaf should be collected including the leaf sheath; it should be folded into an *N* or *V* shape. A young and an adult leaf should always be collected. Depending on their shape and size, inflorescences can be sectioned at the base of the scape (the stalk that supports the inflorescence) and folded as many times as necessary. It is possible to use the blunt side of a machete to mark the points where it will be folded or cut the inflorescence or infructescence in half, if they are globose, similar to the pineapple.

For large bromeliads whose leaves exceed 3 m in height, the leaf is measured, and samples from the tip, middle, and base of the leaf are herborized. These fragments should not exceed the size of the voucher specimen. Ideally, the leaves and inflorescence would be pressed separately.

Small bromeliads are collected and pressed in their entirety. We recommend care during their collection because several bromeliad species have associations with insects. For example, *Aechmea mertensii* (G. Mey.) Schult. & Schult. f. has an association with leafcutter ants. The collector should clean the debris that accumulates at the leaf base. Bromeliad leaves are arranged in spirals and form a microhabitat for certain groups of animals (insects and amphibians) and plants. Therefore, it is useful to record the presence of animals and other plants in the water accumulated by the leaves.

3.2 Woody Vines (Lianas)

Vines or creepers, popularly known as lianas or branches and commonly associated with the families Bignoniaceae, including such plants as cipó-alho (*Arrabidaea* spp.), and Convolvulaceae, including such plants as the Brazilian jalap (batata-de-purga) (*Operculina macrocarpa* (Linn) Urb.), are important among the plants used by traditional communities. Because lianas usually cling to larger plants, they can be confused with the supporting plant; therefore, different aspects of their collection are featured.

Apical and basal branches should be collected along with the respective structures that fixed them to the supporting plant. During pressing, the collector should indicate the position of the collected branch because in certain species, branches are highly diverse. The stem and/or more developed branches should be cross-sectioned because they provide key information for the identification of the family.

Branches are rolled up so that the bigger leaves can be folded or even sectioned to facilitate herborization. If possible, the collector should provide the name of the supporting plant.

The collector should not add sterile and leafless branches to samples with fertile branches to avoid mixing different material because lianas are rarely found in isolation. Usually, several liana species are found tangled together with leafless stems, and their leafy stems are spread in the tree canopy.

3.3 Arecaceae (Palm Trees)

This group includes the known palm trees, the babassu palm (coco-babaçu) (*Orbignya phalerata* Mart.), carnauba wax palm (carnaúba) (*Copernicia prunifera* (Mill.) H.E. Moore), buriti palm (*Mauritia flexuosa* L. f.), and coconut (*Cocos* spp. and *Syagrus* spp.). These species are important because of their multiple uses. For example, in house construction, the trunks are used for the vertical structures, and their leaves are used for covering the house. They also feature edible fruits.

The same method that is used for collecting arboreal species is used for collecting large palm trees.

It is essential that the following features are recorded: the total height of the plant; stipe height; the presence of anchoring roots; the number of leaves in the canopy; leaf and petiole lengths; the number of leaflet pairs; the length of the inflorescence and its axes; and the presence and distribution of thorns.

The sampling should include the leaflet, portions of the middle and tip of the leaf blade, representative parts of the inflorescence, flowers, and fruits.

During pressing, all leaflets on one side of the rachis should be sectioned, leaving their respective vestiges intact; several axes of the inflorescence should also be cut, and the bases should be left to indicate where the insertions were located. Each paper on which plant parts are placed should be marked with the name of the collector and the number, the name of the organ (e.g., leaf or inflorescence) to which the sample belongs, and an indication of the corresponding fragment (Fig. 4). Even if voucher specimens are from different parts of the plant, they receive the same

Fig. 4 Pressing of an Arecaceae specimen, of which different parts received the same collection and registration number

collection and registration number because they are treated as a single sample. For example, F. Vieira (58), Leaf, base; F. Vieira (58), Leaf, middle; and F. Vieira (58), Leaf, tip.

3.4 Plants with Bulky Parts (Cacti, Inflorescences, Fruits, and Tuberous Roots)

Plants with stem tubes, bulbs, tuberous roots, xylopodiums, or even large inflorescences and fruits present problems during herborization because they contain large amounts of water in their tissues, and some parts of the structure are thick.

These bulky parts are serially cut longitudinally and/or cross-sectioned and pressed separately from the more sensitive parts to reduce the loss of their primary characteristics because of creasing or wrinkling.

The blotting paper should be changed daily to avoid infestation, especially by fungi, which damage the material, and to facilitate drying.

Excessively succulent plants, especially certain pseudobulbs of orchids, can be cut at their fleshiest parts, which facilitates water evaporation during drying. A longitudinal cut should also be performed to remove part of the internal tissue.

For cacti, the cladodiums with ribs should have their fertile parts cut longitudinally. To show the number of ribs, a 2-cm-thick cross-section should be added to the sample.

3.5 Aquatic Plants

Water lettuce (*Pistia stratiotes* L.—Araceae) and common water hyacinth (*Eichhornia crassipes* (Mart.) Solms—Pontederiaceae) are good examples of this plant group. Aquatic plants require special care because of their organ structures and the amount of water contained in their tissues. These plants should be collected carefully because most of them are notably delicate, with roots that are strongly attached to the substrate, and could be damaged with any rough movement.

Aquatic plants should be herborized at the sampling site because their flowers wilt easily. We recommend distending the flower parts with a small brush or tweezers, wrapping them in cellophane or bond paper (40 kg) and attaching them to the blotting paper using adhesive tape. The rest of the procedure should follow the general rules for collecting, pressing, and drying plant material.

4 Sample Collection for Ethnobotanical Studies

4.1 Collection of Wood Samples

Timber resources are widely used in several environments, especially as fuel, but also in housing construction and for producing different artifacts [4] (Fig. 5a). Wood samples consist of a $15 \times 8 \times 4$ cm block of heartwood. Usually, this block is cut from a bigger one that has been removed from a tree trunk after slow drying to avoid cracking in the sample. This material can also be used for the analysis of density, moisture, and total ash content in ethnobotanical studies that address phytofuels.

4.2 Sampling of Material for Phytochemical Analysis

Any plant part can be collected. During collection, the samples should be placed in blotting paper bags, and the larger parts that do not fit in the bags should be wrapped in blotting paper. The use of newspaper is not recommended because it can release the chemical substances used for printing, which can damage/contaminate the sample. The researcher responsible for the analysis should determine the amount of material collected. However, certain precautions should be taken during collection. For example, avoid girdling (when the entire bark is removed, and the phloem is reached), which can lead to the death of the plant (Figs. 5b, c).

The reader should remember that certain preservative solutions commonly used to avoid contamination during the drying process, such as 70 % ethanol and camphor, could change the plant's natural properties. In such cases, the plant material should be dried without being treated with a preservative solution.

4.3 Edible Plants

In studies on the nutritional value of plants, the researcher should be careful with the collection, storage/packaging, and preservation of fruits and other parts of interest (e.g., the central portion of the stem, the stele (Fig. 5d) and roots) to maintain them in perfect condition for laboratory analyses. Immediate processing is always recommended; however, because it is not possible most of the time, it is important to store these parts in a cooler or freezer to ensure that the chemical characteristics of the plant are maintained until these components are isolated and subsequently analyzed.

Fig. 5 (**a**) Wood for making artifacts; (**b**) bark extraction for a phytochemical study; (**c**) appropriate amount of dry bark for use in analysis; (**d**) blue columnar cactus (Facheiro) (*Pilosocereus pachycladus* F. Ritter) used to make sweets; (**e**) checklist-interview: dry plants in voucher specimens (field herbarium) and photos shown on a laptop; (**f**) a respondent showing plants that were cited as useful near his house during a guided tour

4.4 Material Sampling for a Checklist-Interview

Visual prompts are increasingly used as tools in ethnobotanical studies (*see* [8–10]). Among the different types of prompts, interviewees can be shown *photos* (printed or on a laptop and of high quality) of the species of interest or even a *field herbarium*, which consists of an album or a block of voucher specimens containing 32×21 cm samples (Fig. 5e). Showing a reasonable number of plants to the interviewee should also be considered to avoid transforming a useful tool into a long and tiresome methodology for the study participants, which may result in the loss of key information about a particular group of plants.

4.5 Material Sampling During a Guided Tour

On a guided tour during the interview, the persons participating in the study show the cited plants, which may occur in their backyard, in a neighboring plot, or even in a distant area. The tour provides a good opportunity to collect the specimens cited as useful because the respondent is nearby and can assist in ensuring that the correct plant is collected (Fig. 5f).

5 Final Considerations

Generally, all of the steps that concern preparations for the sampling of plant species, whether for general knowledge of the flora or for more specific knowledge of a particular area (e.g., catalog of medicinal, timber, or edible plants), until the samples reach the herbarium are important and contribute to the understanding, conservation, and use of these resources by traditional and nontraditional communities.

Acknowledgements

The authors thank Dr. Luiz Vital Fernandes Cruz da Cunha and Simone Santos Lira Silva, M.Sc. for providing a number of the cited studies and Alyson Luiz Santos de Almeida, M.Sc. for Fig. 5a, d.

References

1. Cotton CM (1996) Ethnobotany: principles and applications. Wiley, London
2. Ming LC (1996) Coleta de Plantas Medicinais (sampling of medicinal plants). In: Di Stasi LC (Org). Plantas Medicinais: Arte e Ciência. Um Guia de Estudo Interdisciplinar. Editora da UNESP, São Paulo, pp 69–86
3. Fidalgo O, Bononi VLR (1984) Técnicas de Coleta, Preservação e Herborização de Material. Botânico, São Paulo
4. Mori AS et al (1989) Manual de Manejo do Herbário Fanerogâmico. Centro de Pesquisas do Cacau, Ilhéus, BA
5. IBGE (1992) Manual Técnico da Vegetação Brasileira, Manuais Técnicos em Geociências. IBGE, Brasília
6. Martin GJ (1995) Ethnobotany: a methods manual. Chapman & Hall, London
7. Bezerra P, Fernandes A (1989) Fundamentos de Taxonomia Vegetal. EUFC, Fortaleza, CE
8. Medeiros PM et al (2010) Uso de Estímulos Visuais na Pesquisa Etnobotânica. In: Albuquerque UP, Lucena RFP, Cunha LVFC (Orgs) Métodos e Técnicas na Pesquisa Etnobotânica. Recife, Pernambuco
9. Medeiros PM et al (2008) The role of visual stimuli in ethnobotanical surveys: an overview. In: Albuquerque UP, Alves MA (eds) Current topics in ethnobotany, 1st edn
10. Case RJ et al (2006) Ethnopharmacological evaluation of the informant consensus model on anti-tuberculosis claims among the Manus. J Ethnopharmacol 106(1):82–89

Chapter 12

An Introduction to Zoological Taxonomy and the Collection and Preparation of Zoological Specimens

Kleber Silva Vieira, Washington Luis Silva Vieira, and Rômulo Romeu Nóbrega Alves

Abstract

Humans have an innate capacity to create categories and classify them according to particular characteristics. The attributes we refer to things and organisms have served as the basis for developing many classification systems, although taxonomy is the only one used to study biodiversity. Having access to biological diversity it is not a simple task, and investigators must know a series of prerequisites ranging from the collection and preparation of zoological material to their correct storage in scientific collections. The variety of collection and preparation methods for animal specimens is almost equal to the diversity of known taxonomic groups—although it is possible to consider general methods that can be applied to most taxa that will facilitate access by systematists, taxonomists, and nonspecialists alike to the information contained in those specimens. The objective of the present chapter is to present a short introduction to zoological taxonomy and explain the general methods used in collecting and preparing zoological material, in a manner that can be appreciated by nonspecialists. We also wish to provide sufficient orientation so that zoological material can be correctly deposited in scientific collections and supporting a better use of the information presented in the specimens by researchers from different areas, including ethnobiology.

Key words Systematics, Methods, Zoological collections

1 Introduction

When humans evolved the power of abstraction we became capable of restructuring the manner in which we observe the world, honing an innate capacity to order things that we judged to be, or not similar—including objects and plants and animals, and even ourselves. This ordering capacity bequeathed humans the power to perceive the great diversity of forms that exist in nature— and the set of methods employed in classifying the diversity of living things came to be called taxonomy [1].

Although all human cultures have developed specific manners of classifying things and organisms [2] they have differed in their greater or lesser power of organizing that information. The Greeks

Ulysses Paulino Albuquerque et al. (eds.), *Methods and Techniques in Ethnobiology and Ethnoecology*, Springer Protocols Handbooks, DOI 10.1007/978-1-4614-8636-7_12, © Springer Science+Business Media New York 2014

were credited with perfecting our manner of seeing and classifying the world, and Aristotle fundamental in his logic (inspired, in turn, by Plato) the concept of classes and how they related to others—a system of logically positioned categories that conform to a hierarchy of values [3]. In spite of being quite old, this classification system is as useful today as it was in the time of Aristotle, and is currently used to educate children and body laymen in informal conversations, and is employed in more technical matters by numerous professionals such as economists, ecologists, physiologists, geneticists, doctors, and taxonomists [4].

Aristotle's method served as the basis for the development of traditional taxonomy at the beginning of the eighteenth century when scientists employed the formal concept of classes, and a new field of science emerged from the combined efforts of Peter Artedi and Linnaeus to classify the natural world [5]. Taxonomy, as we understand it today, creates biological groups that share certain characteristics and elaborates names for those groups. The classification system of Linnaeus also employs the concept of categories, which represent another part of its methodological structure. Biological classifications therefore serve as sources of information about biodiversity and reference systems concerning biological organisms [6].

Researchers, taxonomists or not, interested in investigating biological diversity must be aware of a series of different collection techniques and preparation methods for organisms that are prerequisites for their inclusion into scientific collections. The variety of these collection and preparation methods is almost proportional to the diversity of taxonomic groups currently known [7]—but even though this array of techniques is quite numerous, it is still possible to discuss general methods that can be applied to most of the taxa that interest researchers.

The collection and preparation of animal specimens does not constitute a sport, nor do they represent acts of meaningless violence or cruelty, as many nonscientists imagine—but rather a methodology that is indispensable to the investigation of biological diversity and unique in providing taxonomists, systematists, anatomists, physiologists, and ecologists with information that would not be possible in any other manner. Scientific collections are archives containing our accumulated information about nature, and these collections, many of which are hundreds of years old, are open to any and all individuals interested in research or teaching.

The objective of the present chapter is to present a short introduction to zoological taxonomy and explain the general methods used in collecting and preparing zoological material, in a manner that can be appreciated by nonspecialists. We also wish to provide sufficient orientation so that new material can be correctly deposited in scientific collections and the information contained within those specimens preserved and used constructively by researchers in many different areas, including ethnobiology.

2 Zoological Nomenclature

The publication of the tenth edition of the *Systema Naturae* near the beginning of the second half of the eighteenth century can be considered to mark the birth of modern taxonomy [8]. After the death of Peter Artedi, one of the originators of the "Systema Naturae" the Swedish naturalist Linnaeus presented his work describing, so far, the most appropriate way of classifying and naming organisms.

In his proposal, Linnaeus classified and created a series of divisions so inclusive that encompassed from minerals to animals, being also useful in the classification of microorganisms. The nucleus of the classification methodology presented in the *Systema Naturae* [9] was not in itself totally new, as it basically recognized similarities and grouped entities (whether animals, plants or minerals) into groups with exclusive names—as had been done since ancient Greece [10]. Nonetheless, the greatest innovation in the *Systema Naturae* was, without a doubt, the practice of assigning binomials to those entities.

The method applied by Linnaeus borrows the concept of Aristotelian classes in the construction of classification levels, which he denominated "taxa". This conferred to the system a platonic aspect in explaining the observed similarities between objects and in naming categories—and gave Linnaeus the title of essentialist in the history of biology. In addition to its taxonomic structure, the methodology developed by Linnaeus adopted the concept of categories when the objective was to describe the attributes of the specimens or types, and it was here that the binomial rule took on its importance. Within the dimension of a category, a genus (origin, tribe, offspring, genus) is hierarchically more encompassing than an eidos (exterior aspect, form, way of being), but such terms are relative and dependent on references, and a genus could be an eidos when compared to a higher level in the hierarchy. Linnaeus created only five categories within his system: Species > Genus > Or der > Class > Kingdom, all of which were very precise from a logical-ontological point of view.

Zoological nomenclature, as originally developed by Linnaeus, is now regimented by the International Code of Zoological Nomenclature [8] that established a system of rules and recommendations dealing with the correct manner of creating and applying zoological names. There are, of course, other nomenclatural codes in biological taxonomy, but the zoological code is independent and only recognizes its own rules and recommendations [11]. Its objective is to assure stability and universality in the scientific names given to animals, and to guarantee that each is unique. The origin of this code was a consequence of the confusion of names assigned to animals in the scientific literature in the period around the first half of the nineteenth century as a result of the expansion of the Linnaean system and its acceptance in many countries and in many different areas of zoological research. The regimentation of the International Code of Zoological Nomenclature has eight basic principles [8]:

1. The code avoids dealing with taxonomic judgments, which must not be made subject to regulation or restraint
2. The nomenclature used does not determine the inclusion or exclusion of any taxon, nor the order to be adopted for any group of animals, but does furnish the name to be used for a taxon, and this name will indicate its hierarchical level
3. The creation of a name-bearing type permits its application to a taxon without infringing on taxonomic judgments. Every name, according to the Code (except for names that designate "a collective groups" and for taxa above the level of family) must is permanently attached to a name-bearing type. For species and subspecies, this name-bearing type designates at the same time the name of a specimen (or number of specimens) that constitute the name-bearer of the taxon; for genera and subgenera, it is the nominal species; for taxa at the level of family it is the nominal genus. For this reason, when a taxon at some level is created by a taxonomist, this taxon may contain several name-bearing types, each with a name that is available for use at that level. The principle of priority (which can be modified in the interest of stability and universality) is used to determine which of them is the unique valid name
4. Nomenclatural rules are tools that can be used to guarantee maximum stability compatible with taxonomic freedom. As such, the Code recognizes that rigidity in the application of the Priority Principle can, in certain cases, cause problems in cases of widely used or well-known names by the validation of another little-known name, or even a name that has long been forgotten. As such, the rules orient the taxon is to consistently use the Priority Principle, except in situations when its application would be incompatible with stability or universality or cause unnecessary confusion. In these circumstances, the Code contains provisions that modify the automatic application of the Priority Principle and the establishment of the primacy of names, or in assigning name-bearing types, in spelling a name, or in other similar situations
5. To avoid ambiguity, the use of the same name for different taxa is prohibited. This is the Principle of Homonymy
6. The Code provides orientations to zoologists that must establish new names as well as rules to determine if some name, which was previously proposed, is available and with what priority; it also considers cases in which a name needs to be corrected to

be correctly used—the Code in this case allows for the confirmation of the name-bearing type of the taxon it denotes to be ascertained (and, when necessary, its fixation)

7. The Code also allows self-interpretation and self-administration by establishing an operational International Commission of Zoological Nomenclature and provides conditions under which the Code can be modified
8. "Jurisprudence" does not exist in Zoological Nomenclature. Nomenclatural problems are decided by the direct application of the Code, and never by referral to any precedent

Although all these principles are important, the laws of Priority, Homonymy, and Synonymy are consider the most significant by those who are interested in suggesting new proposals for the names of taxa. It is convenient to note here that the Code attempts to reach its goals of stability and universality by way of these principles, definitively prohibiting homonymy at the level of family and genus in all of the animal kingdom. At the species level, homonyms are prohibited within each genus [11]—although the Rule of Homonymy does not apply in cases in which the generic zoological names are identical to names of plants or microorganisms, due to the principle of independence of the Code. Another situation prohibited by the Zoological Nomenclatural Code is utilization of two or more names for the same taxon (synonymy); when an error of this type is encountered it should be immediately corrected.

If the same name is given to two or more taxa (homonymy) or if two or more names are attributed to the same taxon (synonymy), the Code indicates the Rule of Priority to resolve this type of problem. The Rule of Priority is a very direct and simple principle, but one of the most important, and states that if two or more homonyms or synonyms are encountered the oldest published name will be considered valid.

All of the names published in this way must have an author (or authors) and a date, with the author being considered the person who first published a given name. According to the Code, the name of the author and the date do not compose part of the official scientific name of the taxon, in spite of the fact that they are generally cited together (ex. *Elachistocleis ovalis* Schneider, 1799; *Elachistocleis ovalis* Schneider; or only *Elachistocleis ovalis*). Master's and Doctoral theses are not considered publications to this end, as the Code only considers texts that were published in specialized and indexed periodicals.

When a species is transferred to another genus the name of the author and date of publication should be presented in parenthesis (ex. *Ceratophrys aurita* (Raddi 1823)), this represents a special case that recognizes that proposed taxonomical rearrangements are potentially reversible, as they are subject to different interpretations by different authors.

Proposals for new names, especially for species, require that the author(s) designate a type species. This process, which the Code designates as the principle of typification, consists of choosing a specimen (if that is the case) that will fix the name of a given taxon. This type-specimen may be a whole animal or part of one, a colony or part of one, or even the activity of an animal (in the case of fossils) or impressions, molds, or counter molds. In addition to these cases, there are a series of others that are discussed in detail by Bernardi [11, 12] and by Mateus [13].

If the author uses a single specimen for the description, it will be denominated the holotype, which is the most common situation. In case the author needs more specimens he/she can designate a holotype and the others will be called paratypes. Although the author may choose not to designate a holotype, the entire series will serve to fix the new name and each specimen will be called a syntype. In this situation, any author may subsequently choose a single specimen to be the agent that fixes the name—and this specimen will be the lectotype, and the others be called paralectotypes. The difference resides here only in the date of the original publication or those published later. In special situations, in which the holotype or lectotype is lost, an author may propose a neotype, although this is only done in situations of extreme necessity. The formation of a specific name must respect the binomial rule and be formed by more than one letter. Generic names are always singular nominative nouns (ex. *Leptodactylus*) and likewise formed by more than one letter. At the family level, the name that is used to designate the genus loses its singular genitive termination and receives the suffix -idae (ex. Leptodactylidae). The formation of the term that designates a subfamily follows the same principle, although the suffix utilized is -inae (Hylinae). These principles are recommended by the Code, which also recommends the use of the suffix -oidea in the designation of a superfamily (Tyrannosauroidea). Additionally, the suffix -ini (Odontophrynini) is used for tribes and -ina for subtribes.

The development and expansion of the Phylogenetic System in the second half of the twentieth century revealed some problems with the traditional nomenclatural rules when applied to supra-specific monophyletic groups. As a result of these conceptual and practical problems, nomenclatural systems exclusive to the Phylogenetic System were developed [14, 15]. Some of these proposals were not well received by the scientific community (whether due to rivalry between different schools or to the fact they were not functional) and an alternative nomenclatural proposal for the Phylogenetic System known as the PhyloCode has been better accepted and increasingly widely used in the last 5 years [16].

There are a number of similarities between the PhyloCode and traditional zoological nomenclatural rules, as for example their mutual objectives of determining stable, universal, and unambiguous methodologies. Both codes avoid infringing on the personal judgment of the taxonomists in the choice of names, and attempt to apply the Principle of Priority to correct eventual cases of homonymy or synonymy. The PhyloCode is nonetheless independent of the Traditional Code and does not apply a taxonomic hierarchy, recognizing the categories of clade and species only as different biological entities and not as hierarchical levels. A species is therefore treated as a segment of a populational lineage while a clade is considered a monophyletic group of species (or organisms). Both are products of evolution, with a phenomenological purpose, and the fact of being named does not make any difference at all. The act of giving a name, in this case, would be a fortuitous consequence, in which the recognition of diagnostic characters would represent questions to be decided through the use of empirical evidence rather than cases of personal decisions.

3 The Collection and Preparation of Zoological Material

Species descriptions following the international code of zoological nomenclature, taxonomic revisions, and morphological, biogeographical, reproductive or ecological studies, as well as the phylogenetic system of a given species or group of species are all based on specimens collected and deposited in scientific collections. Specimens collected and deposited in both scientific and didactic collections are used in graduate and postgraduate research and laboratory classes, biology courses, veterinary medicine, ecology, oceanography and agronomy, among many others.

Additionally, ethnobiological projects that study the utilization of animals or animal parts by traditional societies can also use information gathered from specimens collected by the researcher himself or from specimens deposited in academic collections. The enormous diversity of animal species, associated with the wide varieties of environments that they occupy, demand corresponding diversification of the collection and preparation methods used to incorporate these organisms in zoological collections [7]. According to Caramaschi [7], a number of basic procedures should be followed in order to guarantee adequate collections with the least possible risk to the integrity of the animal populations and to the collector himself. As such, this chapter presents the general technical principles involved in collecting and preparing vertebrate specimens, the equipment utilized, the types of zoological collections possible, and their importance.

3.1 Collecting

Zoological specimens are collected when there is a real necessity for specimens for scientific and/or educational purposes, and if the conditions for their preparation, transport, and storage are adequate. Random collections or collections without documentation, or those with inadequate preparation and storage facilities should be discouraged. Additionally, the collector must respect the environmental regulation and licensing procedures of the country where they are working in order to avoid judicial complications.

There are basically two types of collecting methods: general and specific (senso Vanzolini and Papavero [17]). A general collecting is undertaken by the systematic exploration of a given area with attention being given to all of the animals encountered. In reality, this collecting method is impractical, as no one can go into any field and collect everything; additionally, the logistics and equipment and materials necessary for this type of collection would be quite impractical. Researchers are generally interested in a particular taxonomic group, even though they may come across a collectible animal species outside of the focus of their work. Nonetheless, the researcher must pay attention to all informative details even if apparently external to the specific group they are interested in.

According to Vanzolini and Papavero [17], specific collecting is characterized by searching for and capturing specific specimens—which might be just a single species or a certain taxonomic group, agricultural pests, or organisms that inhabit a specific environment or special microhabitat (e.g., species that live inside bromeliads, caverns, or fossorial organisms). Specific collecting does not have to be rigid [7], and if the researcher is an area with rare or interesting species, the opportunity to collect some of these individuals should not be wasted (even if they are not the direct object of the study). It must be noted, however, that this particular collecting situation still must obey the criteria established by the collecting permit and must never be applied in cases of rare species that are threatened with extinction.

3.2 Series

Series are composed of individuals of one or more species collected in a given area within a given period of time. Independent of which taxonomic group they belong, animals can present intraspecific variations that are individual or generated by sexual dimorphism or by age. Individuals of the same species from different localities can show accentuated variations when closely examined [7]. As such, studies of a given species should analyze as many series as possible, and preferentially those from various localities, covering the total geographical distribution of the focal animal.

A good series should contain, in principle, sufficient numbers of specimens to cover all or most of the variations encountered in a population. However, this number should not affect the quality, in other words, should be collected as many individuals as possible to prepare and transport in good condition without spoil the collection. Due to the environmental laws that govern the emission of collecting permits for animals, the upper number of specimens to be collected is always restricted. In this case, the series should be as all-inclusive as possible and include specimens of both sexes, and of all sizes, colors, and ages. As such, a good collection is one that does not select only males or females, or the largest and most visible specimens, but rather is performed more or less randomly without any individual or collective choices [7, 17].

3.3 Field Notebooks

Field notebooks should be of high quality and have hard covers and sturdy paper, and be of a size that is both convenient to use and easy to store. The pages should not become loose easily—and for this reason it is best to use notebooks whose pages are sewn together rather than stapled or spiral-bound. Some researchers prefer to use loose but numbered pages that can later be bound in a binder [7].

Field notebooks should be used to record three basic types of information [7, 17]: (1) Trip diaries that register both scientific and human-related incidents, observations of the local inhabitants and local customs, as well as physiognomic and ecological descriptions of the study localities; (2) The catalog numbers of the material collected, including, for each number, all of the data noted for the specimen (or lot of specimens) that it corresponds to. This numbering must be done very carefully to avoid repetitions or numbering errors that could lead to a loss of information; (3) Observations about the animals and the sites. In addition to descriptions of the local environment, information should also be included that can be used to critically define the collected material, such as where it was collected, the date, the beginning and end of the collection period, the number of people collecting, the characteristics that can be lost or altered at death or during the preservation process (such as behavior, coloring, weight, emitted sounds, or luminosity).

3.4 Tagging and Labeling

Any and all specimens or lots that are collected must be well identified so that the information will not be lost and the collecting nullified. The surest procedure is to tag the material in the field during collection. In this context, the tag is physically bound to the collected specimen. The tag may be only numerical, containing the field number followed by the identification of the collector, or it may include written data such as the weight, color, size, sex, and sexual maturity of the specimen [7, 17]. The tag should be made of excellent quality material—especially tags used to identify specimens that will be conserved in humid environments or that will come into contact with formaldehyde or alcohol. For tags that will only be exposed to dry environments, good quality and firm-textured paper will be sufficient.

Tags can be of five types: fiber—which can be previously printed or filled in with Nankin Indian Ink; tin—which is more difficult to prepare, as the letters and numbers must be stamped; cardboard—which can be printed with a sequential numbering stamp using printing ink; plastic—which are made from continuous rolls of plastic tape and a sequential numbering stamp (so that the letters and numbers are easily visible); or tracing paper—which should be of excellent quality (at least 120 weight), and the letters and numbers should be written legibly using good quality Nankin Indian Ink.

An additional label should be included in the recipient that holds a specimen (or lot of specimens) collected in a given area. This label generally contains three basic items of information: (1) the locality where the material was collected (farm, stream, river, lake, beach, etc.), which should be indicated with the greatest precision possible, preferentially georeferenced using a GPS device; (2) the collection date (day, month and year) followed by the name of the district or municipality and then the state and country. To avoid confusion, the month should be written in Roman, not Arabic numerals; and (3) the name(s) of the collector(s), with the abbreviation "Col." or "Leg" before them (from *legit* in Latin, which means collected).

The recipient label should be made of good quality tracing paper (minimum weight of 120). The information must be legible, and Nankin Indian Ink is recommended or inkjet or laser printing (as long as the ink used is not soluble in formaldehyde or alcohol).

3.5 Preparation

This procedure allows a specimen to be stored indefinitely in a collection without any serious distortion or degradation of its parts and should inhibit cell destruction by autolysis as well as impede bacterial and fungal attacks.

Preparation is sometimes confused with fixation, but there is a great difference between these two methods. Fixation stabilizes the animal tissues, while preparation allows the zoological material to be stored for indeterminate periods (as explained in the paragraph above). There are two principal types of preparation: liquid and dry [17].

Liquid preparation uses fixing and conserving solutions. The principal fixer used is formaldehyde (which is a gas in its pure state). Commercially available formaldehyde is generally used for fixing animal tissues and consists of a saturated aqueous solution containing 40 % formaldehyde (commonly called 100 % formaldehyde). The final formaldehyde concentrations most commonly used are 5 and 10 %, appropriately neutralized, with the 5 % solution being favored for watery and delicate specimens.

Alcohol at 75° GL is the principal liquid preservative solution and its concentration should never be below 50° GL. Materials preserved in alcohol should be inspected periodically to avoid significant decreases in alcohol concentrations and/or liquid volumes [7, 17].

Fig. 1 Fish collecting, of the selective type, using throw-nets (Photo: Virgínia Diniz)

3.6 Sampling or Collecting Methods for Vertebrates

There are two basic methods for sampling or collecting vertebrates: (1) active searching (which is also known as general or visual searching) for a set period of time, which consists of searching for animals in their native habitats and microhabitats (sensu Martins and Oliveira [18]); and (2) passive sampling, using different types of traps. There are various techniques used in each of these sampling methods depending on the vertebrate group being collected.

3.6.1 Fishes

The fish capture should be undertaken in all the aquatic environments to be sampled, whether continental or marine. Fishing equipment is generally of two types: selective and nonselective. Selective instruments consist of traditional fishing gear as well as nets with meshes and specific shapes designed to capture fish of certain sizes and with similar lifestyles (Fig. 1), including fishing lines, fishing poles, sieves and dip net, longline fishing (both the open ocean and bottom type), fishing nets such as cast net, trawl net, gillnets and hoop net (also used in capturing other vertebrates) [17, 19]. Nonselective equipments are those used when no importance is given to the size or lifestyle of the fishes to be collected, and includes electrical fishing equipment as well as chemical substances such as rotenone. The latter is very efficient in small pools left by receding rivers or tides, or in rocky streams.

Nonselective methods have many disadvantages and are generally not recommended. The use of chemical substances, for example, is considered an environmental crime in Brazil according to Article 35 of Law No. 9605 of February 12, 1998. Electrical fishing equipment is not only difficult to transport and store but also can occasionally cause serious accidents [19].

The collected specimens are fixed in 10 % formaldehyde. Small specimens can be placed in an aqueous solution of formaldehyde while still alive, to guarantee the fixation of their internal organs and musculature. Medium and large specimens require formaldehyde injections through the mouth and anus (or cloaca), into the musculature, at the base of the swimming fins, and into their internal body cavity [19]. The fish can be tagged by tying a label to the caudal peduncle, by passing a line through the mouth brachial chamber, or by placing a tracing paper label inside the jars or plastic sacks containing formaldehyde used to store them (plastic sacks can be transported inside a cooler or other such container). If prepared tags are not available, labels can be rolled up and stuffed into the mouths of the collected specimens [17]. After these procedures, the animals should be maintained immersed in 10 % formaldehyde in adequately sized recipients and kept there for 2 weeks. After this period they should be washed to remove excess formaldehyde and then placed in recipients containing 75° GL ethyl alcohol.

3.6.2 Amphibians and Reptiles

Amphibians and reptiles are often collected together and denominated herpetofauna. There are essentially two basic methods used for capturing these animals—active or passive capturing. Active capturing or visual searching for defined periods (sensu Martins and Oliveira [18]) can be undertaken during either the day or night, while passive captures use snares and traps.

Active collecting of amphibians and reptiles can be undertaken in two ways: General—consisting of wandering randomly in a predetermined area while examining the greatest number of possible microhabitats (in the vegetation; underneath the leaf litter, fallen trunks or rocks; in crevices or on rock outcrops; in burrows along the margins of streams and lakes, or within them); and Systemized—searches in which the researcher surveys and established transect to encounter these animals. The most commonly used sampling methods in systemized collecting include audio strip-transects to sample anuran amphibians [20], transect sampling [21], and quadrant sampling [22].

The principal equipment utilized in herpetofauna collecting includes: long-sleeved shirts and long pants and a hat, a backpack, canteen, penknife, knife or machete, a field notebook and Nankin Indian Ink pens, hand-held flashlights or head lamps, extra batteries, boots, and snake garters. Rubber boots are recommended for use in swampy areas. Collectors also use slingshots, and large tweezers (35 cm long) to capture animals hiding in holes or burrows, and bamboo pole lassos. Compressed air pistols with 4.5 mm lead shot are used to collect lizards. Lutz lassos and poles with hooks are used to contain and manage snakes. Wooden boxes and cloth and plastic sacks are used to transport the collected specimens.

The principal method of passive collection involves the use of straight-line drift fences and pitfall traps made from 60 L (or larger)

Fig. 2 Interception and fall-traps with directional fences utilized in collecting amphibians and reptiles (Photo: Washington L.S. Vieira)

plastic jars or buckets that are buried to their rims and separated by plastic drift fences at least 5 m long and 1 m high (Fig. 2). These traps must be inspected at least twice a day (in the early morning and before nightfall). This method is very useful in capturing species that are difficult to detect by active searching and aids in standardizing and maximizing herpetofauna collections in any area.

Other devices such as "Glue Trap," "Sticky-trap" and "Trap-a-roach" are largely used to capture arboreal specimens [23–27]. Funnel traps are generally set near bodies of water to capture tadpoles, snakes and turtles and can be associated with pitfall traps to maximize collection efforts [24, 26].

Cage traps ("Tomahawk" and "Sherman") are principally used to capture mammals but are equally useful in collecting large lizards, turtles, and crocodilians. The baits used in these traps must be renewed daily (meat, fruits and/or a mixture of sardines, flour, and peanut butter) and nontarget animals attracted to these baits can often be used as bait for larger animals [28].

The specimens collected can be transported in cloth or plastic sacks to the laboratory where they are anesthetized and killed. In the case of amphibians, the specimens can be euthanized by inhalation of ethyl ether or by injections with high doses of Benzocaine, Lidocaine or Xylocaine, or by drowning in 40° GL alcohol.

Reptiles are generally killed by inhaling ethyl ether or by injecting them with high doses of Xylocaine, Pentobarbital, or Ketamine [29]. All of the specimens are ultimately fixed in 10 % formaldehyde and then conserved in 75° GL alcohol.

3.6.3 Mammals

Collecting mammals in scientific search has three principal objectives: to study their skins, bony parts, and soft organs [17]. The information contained in the collected specimens and deposited in scientific collections can serve as the basis for studies of systematics, evolution, biogeography, genetics, ethnobiology, and ecology.

Mammals are encountered in almost every type of habitat—which makes it difficult to establish fixed methodologies for their capture [17]. Nonetheless, collecting of these animals can follow the sampling methods established for other specific groups, depending on the objectives of the study. Mammals can usually be collected by two distinct techniques: active collecting, also known as direct searching; and by using traps [17].

According to Vanzolini and Papavero [17], active collecting is utilized for certain groups, principally medium-sized animals such as small felines, foxes, small deer, wild pigs, and monkeys. This type of collecting/hunting commonly involves the use of a 36 caliber or larger shotgun, 22 caliber rifles, and shells with small (ns. 8 and 10) or large shot (ns. 3 and 5). To use this equipment the researcher must have a license for owning and carrying a firearm as well as certification for their use. If the firearm belongs to an institution, the researcher must carry a letter from that institution allowing him to use it in the field [30].

Other equipment is also used to collect mammals, and most items have been listed above under the topic of amphibians and reptiles. However, collectors should also take with them into the field (especially for preparing the skins) scissors, tweezers, cotton, sewing thread, arsenic, borax (sodium borate) or taxidermy paste, wire, pliers, brushes, sawdust or cornmeal (used to absorb fat, blood, or other body fluids during the preparation of the skins), pins, and wooden or cork boards that can be used to hold the skins while they are drying. The utility and utilization of these equipments are well described in the publications of Vanzolini and Papavero [17] and Auricchio [31].

Various types of traps are used to collect mammals, and there is considerable variation among the different models. The most commonly used types are Tomahawk and Sherman traps (Fig. 3), although other less-common models are available such as Havahart and Longworth traps [31].

Tomahawk and Havahart traps are made from galvanized wire with openings at one or both ends and are used to capture larger animals such as felines, foxes, bush-dogs, and striped hog-nosed skunks. Sherman traps are made from aluminum or galvanized metal sheets and can be collapsible, or not; they are generally easy

Fig. 3 Sherman-type traps used for collecting mammals (Photo: Washington L.S. Vieira)

to set up and come in many different sizes. Longworth traps are also made from aluminum and have a tunnel entrance with an automatic trap door mechanism and a holding cage coupled to the posterior part of the tunnel when the trap is fully armed [31]. To attract animals, these traps must be baited on a daily basis with a mixture of fruits and sardines or flour and peanut butter; they should be checked early in the morning. Rat and mouse traps can also be used to collect small mammals such as rodents and marsupials, although they have inconvenience of crushing some part of the animal—making its skeleton or skull unusable for study [17, 31].

Intercept and pitfall traps linked by straight-line drift fences [32–34] are also used and can maximize mammal collections (Fig. 3) by capturing small and medium-sized animals such as rodents, marsupials and armadillos (Fig. 4). Details of this method were described earlier, under the topic of amphibian and reptile collections.

Fine-mesh mist nets made from very thin black nylon thread come in many sizes, although the most common models are 6 m long by 3 m tall. These nets are extended and held vertically by aluminum or bamboo poles (Fig. 5); they should not be tightly stretched, and their holding poles should be placed to allow the formation of bulges or overlaps that allow the animals to become entangled as they struggle [17, 30, 31, 33, 35].

There are different ways of erecting these net barriers, and they can be stretched in a line or in the form of a "Y," suspended from trees, placed around abandoned human habitations, over streams and small lakes, at the margins of large rivers, or near burrows or the entrances of caves [35]. Another type of net used to capture

Fig. 4 *Euphractus sexcinctus* captured using interception and trap-falls with directional fences (Photo: Washington L.S. Vieira)

Fig. 5 Capturing bats with mist nets (Photo: Washington L.S. Vieira)

bats (although little-used yet in Brazil) are harp traps, which consist of two aluminum supports holding an array of thin threads. For more details, see Jones et al. [34] and Kunz et al. [35].

Mammal inventories, in addition to using the methods described above, can be complemented by recording animal tracks. These registers are made by photographing the tracks or making plaster molds. "Photographic traps" use an apparatus that triggers a camera when an animal passes near it. Owl pellets—the regurgitated remains of owl meals that contain the bones and skins of small mammals—can be used to complement inventories [31]. Additionally, interviews with local residents of human communities and hunters will often supply interesting information, although the precision of their observations will depend on obtaining supportive elements from them (such as skulls or skins) that will

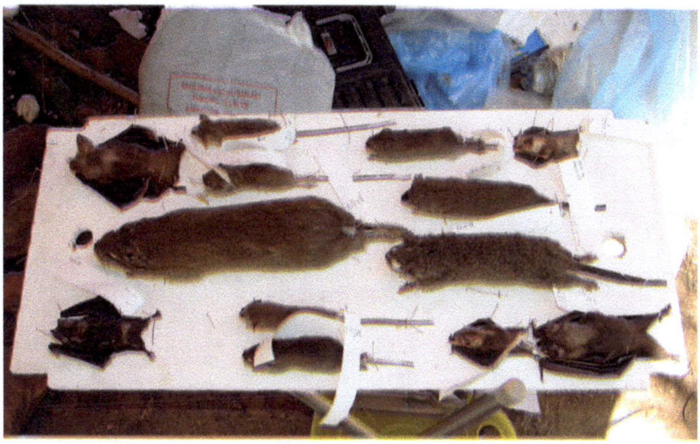

Fig. 6 Specimens preserved by taxidermy and labeled in the field (Photo: Washington L.S. Vieira)

allow more precise identifications of the species and serve as testimonial material.

Local residents also commonly stock products derived from local animals (such as their skulls and skins) used for medicinal or ornamental purposes [36–41]. Once donated, this biological material will allow secure identifications of the species. The practical knowledge of hunters and their hunting techniques can often maximize collection results. For more details, refer to Alves et al. [39].

The specimens collected can be killed using anesthetics commonly employed by veterinarians or with ether or chloroform. Injections with alcohol or 10 % formaldehyde at the base of the cranium are commonly used for small individuals [31], with large doses of cyanide being used for large animals [17]. The specimens can be prepared by liquid or dry techniques. Liquid preparation involves injecting 10 % formaldehyde into the specimens and their subsequent conservation in 75° GL alcohol. Dry preparation is accomplished through taxidermy (Fig. 6).

Packing and transport of the specimens in liquids can be accomplished using jars or large vessels containing 75° GL alcohol, or in boxes (in the case of skins preserved by taxidermy). More details are available in Vanzolini and Papavero [17] and Auricchio [31].

3.6.4 Birds

Collecting of bird specimens for scientific purposes are made with two principal objectives: (1) to study their plumage and other structures of their external morphology, and; (2) to study their skeletons and soft tissues. Additional information is usually gathered concerning their nests, eggs, behavior, and songs [17]. In general, birds can be collected by either active (direct searches) or passive collecting.

Active searches are undertaken during the day, generally in the first hours of the morning, or at night fall, by way of random

Fig. 7 Capturing birds using mist nets (Photo: Washington L.S. Vieira)

searches or along preestablished transects. According to Bibby et al. [42], active searching along transects is more appropriate as it allows the researcher to estimate bird densities at the site and provides more accurate information about habitat and microhabitat use and gradients of species distributions in the area. Long transects can be divided into smaller sections, thus facilitating observations and collecting.

Birds can be brought down with a 36 caliber shotgun using small (n. 8) or large shot (n. 5). The caliber of the shotgun, the sizes of the shells and the shot must all be adequate to the size of the bird, its distance, and to the collecting locality. The use of this equipment requires a hunting permit and a registration permit for the firearm itself. Additionally, if the firearm belongs to an institution, a letter must be presented indicating that the bearer is authorized to use it [30].

Much of the other equipment used for capturing birds is used with other animal groups, although a good pair of binoculars and camera are required as well as a recording machine. If the recorder is not digital, extra tape cassettes are recommended. Additional equipment includes a directional microphone, reserve batteries, banding rings (for banding birds that will not be collected), a needle-nosed pliers, paper used to prepare cones (utilized in transporting the specimens), and cotton. Addition material, principally that used in preparing skins, was listed above in the section dealing with mammals. The purposes and uses of this equipment have been described in good detail by Vanzolini and Papavero [17], and Auricchio [31].

The method most commonly used to collect birds employees mist nets (Fig. 7), which were described earlier in the section dealing with collecting mammals. These nets should be set up before the birds begin their activities (generally between 04:00 and 04:30 hours in the morning, depending on the geographical localization of the sampling area). Once the nets have been installed the researcher should remain nearby in order to inspect them (although the interval between inspections may vary according to the methodology used).

The collector can often maximize the collecting successes of both active searches and mist nets by recording bird calls and then playing them back to attract other birds. This technique is extremely advantageous because it will attract individuals of a particular species seeking to investigate the electron trap called call—facilitating their capture and economizing time [30].

As with mammal collections, the researcher can interview local inhabitants, hunters, or people who commonly trap birds. The information may not be extremely accurate, however, except in cases where whole specimens (or parts) are available for definitive identifications. The knowledge that the hunters and bird trappers have of the area and the techniques they employ may also aid in maximizing scientific collections. For more details, refer to Alves et al. [39, 40, 43].

Captured birds can be killed by compressing their thorax, which impedes their respiration and heartbeat. Other techniques include injecting Ketamine into the thorax or alcohol or formaldehyde into the base of the cranium, or inhalation of ether or chloroform [30]. These specimens may be prepared in 10 % formaldehyde and subsequently conserved in 75° GL alcohol, or prepared in a dry state through taxidermy.

The specimens can be transported and stored in small recipients or in larger containers containing 75° GL alcohol (specimens stored under liquid conditions), or in boxes (skins that have been prepared by taxidermy). For more details, refer to Vanzolini and Papavero [17], and Auricchio [31].

4 Zoological Collections

The material collected during field expeditions should be housed in permanent zoological collections that are responsible for recording and ordering the specimens and preserved body parts for educational and/or scientific purposes [44]. The collections can be used to support research projects in areas such as taxonomy, biogeography, zoogeography, morphology, ecology, and ethnobiology, and are essential to the academic studies of undergraduate and graduate students. More detailed information about zoological collections and their peculiarities can be found in Martins [44] and Franco [45].

Zoological collections can be either didactic or purely research collections. Didactic collections are designed toward teaching purposes, demonstrations, and training. They are generally housed in institutions that offer courses in zoology, and the materials usually have short useful lives, due to frequent handling of the specimens. As such, teaching collections must be renewed periodically. The material incorporated into this type collection is often inappropriate for research purposes due to the lack of basic field collection information or because they were partially damaged [44, 45].

Research collections are also known as scientific collections and have inestimable value to taxonomic and biological studies. They are classified as general, regional, private, or special collections.

General collections maintain biological material from different parts of world, are secular, and generally incorporated into museums (although some may be found in universities or other large research institutions). Regional collections unite diversified zoological material from a specific geographical region or locality and are usually kept by local universities, technical schools, or research institutes. Researchers and dedicated amateurs or even private companies sometimes gather zoological material into private collections. These collections are maintained by private resources and are commonly donated or sold to public institutions after the death of the collector or when he/she no longer can (or cares to) maintain the material. As these collections are usually restricted to the specific interests of the collectors or to a given zoological group, they quite often have excellence specimens and records [44, 45].

Special collections house materials with the purpose of supporting specific studies, often those of economic interest, and are designed to aid in solving problems specific to the institutions with which they are associated. The principal types of special collections are: *collections of economic interest*—which can be medical/public health, agricultural, or related to food production; *reference or identification collections*—which hold collections designed to facilitate the identification of specimens by comparisons; *collections of specific taxonomic groups*—which hold collections of specific animals (such as snakes, amphibians, coleopterans, mollusks, or mammals) and are of relevant importance to scientists working with those specific groups [44, 45].

All collections should be maintained in order so that their data and specimens are reliable and easily consulted, and they should be amplified and conserved for future generations. The storage and maintenance of any collection is extremely important and requires constant attention. This set of activities is the primary responsibility of curators, and should follow the specific instructions and procedures outlined in the works of Martins [44] and Franco [45].

Acknowledgments

To Universidade Federal da Paraíba and Universidade Estadual da Paraíba for all the support. The last author would like to acknowledge to CNPq (Conselho Nacional de Desenvolvimento Científico e Tecnológico) for providing a research fellowship.

References

1. Mayr E (1998) This is biology: the science of the living world. Belknap Press of Harvard University Press, Cambridge
2. Papavero N, Balsa S (1986) Introdução histórica e epistemológica à biologia comparada, com especial ênfase à biogeografia. I. Do gênese ao fim do Império Romano do Ocidente. Biótica e Sociedade Brasileira de Zoologia, Belo Horizonte
3. Abreu YS (1994) O método de Aristóteles para o estudo dos seres vivos. Rev Soc Bras Hist Ciênc 11:35–40
4. Amorim DS (1997) Elementos básicos de sistemática filogenética. Holos Editora, Ribeirão Preto
5. Gardiner B, Morris M (2007) Linnean tercentenary. The Linnean Society of London, London
6. Hennig W (1966) Phylogenetic systematics. University of Illinois Press, Urbana
7. Caramaschi U (1987) Manual de técnicas para preparação de coleções zoológicas. Sociedade Brasileira de Zoologia, Curitiba
8. ICZN (International Commission on Zoological Nomenclature) (1999) International code of zoological nomenclature. International Trust for Zoological Nomenclature, London
9. Lineu C (1756) Systema naturae—sistens regna tria naturae in classes et ordines genera et species. Kessinger Publishing, Whitefish
10. Souza E, Silva MF (2006) História dos animais de Aristóteles I. Biblioteca de Autores Clássicos, Coimbra
11. Bernardi N (1994) Nomenclatura zoológica. In: Papavero N (ed) Fundamentos práticos de taxonomia zoológica. Editora UNESP, São Paulo, pp 169–186
12. Bernardi N (1980) O significado dos tipos nomenclatórios do grupo de espécie em taxonomia animal. Rev Bras Entomol 24:175–179
13. Mateus A (1989) Fundamentos de zoologia sistemática. Fundação Calouste Gulbenkian, Lisboa
14. Papavero N, Llorent-Busquets J (1992) El uso equívoco del concepto de 'gênero' en sistemática filogenética (I). Publ Especiales Mus Zool Univ Nacional Autònoma Mexico 5:31–37
15. Papavero N, Llorent-Busquets J, Abe JM (1992) El uso equívoco del concepto de 'gênero' en sistemática filogenética (II). Publ Especiales Mus Zool Univ Nacional Autònoma Mexico 6:61–82
16. Cantino PD, Queiroz K (2010) PhyloCode—international code of phylogenetic nomenclature. http://www.ohio.edu/phylocode. Accessed 16 Apr 2010
17. Vanzolini PE, Papavero N (1967) Manual de coleta e preparação de animais terrestres e de água doce. Departamento de Zoologia, Secretaria da Agricultura do Estado de São Paulo, São Paulo
18. Martins M, Oliveira ME (1998) Natural history of snakes in forest of the Manus region, central Amazonia, Brazil. Herpetol Nat Hist 6:78–150
19. Souza AM, Auricchio P (2002) Peixes. In: Auricchio P, Salomão MG (eds) Técnicas de coleta e preparação de vertebrados para fins científicos e didáticos. Instituto Pau Brasil de História Natural, São Paulo, pp 17–33
20. Zimmerman BL (1994) Audio strip transect. In: Heyer WR, Donnelly MA, McDiarmid RW, Heyek LC, Foster MS (eds) Measuring and monitoring biological diversity. Standard methods for amphibians. Smithsonian Institution Press, Washington, DC, pp 92–97
21. Jaeger RG (1994) Transect sampling. In: Heyer WR, Donnelly MA, McDiarmid RW, Heyek LC, Foster MS (eds) Measuring and monitoring biological diversity. Standard methods for amphibians. Smithsonian Institution Press, Washington, DC, pp 103–107
22. Jaeger RG, Inger RF (1994) Quadrat sampling. In: Heyer WR, Donnelly MA, McDiarmid RW, Heyek LC, Foster MS (eds) Measuring and monitoring biological diversity. Standard methods for amphibians. Smithsonian Institution Press, Washington, DC, pp 97–102
23. Bauer AM, Sadlier RA (1992) The use of a mouse glue traps to capture lizards. Herpetol Rev 23:112–113

24. Scrocchi G, Kretzschmar S (1996) Guía de métodos de coleta y preparación de anfíbios y reptiles para studios científicos y manejo de colecciones herpetológicas. Fundación Miguel Lillo, Ministerio de Cultura y Educación, San Miguel de Tucumán
25. Whinting MJ (1998) Increasing lizard capture success using baited glue traps. Herpetol Rev 29(1):34
26. Savage JM (2002) The amphibians and reptiles of Costa Rica. A herpetofauna between two continents, between two seas. University Chicago Press, Chicago, IL
27. Franco FL, Salomão MG (2002) Répteis. In: Auricchio P, Salomão MG (eds) Técnicas de coleta e preparação de vertebrados para fins científicos e didáticos. Instituto Pau Brasil de História Natural, São Paulo, pp 77–115
28. Calleffo ME (2002) Anfíbios. In: Auricchio P, Salomão MG (eds) Técnicas de coleta e preparação de vertebrados para fins científicos e didáticos. Instituto Pau Brasil de História Natural, São Paulo, pp 45–73
29. Ross LG, Ross B (2008) Anaesthetic and sedative techniques for aquatic animals, 3rd edn. Blackwell Science, Oxford, p 236
30. Auricchio P (2002) Aves. In: Auricchio P, Salomão MG (eds) Técnicas de coleta e preparação de vertebrados para fins científicos e didáticos. Instituto Pau Brasil de História Natural, São Paulo, pp 127–147
31. Auricchio P (2002) Mamíferos. In: Auricchio P, Salomão MG (eds) Técnicas de coleta e preparação de vertebrados para fins científicos e didáticos. Instituto Pau Brasil de História Natural, São Paulo, pp 151–194
32. Corn PS (1994) Straigght-line drift fences and pifall traps. In: Heyer WR, Donnelly MA, McDiarmid RW, Heyek LC, Foster MS (eds) Measuring and monitoring biological diversity. Standard methods for amphibians. Smithsonian Institution Press, Washington, DC, pp 109–117
33. Jones C, McShea WJ, Conroy MJ, Kunz TH (1996) Capturing mammals. In: Wilson DE, Cole FR, Nichols JD, Rudran R, Foster MS (eds) Measuring and monitoring biological diversity. Standard methods for mammals. Smithsonian Institution Press, Washington, DC, pp 115–155
34. Cechin SZ, Martins M (2000) Eficiência de armadilhas de queda (pitfall traps) em amostragens de anfíbios e répteis no Brasil. Rev Bras Zool 17(3):729–740
35. Kunz TH, Hodgkison R, Weise C (2009) Methods of capturing and handling bats. In: Kunz T, Parsons S (eds) Ecological and behavioral methods for the study of bats. Johns Hopkins University Press, Baltimore, MD, pp 3–35
36. Alves RRN, Rosa IL (2007) Zootherapy goes to town: the use of animal-based remedies in urban areas of NE and N Brazil. J Ethnopharmacol 113:541–555
37. Alves RRN, Rosa IL (2006) From cnidarians to mammals: the use of animals as remedies in fishing communities in NE Brazil. J Ethnopharmacol 107:259–276
38. Alves RRN, Rosa IL, Santana GG (2007) The role of animal-derived remedies as complementary medicine in Brazil. Bioscience 57: 949–955
39. Alves RRN, Mendonça LET, Confessor MVA, Vieira WLS, Lopez LCS (2009) Hunting strategies used in the semi-arid region of northeastern Brazil. J Ethnobiol Ethnomed 5:1–16
40. Alves RRN, Gonçalves MBR, Vieira WLS (2012) Caça, uso e conservação de vertebrados no semiárido Brasileiro. Trop Conser Sci 5(3):394–416
41. Alves RRN, Rosa IL (2013) Animals in traditional folk medicine: implications for conservation, vol 1. Springer, Berlin
42. Bibby CJ, Burgess ND, Hill DA, Mustoe SH (2000) Bird census techniques. Academic, London
43. Alves RRN, Nogueira EGG, Araújo HFP, Brooks SE (2010) Bird-keeping in the Caatinga, NE Brazil. Hum Ecol 38:147–156
44. Martins UR (1994) A coleção taxonômica. In: Papavero N (ed) Fundamentos práticos de taxonomia zoológica. Editora UNESP, São Paulo, pp 19–43
45. Franco FL (2002) Coleções zoológicas. In: Auricchio P, Salomão MG (eds) Técnicas de coleta e preparação de vertebrados para fins científicos e didáticos. Instituto Pau Brasil de História Natural, São Paulo, pp 283–318

Chapter 13

Sampling in Ethnobotanical Studies of Medicinal Plants

Mariano Martinez Espinosa, Isanete G.C. Bieski, and Domingos T.O. Martins

Abstract

Ethnopharmacological and ethnobotanical studies represent one of the strategies of medicinal plants selection most widely used and recognized today. Ethnobotanical surveys may support chemical and pharmacological studies in the search for herbal medicines for the treatment of the most common diseases. Several methods can be employed in carrying out these surveys, with particular emphasis on ethno-directed studies, taking into consideration shorter time and lower cost required for data acquisition and analysis. In this regard, probability and non-probability methods of sampling have been widely employed. Therefore, the aim of this chapter is to present the fundamentals of sampling in the area of ethnobotany. For this purpose, brief presentation of the main probability and non-probability sampling methods will be presented with two practical illustrations from recent researches conducted recently. It is important to note that non-probability methods can be used in ethnobotanical survey of medicinal plants, but its use do not allow for statistical inferences on the population, since they are not based on probability theory. For a better understanding of probability sampling method, we utilized two examples of sample design. It is hoped that this write-up will contribute in a significant way to sampling design in ethnobotanical and ethnopharmacological studies minimizing biases that are common in this type of studies.

Key words Population, Sample, Statistical inference, Ethnopharmacology

1 Introduction

The use of statistics dates back to 4000 B.C., when it was used by warriors in conquering people of different territories. It however came into limelight in the nineteenth century gaining importance due to its application in the different areas of knowledge [1].

Statistics is an important tool employed in collecting, analyzing, summarizing, and drawing conclusions from data, using sound processes and scientific tools. Statistical analysis is used in designing scientific studies to increase consistency, measure uncertainty, and produce robust datasets and information. As such, statistical methods are used extensively throughout the scientific process, from the design of research questions through data analysis and to the final interpretation of data, thus enables drawing of valid conclusions.

Ulysses Paulino Albuquerque et al. (eds.), *Methods and Techniques in Ethnobiology and Ethnoecology*, Springer Protocols Handbooks, DOI 10.1007/978-1-4614-8636-7_13, © Springer Science+Business Media New York 2014

One of the main goals of statistics is to make inferences from information contained in data obtained from one or more samples of a given population. Research planning requires elaboration of sampling plan or design, which includes defining sample size and sampling techniques to be employed in sample selection from the given population, in such a scientific manner that the sample obtained "honestly" represents the entire population.

In addition, sampling is done in order to provide reliable results with minimum bias or error and in such a way that representativeness is guaranteed and thus allows the researcher to draw conclusions for the entire population. It is clearly necessary to define the type and sampling method before hand for adequate data collection [2].

There are broadly two types of sampling techniques based on sample selection design: probability and non-probability. These two main categories encompass different sampling methods.

In probability sampling, all elements in the population have some opportunity of being included in the sample, and the mathematical probability that any one of them will be selected can be calculated. Whereas, in the case of non-probability sampling the samples are gathered in a process that does not give all the individuals in the population equal chances of being selected.

Both types of sampling can be used in ethnobotanical and ethnopharmacological investigations for the selection of medicinal plants [3–9].

However, it should be noted that non-probability sampling has many limitations when compared to probability sampling, especially in terms of accuracy and representativeness. Thus, non-probability sampling may be used only in surveys in which data obtained are not meant to be used to make statistical inferences from the population [2, 6, 10].

1.1 Basic Concepts of Statistical Inference

1.1.1 Population or Universe

Universe is the set of all elements of interest that have one or more common characteristics that can be counted, measured or ordered, and from which inferences can be made [2, 11]. Thus, this set consists of elements and the same can be represented by (1):

$$E = \{1, 2, \ldots, N-1, N\} \quad (1)$$

Population is the entire collection of individuals, objects, or simply elements about which information is desired. Population may be finite (determined number of elements), or infinite, i.e., an indeterminate number of elements.

1.1.2 Element of a Population

According to Bolfarine and Bussab [2], element of a population or basic unit i is any element belonging to the set E ($i \in E$). The basic unit may be single or collective (group or set), although the unit of observation and analysis is the element [11].

1.1.3 Characteristic of Interest

Variables of interest of a population are the common characteristics of the information of each basic unit of that population. A *variable* is a characteristic of interest to be measured for each unit in the sample. In general, variables are usually represented as letters X and Y, other notations may also be used or better still with X_i and Y_i. However, the latter notations are recommended when there are several features or variables.

1.1.4 Population Parameter

A population parameter is the numerical measure used to describe the characteristic of a population of interest. Usually, the parameters are represented by letters of the Greek alphabets (α, β, etc.). The most utilized are the total, average, proportion, and population variance.

1.1.5 Census: Advantages and Disadvantages

A census consists of collecting information of all various elements of a population or universe. The advantage of a census is that it presents the information of all the elements existing in a given population.

The main disadvantages of census are the high cost, vast amount of time needed, paucity of skilled personnel for the collection, inability to observe all elements of the population, and destructive processes, among others. A solution to these problems is to select part of the elements (sample) and from this part inferences about the population under consideration is made.

1.1.6 Sample Sampling Unit

A sample is a subset of the population and consists of n elements [2, 3]. In practice, it usually employs random samples, in which each elements of the population has the same chance of being selected in the sample.

When a survey is carried out, it is recommended to select elements of the population by means of the sampling units. These units or elements in which the measurements are made to obtain data to be analyzed are called sampling units. The sample unit can be single or collective [3, 11, 12].

It is essential to obtain an appropriate sample in order to be able to generalize the findings. This is in addition to ensuring good representation of the population of a particular survey. Thus, one of the important steps in research planning is the data capture, in which sampling design and sample size determination are fundamental, in order to have statistically reliable results.

1.1.7 Variables

A variable to be studied in a population consists of common characteristic that is desired to be measured (or observed) in each unit of a given population. Variables can be qualitative or quantitative.

A *qualitative* variable (or *categorical*) can be classified into *nominal* and *ordinal*. Contrary to the nominal qualitative variables, ordinal variables are presented with a natural ordering [13, 14].

Variables that can be counted or measured are termed quantitative and can be classified into discrete and continuous. Discrete variables are those that assume finite number of values that can be counted or referred. For example, the number of individuals in a community and the number of plants in a forest garden.

Continuous variables are those that can assume an infinite set of values and are generally obtained by measurement. For example, the height of a plant (m) and the weight of a fruit (g).

1.1.8 Data

Numerical measures (or attributes) describe the results obtained (or observed) in each unit constituting the data for each variable studied.

Variations or errors are common occurrence in researches at all levels of measurement involved in data collection of a particular feature of interest.

1.1.9 Error

When measurements of a particular characteristic of a sample or population are repeated under the same conditions, the results of these measurements are not necessarily identical. The unpredictable fluctuation that is observed in measurements or observations of the results is called error [2, 11].

1.2 Different Types of Sampling

Sampling is the process by which inference is made to the whole by examining a part. The sampling method is the scientific procedure of selecting those sampling units which would provide the required estimates with associated margins of uncertainty, arising from examining only a part and not the whole. There are broadly two types of sampling techniques based on design of sample selection: probability and non-probability. These two main categories encompass different sampling methods.

1.2.1 Nonrandom or Non-probability Sampling

In non-probability sampling type, probability mechanism is not utilized in the selection of units or sample elements of interest. The selection of elements usually subjected to the experience and knowledge of the researcher.

In ethnobotanical surveys, non-probability sampling methods most commonly employed are sampling by convenience, quota sampling, and snowball sampling [6–10]. It is important to note that these methods are not recommended when there is the desire to extend the results to the population and to make inferences. However, such methods may be useful in exploratory research.

Sampling by convenience. A convenience sample is simply one in which the researcher uses any subjects that are available to participate in the research study. That is, a population is selected because it is readily available and convenient. This sampling method has a limitation, in that it cannot be generalized to the population [15].

Therefore, this sampling method does not guarantee that the sample is representative, and therefore, the data from this sampling is only valid to the sample itself and cannot be extended to the population. This sampling technique can be used in exploratory research, especially in situations where one desires to have an overall idea and to identify general aspects of the variables considered.

This sampling procedure usually takes shorter time, relatively costs less, and is very easy to apply [6, 15].

Sampling by quotas. This sampling method is quite similar to stratified sampling method. However, in each group or stratum, one nonrandom sample is selected. Although the sample size for each stratum is proportional to the size of the stratum in relation to the total population, the sample may not necessarily be representative of the group or stratum of the population in consideration. In general, the sample in each stratum is chosen using sampling by convenience method. Consequently, the samples obtained by this method also do not permit making inference about the population under studied [15, 16].

Snowball sampling. This sampling method is widely used for intentional selection of expert informants. In this method, the researcher goes to the study site and seeks in the community, an informant that is culturally competent (a local specialist) about the topic of the research, who then indicates others with similar competence, repeating the process from new participants progressively included until all local experts in the community are completely covered [17]. It is worth mentioning that samples obtained by the snowballing method are not representative and cannot be used for making statistical inferences about the population [6].

1.2.2 Random Probability Sampling

In probability sampling, the elements of the population of interest are selected by a probability procedure in which the samples are determined at random [2]. By using this method, all the elements have equal chance of being included in the sample. It guarantees representativeness of the sample with increased robustness [6, 17].

There are several procedures of implementing probability sampling. There are several probability sampling methods.

The four probability sampling methods most commonly used in ethnobotanical surveys of medicinal plants are simple random, stratified, systematic, and cluster or complex sampling [2, 9, 18].

Simple random sampling. Simple random sampling is a sampling method whereby a sample of size n of a random variable X, with a given distribution, is selected from a finite population of N units, such that each sample has the same probability of being selected and all members of the population have the same probability of being included in the sample [2, 6, 10, 14, 19].

The researcher, before commencing research and selecting samples must determine the appropriate sample size needed to achieve the desired accuracy. The accuracy of an interval represents the range permitted between the true value of the parameter of interest and the point estimate.

In order to determine approximate sample size, the parameter of interest must be known. This is usually a mean or a proportion. It is also important to ascertain whether the population size (N) is known, since the expressions utilized in the determination of sample size depend on availability or otherwise of these information. It is also important to choose confidence level and the limit to the error of estimation.

In this chapter, only expressions used in determining a sample size for estimating proportion (p), with a confidence level of $1-\alpha$. This is because most of the variables considered in the area of ethnobotany are generally categorical and are often used in associations and prevalence ratios.

Thus, if the population size (N) is unknown or if it is too large to be determined, the approximate sample size of the sample (n), the expression (2) are used for the estimation of the value with a confidence level $1-\alpha$ [18].

$$n = \frac{\left(z_{\alpha/2}\right)^2 p(1-p)}{d^2} \tag{2}$$

where

n = approximate sample size;
p = proportion of population considered;
d = limit to error estimate;
α = level of significance considered;
$Z_{\alpha/2}$ = value obtained from the table of standard normal distribution [18].

If the population size (N) is known, expression (3) is used in place of (2) to determine the approximate sample size (n) and to estimate the value of p with a confidence level $1-\alpha$ [2, 6, 18].

$$n = \frac{Np(1-p)}{(N-1)\left(d / z_{\alpha/2}\right)^2 + p(1-p)} \tag{3}$$

where

n = approximate sample size;
N = population size considered;
p = population proportion considered;
d = limit to error estimate;

α = level of significance considered;

$Z_{\alpha/2}$ = value obtained from the table of standard normal distribution [18].

Stratified random sampling. A simple random sampling is more representative of the population if the elements are homogeneous, otherwise the population to be studied should be divided into subgroups or strata of elements in order to be as homogeneous as possible.

A stratified random sample of the elements is the division of the population into strata or groups, according to some characteristic of interest and subsequent selection of a simple random sample from each stratum [2, 10, 19].

To determine the approximate size of a stratified sample, it is necessary to know the parameter of interest, which by and large is usually the mean or proportion. It is also important to know whether the population size (N) is known, since the expressions used to determine sample size depend on availability or otherwise of this information, besides from the value established for error estimation and the level of significance. In this case also, only expressions used in determining sample size for calculating proportion (p) or prevalence will be considered.

Firstly, the sample size is determined by considering the same value of proportion for all strata. In this case, the sample size is determined using expression [2]. If the population in each stratum is given as N_k, then the population size (N) is given by the summation of the size of each stratum [6], which can be obtained by the expression (4):

$$N = N_1 + N_2 + \ldots + N_k = \sum_{i=1}^{K} N_k \tag{4}$$

To determine the sample size for each stratum, it is necessary to first determine the fraction of population (W_k) for each stratum, which is obtained by the expression (5):

$$W_k = \frac{N_k}{N} \tag{5}$$

where

N = total size of the population;

N_k = population size of each stratum;

W_k = population fraction of each stratum.

Thus, the sample size for each stratum (n_k) is obtained by expression (6), multiplying the population fraction in each stratum by the sample size (n):

$$n_k = W_k \quad n, \tag{6}$$

for $k = 1,2,\ldots, K$

where

n_k = approximate sample size of each stratum;
n = approximate total sample size;
W_k = population fraction for each stratum;
K = number of strata.

If the proportions are different for the strata, the expression (7) can be used to estimate the approximate sample size (n) required to estimate p [10, 12, 19]

$$n = \frac{\sum_{i=1}^{K} N_i^2 p_i (1 - p_i) / W_i}{N^2 d^* + \sum_{i=1}^{L} N_i p_i (1 - p_i)} \qquad (7)$$

where

K = number of strata in the population;
N_i = number of units in the stratum i;
N = number of units in the population, i.e., $N = N_1 + N_2 + \ldots + N_K$;
W_i = fraction of observations assigned to the stratum i, i.e., $W_i = N_i / N$;
p_i = population proportion of stratum i;
$d^* = d^2/(z_{\alpha/2})^2$, where d is the limit to error estimate;
$Z_{\alpha/2}$ = value obtained from the table of standard normal distribution [18].

In order to estimate the value of p_i when it is unknown, values from similar surveys are often used, or when this is also lacking, a pilot sampling may be conducted. However, when both options are not available the value $p_i = 0,5$ may be used as an estimate. Justifications for this value can be found in specialized books on sampling and its mathematical and statistical proofs are readily found in the books of statistical inference [2, 3, 14].

Systematic random sampling. An alternative to simple random sampling is systematic sampling. Systematic sampling is a sampling method that allows random selection of an element of the first k elements of a reference system and subsequently the selection of every kth element [12].

To systematically select a sample it is necessary to:

1. Have a complete list of elements of the population
2. Number the elements of the population from 1 to N
3. Know the number of elements to be selected (n)
4. Define the interval or range of selection: $k = N/n$, to determine the value of k, may require a rounding off

5. Choose a random number r between *1* and k; the first randomly selected number would be r, the second number selected would be $r+k$, the third would be: $r+2\times k$, the fourth number would be $r+3\times k$, etc.
6. Use the numbers selected to identify elements of the list selected as samples
7. Next should be used so the numbers selected to identify the elements of the list selected in the sample

Random cluster sampling. In simple random probability sampling methods, the sampling unit is consisted of only one element of the population. When units of the population are grouped into sets or clusters and, in turn, some of these are randomly selected to form a sample, the method is called a cluster sampling. Thus, a cluster sampling is a random sample in which each sampling unit is a set or cluster of units [2, 12, 14].

A cluster sampling is especially useful when the population in question is large and spread over wide geographical area. In this sampling method, the first step is to specify an appropriate number of clusters. In general, the units within a cluster tend to have similar characteristics. On the other hand, the clusters should be as heterogeneous as possible between them, to be able to take economic advantage of cluster sampling.

Numerous small clusters are advantageous to control variability, but few large conglomerates are economically advantageous. Therefore, one should find a balance between the number and size of clusters. The fact is that, there are no "magic" rules that always work best to make this decision. Each problem must be studied separately, but pilot samples may be undertaken to help the researcher in making the right choice [2, 12, 19].

There are two types of cluster sampling methods, namely: single-stage cluster sampling and multiple-stage cluster sampling. In single-stage cluster sampling, all of the elements within selected clusters are included in the sample, whereas multiple-stage cluster sampling is obtained when the researcher only selects a number of subjects from each cluster-either through simple random sampling or through systematic random sampling, as in the cases of municipalities, census sectors, households, and individuals [2, 3, 12]. According to Albuquerque and Hanazaki [3], the expression (8) can be employed to obtain the number of clusters (m):

$$m = \frac{N}{n} \tag{8}$$

where

m = number of randomly selected clusters;
n = approximate sample size;
N = total size of the population.

It should be noted that if there are strata, the approximate number of conglomerate to be selected in each stratum is denoted by m_i, which is obtained by using the expression (9):

$$m_i = \frac{m \times M_i}{M} \quad for\ i = 1,\ldots,K \tag{9}$$

where

m = number of randomly selected clusters;
M_i = number of clusters in each stratum;
K = number of strata;
M = total number of clusters in the population.

In order to determine the number of individual in each randomly selected cluster and region (n_{ci}), the expression (10) can be used:

$$n_{ci} = \frac{N_{cs}}{N_{es}} \times n_k \tag{10}$$

where

N_{cs}: population of the randomly selected cluster;
N_{es}: total population of the stratum selected;
n_k = approximate size of the sample in each stratum.

2 Illustrated Examples

For this section, two examples of researches carried out in the state of Mato Grosso, Brazil will be used as case studies.

2.1 Example of Stratified Sampling

The data obtained from an ethnobotanical survey conducted among the residents of the communities of Nossa Senhora Aparecida do Chumbo District (NSACD) located in Poconé municipal, Mato Grosso, Brazil in 2009 [20] was used. NSACD is comprised of 37 communities, distributed in 16 micro areas, totaling 3,652 individuals belonging to 1,179 families. Micro areas comprise of groupings of communities having large and small number of families. It is an official term employed by the Ministry of Health for logistics purposes to facilitate provision of health-related services to families in these communities who attend the local Family Health Units [21] as presented in Table 1.

In this particular study, a population consisted of an individual (informant) per family (NSACD) with at least 40 years of age and has been residing for at least 5 years in the district. Thus, the sampling unit in this example was the family unit. This was because the district is located in a rural area and moreover it is much easier locating an informant within a family [6, 20, 22].

Table 1
Distribution of 37 communities of Nossa Senhora Aparecida do Chumbo district (NSACD), Poconé, Mato Grosso, Brazil, 2009, by micro areas, with total number of individuals and families in each micro area, including sample fraction and sample size

Communities	MA	Total of individuals	N_k	W_k	n_k
Chumbo	1	311	71	0.0602	17
	2	397	78	0.0662	19
	3	238	67	0.0568	16
Canto do Agostinho, Santa Helena Os Cagados, Várzea Bonita	4	179	52	0.0441	15
Furnas II, Salobra, Zé Alves	5	165	59	0.0500	15
Campina II, Furnas I, Mundo Novo, Rodeio	6	279	81	0.0687	20
Campina de Pedra, Imbé	7	188	67	0.0568	16
Barreirinho, Coetinho, Figueira	8	253	95	0.0806	23
Bahia de Campo	9	257	74	0.0628	18
Agrovila, São Benedito	10	184	66	0.0560	16
Agroana	11	186	81	0.0687	20
	12	186	97	0.0823	24
Bandeira, Minadouro	13	248	82	0.0696	20
Carretão, Deus Ajuda, Sangradouro Pesqueiro, Varzearia	14	216	77	0.0653	19
Chafariz, Ramos, Sete Porcos, Urubamba	15	208	67	0.0568	16
Céu Azul, Capão Verde, Morro Cortado Passagem de Carro, Varal	16	157	65	0.0551	16
Total		3,652	1,179	1.000	290

Source: Bieski (2010) [14]
MA micro area, N_k total of families, W_k sample faction, n_k sample size

Thus, using the expression (3) and the total number of families from Table 1 (N= 1.179), considering a p value of 0.5 (p= 0.5) with a confidence level of 95 % ($Z_{\alpha/2}$ =1.96) and a limit to error estimate of 0.05 (d= 0.05), a sample size of 290 or more families ($n \geq 290$) for the 16 strata was obtained. The value of p= 0.5 was attributed because there is no prior information from previous ethnobotanical surveys of medicinal plants in Mato Grosso state.

In order to determine the approximate sample size of each micro area, expressions (5) and (6) were used, which corresponded to the total number of families and sample fraction respectively. The population fraction (W_k) of each micro area was first determined, by dividing the total number of families in each micro area by the total number of families of the population (N= 1,179), as illustrated in Table 1.

Using expression (6), the sample size for each micro area was calculated by multiplying population fraction of each one by 290. This can be seen in the last column of Table 1.

2.2 Cluster Sampling

In this second example data from a population study of elderly units of the Family Health Program (FHP) [23] was considered. The data are kindly furnished by Cuiabá City Council. The data consisted of a register elaborated by the Office of Planning and Urban Development/IPDU (2010) of the Council, as seen in Table 2.

Using the expression (3) and the total number of the elderly in Table 2 (N=7,187), considering p value of 0.10 (p=0.10), established on the basis of a population study in São Paulo, where the prevalence of use of psychotropic medications for this population was 10 % [24], with a confidence level of 95 % ($Z_{\alpha/2}$ = 1,96) and a limit to the estimation error of 0.025 (d = 0,025), yielded a sample size of 514 or more of the elderly ($n \geq 514$).

Since this is a cluster sampling, a design effect of 1.5 to correct the sample size was taken into consideration, taking into account that the population comes from various FHPs units [25]. With this fix, the sample size of 771 of the elderly was obtained ($514 \times 1.5 = 771$).

Also anticipating possible sample loss (elderly), for various reasons such as unanswered or invalid questionnaires, refusals, death, among others, and considering studies reported in the literature and the characteristics of population [24], a loss of 20 % was expected. Thus, the corrected sample size rose from 771 to 925 of the elderly ($n = 771 + 771 \times 0.20 = 925$).

In order to determine the sample size per stratum (region), we considered the expression (6). These values are shown in Table 3.

Notice that in the four regions considered, the elderly are distributed in 30 units of FHPs that are called clusters (M= 30). Thus a two-stage cluster sampling was employed to reduce the variance [26]. Thus, by applying expression (8) 8 conglomerates (m) was obtained (m=7,187/925 = 8). Using this number (m) and applying expression (9), the number of clusters by region (m_i) was obtained, as shown in the third column of Table 4. The elderly enrolled in the units of the FHP by region, municipality.

To select clusters of each region, we performed a systematic random drawing, with the aid of MINITAB, version 15.0. And subsequently, to determine the number of the elderly in each cluster region drawn (n_{ci}), we considered expression (10), as shown in Table 5.

Table 2
Register of the Family Health Program (FHP) units, Cuiabá municipality, Mato Grosso, Brazil, showing the total number of elderly, as male and female, by unit area and region, 2009

Order	Region	Family Health Program	Feminine	Masculine	Total
1	L	Altos da Serra I	68	79	147
2	L	Altos da Serra II	73	95	168
3	L	Dr. Fábio I	95	116	211
4	L	Dr. Fábio II	105	120	225
5	L	Novo Mato Grosso	139	121	260
6	N	1° de Março	115	109	224
7	N	Jardim Florianópolis	99	87	186
8	N	Jardim Umuarama	126	112	238
9	N	Jardim União	80	91	171
10	N	Jardim Vitória I	76	57	133
11	N	Jardim Vitória II	114	102	216
12	N	Jardim Vitória III	98	85	183
13	N	João Bosco Pinheiro	105	114	219
14	N	Novo Horizonte	140	135	275
15	N	Novo Paraíso I	149	181	330
16	N	Novo Paraíso II	95	93	188
17	N	Ouro Fino	149	148	297
18	N	Três Barras	135	115	250
19	O	Despraiado	141	122	263
20	O	Jardim Araça	180	145	325
21	O	Jardim Santa Amália	212	173	385
22	O	Novo Colorado I	91	103	194
23	O	Novo Colorado II	69	76	145
24	O	Ribeirão da Ponte	158	154	312
25	O	Santa Isabel I	129	100	229
26	O	Santa Isabel II	199	205	404
27	O	Santa Isabel III	152	141	293
28	S	Residencial Coxipó I	154	114	268
29	S	Residencial Coxipó II	148	140	288
30	S	Residencial Coxipó III	90	70	160

Source: Cuiabá [27]

Table 3

Number, sample fraction, and sample size of the elderly registered at the FHP units by region, in Cuiabá municipality, Mato Grosso, Brazil, 2009

Region	N_k	W_k	n_k
North	2910	0.4049	375
South	716	0.0996	92
West	1011	0.1407	130
East	2550	0.3548	328
Total	7187	1.00	925

Source: Cuiabá [27]

N_k number of the elderly by region, W_k sample fraction of the elderly, n_k sample size by region

Table 4

Number of units in the FHP (clusters) by region and sample size of clusters to be selected by region, in the municipality of Cuiabá, Mato Grosso, Brazil, 2009

Region	M_i	m_i
North	13	4
South	3	1
West	5	1
East	9	2
Total	30	8

M_i number of clusters in each region, m_i number of clusters selected in each stratum

Table 5

Sample distribution of the elderly, by conglomerate units in the FHP, in Cuiabá city regions, of Mato Grosso state, Brazil, 2009

Stratum	Region	Cluster	Family Health Program	N_{cs}	N_{es}	n_k	n_{ci}
1	South	01	Residencial Coxipó II	288	288	92	92
2	North	02	Novo Paraíso 2	188			97
		03	Jardim Vitória 2	216			100
		04	Jardim Florianópolis	186	809	375	96
		05	João Bosco Pinheiro	219			113
3	East	6	Despraiado 1	263	575	328	150
		07	Ribeirão da Ponte II	312			178
4	West	08	Dr. Fábio II	225	255	130	130
Total				1833			925

N_{cs} population of the cluster drawn, N_{es} population of the stratum selected, n_k approximate sample size for each stratum, n_{ci} number of the elderly in each cluster and region selected

3 Conclusion

It is hoped that this chapter will contribute substantially to sample planning, especially when conducting ethnobotanical and ethnopharmacological surveys of medicinal plants, thereby reducing or eliminating major sources of bias that can affect the statistical inferences and thus the analysis of the data obtained. It also hoped to facilitates comparisons between ethnobotanical and ethnopharmacological studies that may employ the methods presented, which will enable analysis with greater precision and reliability.

References

1. Ignácio AS (2010) Importância da estatística para o processo de conhecimento e tomada de decisão. Rev Paranaense Desenvolv 118:175–192
2. Bolfarine H, Bussab WO (2005) Elementos de amostragem, 1st edn. Editora Edgar Blücher LTDA, São Paulo
3. Albuquerque UP, Hanazaki N (2006) As pesquisas etnodirigidas na descoberta de novos fármacos de interesse médico e farmacêutico: fragilidades e pespectivas. Rev Bras Farmacogn 16:678–689
4. Jesus NZT, Lima JCS, Silva RM, Espinosa MM, Martins DTO (2009) Levantamento etnobotânico de plantas popularmente utilizadas como antiúlceras e antiinflamatórias pela comunidade de Pirizal, Nossa Senhora do Livramento-MT. Brasil Rev Bras Farmacogn 19(1A):130–139
5. Espinosa MM, Bieski IGC, Martins DTO (2012) Probability sampling design in ethnobotanical surveys of medicinal plants. Rev Bras Farmacogn 23(5):1362–1367
6. Cavéchia LA, Proença CEB (2007) Resgate cultural de plantas de usos de plantas nativas do Cerrado pela população tradicional da região do atual Distrito Federal. Bras Heringeriana 1:11–24
7. Poss J, Pierce R, Prieto V (2005) Herbal remedies used by selected migrant farmworkers in El Paso, Texas. J Rural Health 21:187–191
8. Silva MAB, Melo LVL, Ribeiro RV, de Souza JPM, Lima JCS, Martins DT, Silva RM (2010) Levantamento etnobotânico de plantas utilizadas como anti-hiperlipidêmicas e anorexígenas pela população de Nova Xavantina-MT, Brasil. Rev Bras Farmacogn 20:549–562
9. Viganó J, Viganó JA, Silva DS, CTA (2007) Utilização de plantas medicinais pela população da região urbana de Três Barras do Paraná. Acta Sci Health Sci 29:51–58
10. Cochran WG (1977) Sampling techniques, 3rd edn. Wiley, New York
11. Silva NN (2001) Amostragem probabilística: um curso introdutório, 2 revisth edn. EDUSP, São Paulo
12. Kish L (1965) Survey Sampling. Wiley, New York
13. Magalhães NN, Lima ACP (1999) Noções de probabilidade e estatística. IME-USP, São Paulo
14. Bussab W, Morettin PA (2003) Estatística Básica. Editora Saraiva, São Paulo
15. Mattar FN (1996) Pesquisa de marketing. Ed. Atlas (V1), São Paulo
16. Hill M, Hill A (2002) Investigação por questionário. Edições Sílabo Lda, Lisboa
17. Albuquerque UP, Lucena RFP, Cunha LVFC (2010) Modern physics. Métodos e Técnicas na Pesquisa Etnobiológica e Etnoecológica. In: Albuquerque UP, Lucena RFP, Alencar NL (eds) (org) Métodos e técnicas para coleta de dados etnobiológicos, NUPEEA, Recife, pp 41–64
18. Levy PS, Lemeshow S (2008) Sampling of populations: Methods and Applications. Wiley, New York
19. Scheaffer RL, Mendenhall W, OTT RL, Gerow K (2006) Elementary Survey Sampling, 7th edn. Brooks/Cole Cengage Learning, Boston
20. Bieski IGC (2010) Conhecimento etnofarmacobotânico de plantas medicinais utilizadas por comunidades tradicionais do Distrito Nossa Senhora Aparecida Chumbo, Poconé, Mato Grosso, Brasil (Cuiabá). Dissertação de Mestrado 269, Programa de Pós-graduação em Ciências da Saúde, Universidade Federal de Mato Grosso
21. Poconé (2008) Plano Municipal de Saúde de Poconé. Prefeitura Municipal de Poconé. Secretaria municipal de Saúde, Poconé
22. Bieski IGC, Rios Santos F, Oliveira RM, Espinosa MM, Macedo M, Albuquerque UP, Martins DTO (2012) Ethnopharmacology of medicinal plants of the Pantanal region

(Mato Grosso, Brazil). Evid Based Complement Alternat Med 2012:1–36
23. Lopes, SFM (2010) Estudo do consumo de medicamento de psicotrópicos por idosos atendidos nos Programas de Saúde da Família do município de Cuiabá/MT, 2008–2009. Universidade Federal de Mato Grosso e à Faculdade de Ciências Médicas (Dissertation), Mato Grosso
24. Rozenfeld S (2003) Prevalência, fatores associados e mau uso de medicamentos entre os idosos: uma revisão. Cad Saude Publica 19(3):717–724
25. Luiz RR, Magnanin MMF (2000) A lógica da determinação do tamanho da amostra em investigações epidemiológicas. Cad Saúde Coletiva 8(2):9–28
26. Carlini-Contrini B, Barbosa MTS (1993) Pesquisas epidemiológicas sobre o uso de drogas entre estudantes um manual de orientações gerais. Centro Brasileiro de Informações sobre Drogas Psicotrópicas, Departamento de Psicobiologia da Escola Paulista de Medicina, p 56
27. Cuiabá (2008) Equipes de PSF's já estão atuando mesmo antes das novas unidades serem inauguradas. Mato Grosso. Disponível em: http://www.cuiabasmsaude.com.br/index.php?option=com_contentetask=vieweid=467eItemid=97/. Accessed 15 Feb 2009

Chapter 14

Indigenous Populations: Some Peculiarities

Fábio de Oliveira Freitas

Abstract

In this chapter, we consider a number of characteristics inherent to interacting and working with indigenous populations. Certain aspects of life in an indigenous village may present difficulties and require a researcher to adapt his or her work objectives to fit the reality of the environment. Our aim in this chapter is to present research strategies based on data, previous reports, and field experiences that we consider useful for professionals who work with indigenous populations.

Key words Traditional populations, Indigenous peoples, Cultural dynamics, Cultural diversity, Work methodology

1 Introduction

A frequent question for those who are beginning to work with traditional and indigenous communities is what methods to adopt when undertaking that work. The answer to such a simple question is highly complex. There is no single answer, and what works in one situation may be disastrous in another. Furthermore, the skills of the person who is working with these populations will dictate the best approach.

To assist groups who are starting to work with such communities, our aim here is to bring together a number of points we believe to be important in developing the skills of the professional who is planning to work with indigenous peoples.

We will not present a model of fieldwork in this paper but, rather, describe the wide variety of factors and peculiarities that permeate and/or are intrinsic to both traditional and indigenous communities and should be accounted for when preparing fieldwork methodologies so that both parties reach their goals and achieve the best results.

We focus on indigenous communities because we have more experience in this area; however, much of what is presented can be applied to other traditional communities. We use Brazilian examples in several instances to contextualize our findings.

Ulysses Paulino Albuquerque et al. (eds.), *Methods and Techniques in Ethnobiology and Ethnoecology*,
Springer Protocols Handbooks, DOI 10.1007/978-1-4614-8636-7_14, © Springer Science+Business Media New York 2014

2 Traditional and Indigenous Brazilian Populations

A recent change in Brazilian law, through the publication of Decree 6040 on 7 February 2007 in the Official Gazette of the Union (Diário Oficial da União), mandated the official recognition of a number of traditional Brazilian groups and established the National Policy for the Sustainable Development of Traditional Peoples and Communities (1).

In addition to both the indigenous and Maroon ("quilombola") populations that have already been recognized, a number of other populations were incorporated into this list (1), each with its own cultural and historical peculiarities. As a result, these peculiarities can be included in differentiated public policies, and projects and studies can be designed with respect to these peculiarities.

The exact number of people living within traditional Brazilian populations is not yet known. It is estimated that traditional Brazilian populations represent more than 25 million people in total, or approximately 1/8th of the Brazilian population (2). This figure alone indicates the difficulty or even impossibility of trying to develop a standardized methodology.

For indigenous peoples, there are more reliable official numbers. According to the National Indian Foundation (Fundação Nacional do Índio—FUNAI) (3), 225 ethnic groups or indigenous societies are currently recognized in Brazil, and these groups speak 180 different languages. It is estimated that there may be approximately 63 additional societies that are still isolated and have not yet been contacted.

The general population of these societies is approximately 600,000 people, with two-thirds of them living in villages and the remainder in cities. It is estimated that in the year 1500, when the first European settlers arrived, the total indigenous population was between two and four million people (3).

Thus, there was a drastic reduction in the number of indigenous people, and, consequently, a major human-cultural loss occurred, and the remaining populations are descendants of only a part of what used to exist. Moreover, the culture of the surviving peoples may be a mishmash of cultures that survived over time due to both losses incurred through the extinction of many human groups and changes in the innate culture of the nations—whether those changes occurred due to internal or external factors.

The first point that must be raised within the scope of the present study (i.e., focusing on the indigenous communities) is the great cultural diversity that exists among these populations.

Each of these ethnicities has a different history intrinsic to its culture (traditions, myths, and rites), as well as a history of contact with other communities and with "Western" society in particular. As noted above, there are both isolated groups with virtually no influence from or knowledge of Western society and groups that have been so influenced in terms of behavior and everyday life that they are more closely aligned with Western society than the more culturally preserved indigenous populations.

For example, we have observed the "revival" of certain indigenous groups/ethnicities in recent years, mostly in the northeastern region of Brazil, where the historic contact with colonizers occurred longer ago and was often not particularly amicable. Thus, some groups are living a process of cultural "rescue" or even neogenesis. Some members of the latter context, in cases where the traditional culture of the groups, including language, rites, and myths, has been almost lost, structure themselves by "borrowing" the cultural characteristics of other indigenous groups and recorded historical data.

It is clear that although there are traditional cultural groups that have been little if at all influenced by Western society, there are also very acculturated groups for whom only traditional (genetic, cultural) traces remain. Therefore, the way to interact with each of these groups differs as, perhaps, do the objectives.

Below, we discuss and expand upon the peculiarities inherent in the culture and history of indigenous people that we believe provides helpful tips to inform the development and elaboration of a study concept and during the implementation of a study.

3 Relationship

As stated previously, the time of contact a population had with Western society also greatly influences the cultural preservation of a given indigenous people. Moreover, the way that contact was established with these communities also strongly impacts the evolution of their society and its relationship with the environment.

There are cases in which contact has been peaceful from the onset (e.g., with most of the Indians in the region known as Alto Xingu in the state of Mato Grosso) and examples of hostile contacts, many of which resulted in the massacre and extinction of certain indigenous groups.

These different forms of contact had different impacts on the culture of these people. In friendly contacts, there was a cultural exchange, which may have varied in intensity according to the interests of each population, whereas for non-friendly contacts, the relationship is/was imposed.

Even during apparently friendly contacts, Western culture may have been imposed upon indigenous culture. What often occurred, for example, is that groups of missionaries believed they needed to provide "spiritual comfort" to indigenous peoples without knowing what the community's own beliefs were.

Historically, indigenous groups in Brazil have been contacted by different groups from Western society, including settlers, explorers, farmers, rubber tappers, miners/prospectors, lumberjacks, missionaries, governments, militaries, universities, NGOs, and companies.

An example of the different influences each of these groups may have on traditional populations can be observed in the case of the Yanomami Indians. This community was first contacted by prospectors. The results of this contact included health problems (epidemics), environmental problems (the pollution of rivers and deforestation), and, most drastically, conflict and extermination.

Later, beginning in the mid-twentieth century, missionary groups (both Catholic and Evangelical) began to settle among these people (we will not discuss the merits of these missions but simply register the cultural impact that they entailed).

During this missionary work, some Yanomami communities were "colonised" by Evangelical missionaries, while other communities were colonized by Catholics. This divide damaged the family relationships of those who found themselves in different locations with different missionary influences. That is, cultural barriers were created between different Yanomami groups for reasons outside their own culture. Thus, if a person was working with one group, it would be difficult to also work with the other because of the stigma of having gone over to "the other side".

4 Land

Another reality of indigenous communities is their agrarian situation.

Some communities are in an ideal situation, in which they inhabit their traditional territory and it is demarcated for their use (for example, the Indians of the Alto Xingu). Others have demarcated territory that is not their traditional territory (for example, Xavante Indians). Others have partially demarcated territory (whether traditional or not). Finally, there are cases of peoples whose territory is not demarcated (e.g., Guarani—Caiuá in the state of Mato Grosso do Sul).

This demarcation has an influence on cultural dynamics because, if people reside in territories other than those they originally inhabited, there is or was a need to adapt to the new environment. For example, in terms of availability of genetic resources (species found in the environment that people use for food, crafts, construction, tools, and rites), the new territory may lack certain species or products that the new residents have been accustomed to and traditionally used in everyday life (e.g., brazil nuts or prey species) or even sporadically for a specific function (medicine, wood for bow making, oars, etc.).

The time these peoples have already spent in their new territory also has an influence because it allows for more or less time to have adapted to the new environment. Furthermore, whether the biome of the new environment resembles their traditional area will strongly influence their adaptation and relationship to this environment.

There are also cases in which people were removed/driven out/fled their traditional lands to stay in another area for a time but later returned to their traditional territory. This phenomenon was observed in the case of the Panará (or Krenakore) Indians. Due to the arrival of pioneers, the Panará were displaced from the region they inhabited in southern Pará in the 1970s to the Xingu Indigenous Park in the northeastern portion of the state of Mato Grosso. However, they were unable to adapt to this new environment and began to migrate, forming short-term villages in a given place and then moving again, until they eventually returned to the traditional home of their ancestors. This return led to the demand for a reserve to be created for them, which was subsequently granted.

As an aside to the general text and a means of recording this part of their history, the Panará Indians encountered a number of difficulties, especially hunger, during this journey back to their homeland given that they did not have established farm land in any new area they moved to. Sometimes, members of the Panará even moved to a new location before reaping what they had sown. Furthermore, there were no children born during this migration. They simply chose not to conceive children until their return journey had concluded. It was therefore a period of intense pressure on their cultural structure and population.

Returning to the issue of land, a case similar to that of the Panará occurred with the Xavantes Indians of Maraiwatsede, which is one of several distinct territories among those reserved for the many Xavante "lines". In the mid-twentieth century, these Indians were taken from their land and moved to another territory in Água Branca, which was already occupied by Xavante Indians. This conflict created warring factions, and many problems were generated by the interactions of the two groups.

Decades later, these Indians won back the right to their land. However, the interior of their land had been overrun by dozens of farms. This situation both hindered the repossession due to the numerous legal injunctions that had to be processed through the legal system and caused more serious problems that could only be resolved in the medium to long term.

Upon returning to and occupying a part of their land, the Xavante Indians found it to have been devastated. There was no native vegetation on much of the land because much of it had been removed for use as wood or to convert the land into pastures or for planting. These Indians were a group with a traditional food culture based on hunting that found itself in an area with no vegetation, which meant that hunting was severely limited compared to what they had experienced when they lived there the previous decades; thus, the decimation of the native ecosystem had a strong cultural impact on the returned Xavante Indians.

5 Cultural Differences, Language, Behavior, and Gender

A point to bear in mind is that, as a general rule, we will always be foreigners when working with a culture different from our own. No matter how long the relationship lasts and how integrated we become into everyday society by becoming increasingly knowledgeable about the details of another culture, a Western researcher will never become a member of the indigenous people he or she is researching! He or she will only be able to increase the level of respect and trust between parties. The reverse is also true. However, this fact is not a problem, it is just a fact.

Because we are from different cultures, we need to adapt our ways of interacting and passing information to the peculiarities of the people with whom we are working. This applies in relation to our behavior, thinking, interpretations, and cultural background.

An example that nicely illustrates this point comes from a group of people who went to work in an area in Africa in which there were many cases of the disease commonly known as "sleeping sickness". This disease, also known as African trypanosomiasis, is an often fatal disease caused by the unicellular parasite *Trypanosoma brucei*, which is transmitted by the Tsetse fly (4).

The group was trying to introduce ways to avoid being bitten by the insect to the local people. To do so, they prepared a presentation for the community with a number of illustrations and enlarged photos of the parasite, the fly, and the disease cycle.

A few days after the presentation, they perceived that the community had not adopted the group's recommendations. When they tried to ascertain whether the community had understood the dangers of the disease and how to protect itself, they were immediately told "Yes". However, the locals said they were not doing anything about the disease presented by the group, nor were they concerned about it because the enormous fly from the photos did not exist in that region. In other words, the community did not realize that the images shown were enlargements of flies that existed there because they had no knowledge of photography!

The moral of this story is that differences in cultures, ways of both thinking and interpreting, and knowledge sometimes mean that small details undermine the entire effort despite both good intentions and preparation. The information that is interpreted and assimilated may be very different depending on experience and cultural background.

Each culture's means of classifying things is different. For example, direct cousins are considered to be brothers by the Indians of the Alto Xingu. Therefore, if we are researching kinship, a series of questions may arise regarding the genealogy of local families. This variation among cultures' classification schemes also includes the classification of animals, plants, and soil given that the local criteria distinguishing these can be very different from the classifications adopted by current science.

Another issue that can interfere with the relationship is gender. The division of knowledge, tasks, and rights between men and women differs greatly in each community. Sometimes, no matter how good the relationship with and acceptance of our work by a particular group may be, there are limitations to access and information sharing simply because the person is the opposite sex from the one that particular aspect of culture revolves around.

For example, there are certain sacred flutes that are used by many tribes in the Alto Xingu in both festivals and rituals and are stored inside the men's house in the central courtyard of the village. Women are not allowed to see these flutes, and breaching this rule is punishable.

Thus, no matter how good the interaction between the men of these tribes and a woman who is working with them, she generally cannot see or even know the details of these flutes. This applies to many aspects of daily life in many communities, and even Western society, which, despite the significant decrease in gender division achieved over the years, still has cultural aspects related more to one or the other gender.

Certain cultural peculiarities may even seem shocking at first. For example, adult Xavante Indians are the first to be served at mealtime, and the children eat what is left over, which is sometimes not very much! This notion runs contrary to the general practices of Western society.

When asked about this, the Xavante explained that, in their view, the working and hunting adult has to be as healthy as possible so that he is strong and prepared to bring food for the rest of the household. If they have little food and feed the children, leaving the adults weak, they might all soon end up dying.

This is a peculiar way of thinking moulded by years of struggle for survival and food restrictions that cannot simply be criticized by another society that thinks differently; it is simply the way it is! This is an important point for socializing and learning. If we want to know and respect another culture, we must be open to learning other values and understanding why they do something differently than how we expect people to "naturally" behave.

We must simply learn that people think differently! Strange as it may seem, and regardless of whether we agree, we must not attempt to change other peoples' cultures. Watch yourself. Centuries of catechesis were imposed on these peoples through several conflicts with the dominant society. As beneficial as we believe our proposed change is to their culture, why do we always have the arrogance to think that we are right? Thousands of wars between people with different ideas have been fought throughout the history of our planet based exactly on the premise that what I think is right and should therefore be imposed on others.

At the same time, the stereotype of the noble savage, whose every deed is right and who is in harmony with nature, is also a myth created in a time of paranoia when everything had to be politically correct. This stereotype is not true, and many examples can easily be enumerated, some of which will be mentioned in the course of this work. Therefore, there must be "tweaks and adaptations" for specific projects according to common sense and accounting for the previously described arguments in relation to respecting other cultures. It is always paramount to take extreme care with the "cultural security" of a community with which we will work. Having respect and humility while not thinking that we know everything are fundamental tenets for any work we do with other cultures.

6 Improvements in the Villages and Being Sedentary

Another aspect of behavioral change in many traditional communities today is that improvements are being made within their territories and villages.

For example, masonry buildings such as schools, health clinics, and even homes as well as the construction of wells with water tanks and plumbing systems for houses or huts ("maloca") tend to make a village more sedentary. Communities with a history of frequent changes in their location change much less today, and when they do move, the shift is often to a location very close to the one they already inhabited so they can continue to enjoy much of the existing infrastructure.

This trend not only causes behavioral changes but may also result in a greater impact on the environment in which they live. The exploitation of a particular resource, be it food or an item for everyday use such as the species used during the construction of longhouses ("oca"), becomes greater for those communities that remain in one place for longer periods, which exerts pressure on certain resources. Moving to a remote area gives the environment a better opportunity to recover.

Thus, what has begun to occur in many communities is that a tradition/behavior that had a certain impact on the environment that was previously less apparent because the community relocated causes a much greater impact today to the point of creating problems in the environment and cultural balance of the community.

One example is the Xavante Indians. This group has been traditionally semi-nomadic, with hunting and gathering being their primary source of food and agriculture being more incipient and used in a complementary way.

This people farmed small fields, generally based on tubers, which require less farming and can also be harvested at different times (a cassava can be reaped after a year or a year and a half or more—the difference is that it becomes more fibrous, but it can still be used).

The Xavante Indians migrated through an extensive territory while hunting and using different plots left at strategic points. Game was always abundant because they did not stay long in any one place. Today, the Xavante Indians are more sedentary and they were settled in the poorest regions of the Cerrado (Brazilian savannah), where the surrounding farms eliminated much of the native vegetation, which reduced the land's capacity to support animal species.

As they have become sedentary but continue their hunting culture, the pressure on the various animal species has increased. This is especially true with regard to their traditional hunting technique, known as the circle of fire.

The circle of fire, put simply, is made by first placing a fire in a small area in the middle of vegetation. Thereafter, hunters will place fires in the surrounding vegetation over a larger perimeter around the previously burned area, which causes the animals to flee and run towards the previously burned central region, where the Indians lie in wait to kill them.

This hunting method is very efficient for attracting prey to a particular site; however, with repeated use in a given region, the vegetation begins to suffer, and the regeneration of animal species in particular becomes increasingly difficult.

Thus, these people need to adapt their cultural tradition to reflect this new sedentary lifestyle. How they should do this, and how we can help them without substantially interfering in their culture, presents a major challenge to our work.

7 Increasing Population, Increased Pressure on Genetic Resources in the Territory, and Pollution

Another reality that has been emerging in recent decades and relates to both a more sedentary lifestyle and improvements in health and rights (especially land, minimizing conflicts) is that the growth rate of indigenous peoples is generally increasing (today's growth rate of indigenous populations is around four times the average growth rate of the national population).

This trend, coupled with a more sedentary lifestyle, impacts the environment within the reserve. Furthermore, waste problems have begun to develop with the acquisition of more and more industrial products from the city, whereas all of the previously generated waste was organic. In addition, this "internal" generation is augmented by pollution coming from the outside, such as pesticides from nearby farms, the pollution of water from rivers that pass through their territories, and deforestation among other sources.

These developments mean that these people are facing problems that were not a part of their culture; therefore, if we propose work that affects their area, we should account for its newness, and the manner the work is performed must be considered and discussed to involve and account for the local population.

8 Clash of Generations

Significant changes have occurred in the cultural dynamics of many indigenous communities within only a few generations and have created a significant difference between the older and younger members of society.

Whereas many of the elderly in many ethnic groups do not speak Portuguese and have never left their territories, the young all speak fluent Portuguese and frequently visit the city or have lived there for a long period to either study or work. Therefore, many such younger individuals question the old traditions upon returning to their village and tend to disregard what their elders say is the right thing to do.

This generational clash is common in any society, and there have been many generational rifts recorded throughout world history. It is up to each group to ascertain what they want to change and to what degree. However, it can be observed that the direction and intensity of change is not only influenced from within the indigenous community but also heavily influenced, and in some cases imposed, by Western society.

For example, as a result of our laws, indigenous people are forced to organize themselves into associations with a president and treasurer to receive project support, sign work permits, and obtain work as teachers or health workers among other requirements. Naturally, the younger people find it easier to organize themselves in this way and assume these posts because they have a better relationship with the city. This trend causes a power/hierarchy to emerge parallel to that of the traditional chief. Someone who brings money into the village may question an order from the chief.

Western society has also impacted these people through the air. For example, in the Alto Xingu in the mid-1990s, it was very rare to find a television in the village. There was an old one in the FUNAI post and utmost one per village! Even these televisions were often only connected infrequently due to a shortage of fuel for the generator because it need to be fixed or a lack of interest. Watching football was the primary use of these televisions.

Within 10 years, the number of televisions has exploded, and they have become increasingly modern. It is no exaggeration to say that the percentage of homes with flat-screen TVs in these villages was, at one point, greater than in the Brazilian cities.

What is striking is the degree of hypnotism that the device has brought. As the local signal is received by satellite (dish), it is very common that the signal disappears during the space for regional advertisements, and a dark screen appears until the next national advertisement or the programming returns. Amazingly, the audience has been observed to continue to gaze at the screen even during these image vacuums and did not look away or even talk.

This observation leads us to another point of great cultural change. By tradition, the end of the day always signalled a time when the men would meet in the village center and be surrounded by youngsters and children watching and listening to the conversations of their elders. In essence, the children were learning and integrating the functions that they might one day adopt.

What we see today, however, is that these meetings have become increasingly short and have fewer participants, and the youngsters and children stay in their homes to watch soap operas. In other words, even though these people have the right to choose whether to watch television, we are ultimately transferring to them our own society's concepts through what we produce and transmit, which greatly influences them even in terms of their sense of time, which had always been treated more fluidly.

Our influence even affects certain time-related traditions that are deeply rooted in these communities. We will discuss two examples that are again from the Indians of the Alto Xingu.

The first example relates to the length of time children are confined during puberty. There is a rite of passage from childhood to adulthood during which the young are confined to their homes for a given period and are unable to leave or be seen. It is during this period that parents and grandparents teach them their main traditions, stories, knowledge, and teachings. Rituals are carried out and effusions are drunk to strengthen the body and spirit of these children so that they can complete their development.

There are reports of young people being confined for over a year, especially in the chief's family. Today, however, these detentions are increasingly shorter. The young no longer want to stay confined and think it is a waste of time.

Another time of confinement is when a family member dies. The family stays in mourning for a certain time, and many activities must be summarily suspended. However, due to general commitments outside the village, there have been many cases of people who should be in mourning travelling to meetings in the cities and then returning. This would have been quite unthinkable until recently.

9 Greater Flow of Financial Resources to the Villages

One factor that has changed substantially in recent years and is related to the previous factors is the easy access to consumer goods in the villages. This access has several sources that vary from ethnic group to ethnic group. In the northeast, where the indigenous people have been integrated with Western society longer, many have income sources from jobs in the cities and countryside, access to both government grants and benefits (family grants, scholarships), and pensions. The latter form of integration can also be observed among the Xavante Indians in central Brazil.

For other ethnic groups, ecotourism has experienced rapid growth, especially where the people still retain much of their traditional culture. Many tourists pay large sums of money to attend traditional celebrations; others pay to spend time living in one of these societies. There are also those who have begun to explore tourism for specific audiences, such as fishermen, who pay to fish the better preserved territories, which have more abundant supplies of fish.

As this interaction grows, the concomitant cultural exchanges affect traditions. We recently (2012) observed one such case when travelling down the Xingu River by boat. In some sections of the river, we observed signs that had been placed on certain beaches by the Indians. When we inquired what they said, our boatman and the chief of the Ilha Grande village told us they were markers showing tourists the limit of where they could go. They were warnings that any tourist who crossed the line would have to pay for visiting an adjoining village's territory.

Thus, we can observe the concept of ownership, which has hitherto always been very fluid in indigenous cultures, beginning to take on a capitalist character.

Another issue is the wealth of large companies and institutions that pay to link images of Indians to their products, especially those they want to connect to something more natural and environmentally sustainable. Large domestic and foreign broadcasters have also taken advantage of the visual and cultural beauty of indigenous communities by making documentaries on their cultures (which, in a way, is good because it records and encourages the preservation of their traditions while providing resources). More recently, there have even been "reality show" style programs with actors and characters from Western society interacting with indigenous people, which borders on the artificiality inherent in these programs; however, that is a subject for another discussion.

Another source of resources is money from the so-called compensation funds for environmental damage. For example, in

Maranhão, various ethnic groups receive large payments in return for their land having been "crossed" by transmission lines for the power grid between the power plants and cities. More recently, populations have received compensation for both the construction of hydroelectric dams and flooding caused by dams.

The fact is that access to money has significantly increased and is impacting indigenous cultures. For example, there is no longer a great need to farm for food and these peoples now have access to the city and money to buy food. This change leads not only to the risk of cultural loss and the loss of traditional species and varieties no longer being planted but also to the risk of increased health problems, such as diabetes and high blood pressure, through the excessive consumption of salt, pasta, sugar, and other substances.

For example, the Indians of the Xingu region traditionally obtained salt by extraction from the water hyacinth plant. The salt of this plant is composed of KCl and not NaCl as it is the sea salt. In addition to having a different taste, the former does not increase blood pressure, whereas the latter does.

Returning, the methodology used by a researcher has a significant influence on achieving the research goals. We can often list what we aim to achieve by working with these populations when diagnosing the problems to be solved, such as improving the conservation of a particular traditional agricultural species or recovering another. However, there can be vast differences between diagnosis and implementation, as people are often no longer willing to rescue or preserve a given material or tradition that they can replace with something they can buy, and they often have other priorities and commitments such as those that provide them with a source of income.

An indigenous group may be fully aware that a traditional product is disappearing; however, there is sometimes no interest in maintaining it. An example is the fava bean—*Phaseolus lunatus*—in the Alto Xingu. This plant is very much perceived as a cultural marker that survived only due to the customs of the older generation, who alone continue to plant it. Young people, in general, do not appreciate this type of food and have no interest in either eating or planting it. Therefore, the fava bean is slowly disappearing.

Thus, demands can be identified in both conversations and interviews; however, it is difficult to meet these demands effectively. The population must believe their efforts to meet those demands will be compensated, either through realizing their importance or, in many cases, by attaching other benefits during the execution of the work, such as payment for forestry services, being an informant, providing equipment, or fuel they want, or other forms of reimbursement. Unfortunately, in most cases, cultural preservation is no longer a convincing enough argument.

10 The Contemporary Indian and Cultural Echoes of the Past

In the daily life of an indigenous society, one lives with different "levels" of cultural tradition, which can be roughly divided among the contemporary, traditional or historic, and prehistoric Indian cultures.

What we refer to here as a "contemporary" Indian is one who adapts his or her traditional culture to a new reality that he or she has received, often imposed from the outside. This new reality may include legislation, limited territorial mobility, pollution problems, discussions about the use of forest reserves for selling carbon credits, wages and monetary benefits, and other factors.

The historical or traditional Indian is what is stereotypically identified and is the heart of their ancient culture. This state is often the standard envisioned for achievement through the work place in those societies. In many cases, the approach to work is for the preservation of or a return to traditional ways. As a brief aside, one should be very careful not to impose a cultural "freezing" upon these communities by thinking they must be like their parents and grandparents and cannot evolve or change. By definition, culture is dynamic.

It is in the context of the historical Indian that the traditions, myths, rituals, management of both plant and animal use, crafts, construction, and other cultural attributes can be found. It is also here that the impact of Western society and cultural dynamics has been the greatest. Many projects, often demanded by the Indians themselves, attempt to return some aspects of the culture of a given community from being contemporary to a more traditional level.

This effort reflected in projects that revalue the work of indigenous midwives instead of sending pregnant women to the city to have their children, and the revaluation of a healer with his herbs and rituals to complement modern medicine. There are also projects attempting to recover traditionally used species and varieties of plants that have been lost for various reasons. In general, researchers will find ourselves concurrently interacting with these two antagonistic and complementary worlds during any work they do.

The final tradition, that of the prehistoric Indian, permeates the second and is concerned with a history that may have been lost in time. This tradition is the source of much of the ethnobiological information permeating these traditions and entails their stories for the origins of the world, plants, animals, and where the food that they use comes from. In short, everything that gave rise to what they are arrives today via oral tradition.

This tradition refers to what Indians themselves define as the time of their ancestors. The focus is more abstract, with data that often vary widely, and we therefore need to collect these data from a larger number of informants to develop a better idea of the major traditions and their nuances. This area of study includes works focused on archaeology and history, which are often able to rescue some of the history of the origin and evolution of contemporary indigenous people and their traditional materials through specific investigations.

11 Legislation

As with the societal culture, legislation is also dynamic (unfortunately, in many cases, it is less dynamic than it should be!). In recent years, this line of work has changed considerably, which has brought the need to increase legislative knowledge and adapt the working environment to the limits imposed by new legal guidelines.

The most striking law in this line of work was introduced by an interim measure in 2001 that created the Board of Genetic Heritage (Conselho de Patrimônio Genético). This board aims to regulate works involving traditional peoples and check whether the rules on accessing and using traditional knowledge associated with genetic resources have been adhered to. This law, which was generally welcomed in essence, the law was written in a very short time, without discussion, to a certain extent because it seemed to be related to the risk of biopiracy at the time. But it has been generally difficult to implement it in practice.

What the law achieved has been to help standardize work a little more; however, it has also made the activities of honest workers more difficult, whereas those who are intent on biopiracy carry on regardless of whether there is a law against it or not.

The largest observable impact for traditional populations was that they had once again been duped. They were subjected to several campaigns to convince them that they had a strong heritage in both material (their genetic resources) and intangible (their associations with these resources) terms, which made them believe they were sitting on a fortune. For this reason, they were warned not to provide these materials and knowledge to anyone else. In reality, there are few resources and knowledge that have the potential to generate substantial economic dividends.

An adverse side effect of this failure has been felt in many communities, e.g., in the Xingu Indigenous Park. One example relates to the seed exchange network. The driving force for spreading agricultural products around the world and generating variability was seed exchange among farmers. This diffuse agricultural system has always served as a type of buffer against losses given that the exchange of materials would allow a farmer to turn to a relative, neighbour, or friend to obtain a similar seed. This is the heart of the agriculture cultural dynamic.

However, we have observed this exchange network decrease in recent years because the person holding the material is unwilling to pass it on to others in an effort to increase the benefit from the advantage it affords him. We have observed many examples of this among the Kayabi Indians in Xingu, where cases have been reported of a relative only sharing a seed after roasting it, that is, giving it to the other person for food but not planting. This trend is worrying for its impact on species conservation and the generation of diversity, especially considering that there are very few cases in practice where any real gain occurs from using samples collected from these communities.

At the same time, this pattern is a new factor permeating our working relationship with the communities.

Authorizations are an important factor, and many communities today are somewhat "allotted". By law, the Brazilian state regulates and oversees all work involving indigenous communities, and, in theory, the State should authorize any work and people's access to the reservations. However, the situation is notably different in practice.

The State has been both devalued and weakened, and its space is occupied by NGOs, religious groups, businesses, and others. Thus, depending on the area where one works, the rules and necessary authorizations are not official but are imposed either overtly or covertly by the group operating there. Thus, depending on who is in control of a particular region, be it a group, agency, or company, you may find it either easier or harder to gain access and develop your work.

We have primarily focused up to this point on aspects of the culture and history of indigenous peoples and how they may influence the goals and progress of our work with them. We will now shift our focus to issues we should observe at all times to ensure our work is as professional, fruitful, and respectful as possible based on the diverse aspects we might encounter.

The questions and guidelines below are those that we should consider throughout each project, not just at the beginning. Many of these guidelines should be revisited over the course of our work to determine whether we are still on track or need to correct our course.

12 Ask the Right Questions for What You Are Trying to Achieve. Are Your Goals Clear?

The key to success for any study is to clearly determine the goals we want to achieve.

Thereafter, we should adapt the methodology and schedule according to the reality of where we work.

Thus, we must always take time to clearly define the objective. We must clarify what question(s) we want answered. We must clearly identify what is driving the focus of the work and accordingly define both the methodology and form of execution, as this will lead to fewer problems during the execution of the work.

13 How Do We Choose Where and with Whom to Work?

It is common to hear someone who is looking for work say, "I want to work with this or that ethnic group". The reasons for this are many and varied; however, what we often see is that the person has no clarity about the intent of their work and just wants to be there. It is essential to have empathy towards people with whom we are going to work, but at the same time, we must put guidelines in place regarding what we want to do to articulate our intentions with the community and not appear like tourists.

In contrast, there are those who have a specific line of research on which they want to work but still need to find the right place to do their work. This search can sometimes be more difficult because they must match the need/interest of the community to the work they want to do by maintaining empathy between the parties and accounting for the working environment, cost, availability, and so forth.

14 How Do We Become Accepted in the Community?

An essential step for our work is to become accepted by the community. This step is much easier when we are introduced to the community through someone who already has a good relationship with the people.

However, however easily we might be introduced into the community, we will always be under observation, and our immersion in the intricacies of that culture will only be expanded with the gradual passage of time, experience, and growing confidence. Even then, there will be limits to the aspects of the community's culture we have access to, and these limits should be fully respected.

15 Does the Work Focus on Continuing in the Right Direction?

Even with both personal and professional preparations for the work being performed, it is quite common to see people getting distracted and losing their focus. Indeed, indigenous communities have many charms, and we must learn from them, especially in matters of moral growth. At the same time, they make many demands on us that we try to satisfy.

However, if we want to be good professionals, we must know how to separate what is work from what is not. We must regularly monitor ourselves to see if what we are doing is within the proposed objectives. There are activities outside the scope of our work in which we can participate, which is important for better integrating with the community and its culture. However, if these extracurricular activities begin to escalate to the point that work takes a backseat, then this is not right, at least from a professional point of view.

We find it very common that people (with the best of intentions) end up wanting to help meet the demands of the community members or even things that they perceive as demands as a result of these distractions, and before they know it, they end up repeating the mistakes of the past, such as "assistentialism" or worse, attempting to change the habits of the community into a new form of catechism.

In this sense, there are situations that occur in the everyday life of the community that may unfold in very distinct ways from what we would expect based on our vision derived from Western culture. As stated above, any given culture has values and interpretations that differ from the next. Thus, we should not judge too much or take sides on internal cultural issues. What may seem wrong in our culture may be normal in theirs and vice versa. We are just passing through, and there is so much we can learn, including diversity in thought.

We have to work together and not offer assistentialism or catechesis!!

16 What Is the Methodology to be Adopted?

Methodology selection is one of the key points for successful field work; however, it is perhaps also one of the most difficult to determine. There is a wide range of methodologies available, the choice of which depends largely on the approach we take, the community we are working with, and especially our own profile.

We should have the sensitivity to adjust the methodology according to these factors. Moreover, we should be aware that there may be a need to change our methodology as the process develops; the project unfolds; or the situation, relationships, and dynamics become clearer. What is true at any given point may not be true later.

We must adopt more than one methodology even within a single village based on which group of people and focus we are working with.

17 Who Are the Informants/Key People in the Community?

Identifying the key people in the community is an important step in the evolution of the work. There are roles within each society, and identifying the role-players is essential to deepening the scope of the work.

Moreover, there are people who are prominent either because of what they know, the influence they have within the society, or even a mutual affinity. It is essential and fruitful to identify these individuals and foster a productive relationship with them.

At the same time, some peculiarities are interesting. One example is the case we experienced in the village of Ilha Grande involving the Kayabi ethnic group in Xingu. There, one focus of the work was peanuts. Our closest relationship was always with the chief, who was an accomplished farmer. It turns out that, for cultural reasons, he cannot plant peanuts and must "hire" someone to do it. So, even with the close relationship and openness we established with him, some answers about that crop, we only could achieve with another farmer.

18 What Will the Work Schedule Be?

Perhaps the most difficult aspect of doing work with indigenous populations is maintaining a schedule. There are literally a world of factors that may affect the schedule and force us to change or adapt it.

To begin, Indian time is different from Western time. Tomorrow might actually mean tomorrow, or it may mean the day after tomorrow, a week from now, or never! This trait is neither right nor wrong; it is just the way things are. The Indian puts his commitments on a priority "list" that is very dynamic. It is not that he has not committed, he simply found something more interesting or important to do and left our meeting for tomorrow!

If this trend is true when we are present in the field, it is even more so when we ask them to perform actions when we are not present. Unless the indigenous populations are engaged in the project, feel they are really a part of it, and believe it has important benefits for them, we generally find that the task is unlikely to have been conducted according to our protocol when we return.

For example, we might ask indigenous people to record what they ate over a time period so we can analyze their diet, or to keep the bones of the animals they hunted so that we can obtain an idea of their hunting cycle, or to make a note of the actions they take and on which days they undertake them, or we might request that they perform a task necessary for our activities in advance so we can immediately start working when we arrive. However, all of these things are unlikely to occur, especially during the early stages of our relationship with them.

Even when they do provide this information, scientifically speaking, we must be careful when including it in our study because the reliability of these data can be easily compromised unless, like any researcher, they have received appropriate guidance for the work.

Returning to the schedule, just as unforeseen events occur within Western society (e.g., delays in gaining authorization or the release of funds), so too does this occur in indigenous cultures. It may be that a car or boat motor broke, which prevented us from getting to the village; it may be that someone died and the village has consequently gone into mourning and our planned activity is suspended; maybe our resources were released later than expected and the trip to obtain data about planting the fields can only be taken during the harvest; or they discovered that a hydroelectric dam is about to be built in the region and meetings, protests, and trips have been scheduled for the same time we scheduled our field trip, so most of the villagers would not be there; ultimately, these are all unforeseen events.

We therefore have to be focused on our goals rather than our schedule and try to obtain as much information as we can in every moment—taking care to always maintain respect for the culture that we are working with because they have their own day-to-day dynamics and are not always available when we want them. We are foreigners; we are authorized to experience and interact with them within limits that have been earned and must be followed through effort and respect.

Ultimately, when something completely ruins everything we had planned to do, it is best to relax and try to build a better relationship with the community!

19 What Is the Cost and Availability of Resources?

The cost of fieldwork, depending on where it is, can be extremely high, especially with regard to travel, equipment, and field maintenance costs. Thus, the adage on taking a step according to the size of one's leg is quite valid. It is of no use to have the resources to start a project and then not be able to proceed.

It is often worse in this case than if we undertook a smaller project because arranging to work with these communities and then not fulfilling our commitment reinforces what they have learned most about our society: that promises are unfulfilled.

Thus, a very useful tip for building trust with indigenous cultures is to only promise what we are really certain of being able to accomplish. Making promises just to give the impression that everything is fine only leads to a fragile relationship.

20 What Time Do We Have Available for Fieldwork?

As noted above, schedules are very changeable; thus, we usually spend more time than originally planned completing a project. As some projects are tied to scholarships and funding, it is always good to try to complete the work within the scheduled time, without relying on a renewal or extension, so as not to risk having to terminate the work without having accomplished what we had hoped to achieve, which would be frustrating for both ourselves and the community.

It is always good to set levels of goal achievement, with the first for an ideal situation in which everything goes according to plan, another at a medium level, and a third assuming the most pessimistic scenario. For each level of expectations, we must focus on goals and objectives for consistent results. Setting these goals is often not easy in practice, but adequate preparation at this point is in the best interest of the project and can be the difference between having a little, good data or having a lot of incomplete data.

Perhaps it could be said that, through experience, we find that the best approach is often to think "small" and develop parallel actions as the work progresses to expand it while maintaining focus to not run before we can walk. This point is extremely important because no matter how thorough our planning process might be, we can never predict all contingencies and thus can only actually see how things will unfold in practice as we go along.

21 Do I Have the Skills for the Proposed Work?

Something that is likely to occur more often than we might imagine is that a person wants to work with a traditional community but does not have the skills to do so. When we watch a documentary about working in an Indian village from the comfort of our living room, it seems quite enchanting. Beautiful images! Background music! An exotic, pure world to be experienced and "explored" by our adventurous spirit. However, to live this experience, we must be prepared to be exposed to bad weather conditions and attacked by mosquitoes (some of which transmit diseases), which are often present in that same beautiful landscape we observed in the documentary; and the sound in the background is no longer a nice song but a mosquito in our ear that we cannot get rid of.

A person with fieldwork skills must be prepared to feel hot or cold and often hunger and thirst; he or she must be flexible in terms of diet, able to settle into often uncomfortable facilities, make long journeys on foot or even more precarious and uncomfortable forms of transport, relinquish part of his or her privacy (as there is no way of hiding what you are doing in a village), adapt his or her hygiene habits to each local situation, both pack and unpack baggage, carry weight, and endure physical pain; in short, he or she must be able to adapt.

One of these adaptations is time itself. Thirty days goes by very fast in our everyday life in the city; however, in a village, the same length of time can drag so much it sometimes hurts. The days are "longer" in the field, and a person who is not mentally prepared to work with that reality and integrate with the community will tend to waste away and lose momentum as time passes and food is depleted. What the person once considered exotic will now feel like suffering (comfort, warmth, hygiene, and food).

Even the form of relationships with the research team is something that should be considered. It is often said that we only truly get to know someone when we have gone on a trip and lived with him or her full time for a given period. A person's characteristics, which we often overlook or do not notice on day-to-day basis, can become very vexing after a long period of coexistence in the field. The more difficult the situation becomes and the more limited the resources, the more quickly and intensely these personal characteristics emerge.

Thus, knowing how to address the diversity of people's behaviors in the field, as well as knowing how to integrate ourselves into a group, is an important requirement. Especially for the first trip, therefore, it is advisable not to plan too long a period in the field, and we should try to go with someone with whom we have a peaceful relationship, if possible.

22 What Is the Level of Community Involvement with Our Project?

Behind a desk, it is easy to develop a series of projects, activities, scenario projections, and solutions to problems (real or imagined). Everything works well on paper! However, what is really achievable (culturally and practically), and most importantly, what is required?

In many cases, the results of a plan fall far short of its targets because it was created according to the interpretation in our minds without regard to whether the project was really needed by the community, or at least part of it, in the first place or whether it fits into the cultural reality of the community.

In a simple visit to a village, we can list a multitude of activities and projects that we want to undertake using our cosmopolitan outlook. These activities and projects may be structural, environmental, cultural, food oriented, political, economic, rescue based, legal, ecotourism oriented, or cinematic, among many others.

However, what is really demanded? What does the population want, and what are just actions that we ourselves think are necessary based on our worldview?

Is introducing a more productive crop really necessary? Will this introduction not affect other related components (soil depletion, pest introduction, traditional management methods, traditional preparation methods, resistance to abiotic and biotic factors—such as pests—and drought resistance)?

For example, the Indians of the Alto Xingu always planted a type of cassava containing high levels of hydrocyanic acid ("bitter" type), which is extremely toxic and can cause death to those who consume it without proper preparation. This preparation is something that has been traditionally incorporated into the everyday life of these people for millennia. The stage where they withdraw the hydrocyanic acid is laborious but very efficient. However, in the mid-1980s, to minimize the need for cassava processing by the Indians, well-meaning people introduced cultivated varieties of cassava with low acid content, so that they could consume it more easily. This new variety was an immediate success given that it not only reduced their workload but that they could also consume the cassava in different ways, which diversified the food dishes available.

However, there were soon cases of people consuming the "bitter" cassava without processing it, thinking it was the low acid type, and becoming poisoned. In some cases, a part of the entire household died because of this poison. Leaders quickly got together and banned the low acid type cassava from the Xingu. Today, as a rule, all cassava is "bitter" and should follow the traditional processing. There may still be low acid types there, but when in doubt, they are treated as if they are the "bitter" type. Thus, even when we have the best of intentions, it is very easy to destabilize some components of the cultural fabric of a community.

It is therefore necessary to know how to filter and analyze demands from within the perspective of the cultural community. Trying to project the consequences of our direct and indirect actions is a practice that we must constantly undertake.

To minimize and avoid liability for any unforeseen or unintended consequences, discussions regarding each action should be widely conducted with the community, and the final decision of whether the action is taken should come from them and not be imposed. Furthermore, discussion with the community is a basic premise of the success of any project. If they do not feel like a part of the discussion, the project tends to falter and sometimes becomes fruitless.

Conversely, if the project arises from an idea or actual demand from the community and we are only assisting with methodologies, technical support, and resources and some members of the community are participatory actors, the project has a much better chance of surviving and achieving results.

However, even when the idea comes from the community, these projects must be filtered to check what their impact will be on that community, and we must ascertain whether the whole community agrees with it or whether just some of the people put the idea forward. Furthermore, even if the same projects are endorsed by the community, they can become prejudicial if the projects are not properly conducted.

For example, some time ago, a village on the banks of a lagoon in Xingu managed to capture a resource to promote fish farming in hapas nets in the lagoon. This project was demanded by them that they could capitalize on it; however, it was predicted to lead to the introduction of exotic species into the environment. This approach is environmentally risky because it can shrink the food supply rather than expand it by destabilizing the balance of other native species.

23 The Project or the Project?

The title of this section may seem wrong, but it is correct. The reason for this title is the fact that the concept of the project can be interpreted very differently depending on whether we are on the performing or receiving end of the action.

Generally, the technician has a firmly closed, bureaucratic idea of what a project is: it has an aim, materials and methods, goals, a schedule, a budget, and a report. Depending on the funding source, the technician may have a little more or less freedom to make adjustments to this, and the limitations are often greater regarding the allocation of financial resources (rules for what we can and cannot buy or pay for with these resources).

However, for the indigenous communities, it is very common for a project to be interpreted as a resource umbrella under which they can make various demands, whether they are directly linked to the project itself or simply demands the community wants met, which sometimes have nothing to do with the official objective of the work.

Thus, it is essential when discussing the project with the community to make it very clear not only what we want to accomplish but also how the money raised will be used (including the compensation that the community has demanded to authorize the implementation of various actions—for example, a boat; a generator; fuel; use of the village's property and equipment; and payment amounts, if necessary, to community members who participate as informants, woodsmen, and boatmen). It is important that we make these points explicit to ensure that we are not charged for them later.

24 Is This All Within Reason?

Finally, the basic guideline that is essential for anyone working with traditional peoples is to use common sense. There will be many situations in which the course of action at any given moment will be guided more by common sense than any rule, instruction, or knowledge acquired through years of education. What applies in one case may not be true in another because of differing contexts.

Moreover, being given permission to conduct a project does not give us the right to do whatever we want. An example of this restriction is when we receive authorization to photograph and record the everyday moments of the community. Even when we have received authorization (formal or oral) to make such records, there are some situations that should not be recorded because they are either embarrassing or cross boundaries of intimacy or because we were simply asked not to record at a given time. We should not insist—we must be respectful.

In situations where we do not know what protocol to follow, especially during parties or collective rituals, we must wait and see how the community members behave. If there is still doubt, we should ask what to do, such as where we should stand and what we can and cannot do. We must observe and listen more. We should not try to impose our dynamic and not try to avidly grasp at everything all the time. As an old Indian once said, "stop a little before the end of a walk so your soul can catch up with your body. Then, we can take the next step towards the destination".

As we reach the end of this chapter, we can see that this line of research has more questions, doubts, and uncertainties than answers. Even where such answers exist, they are hardly unique and conclusive and vary depending on each person.

Perhaps the greatest certainty we have is that we need to constantly reshape and adapt to each new situation and scenario. This ability to adapt is a significant challenge; however, if we get it right, it is enriching not only professionally but also for our human development.

These tips are intended to help those who wish to follow this line of work and even those who already operate in the field, as they are reflections gathered and accumulated with both time and experience in what is sometimes a surreal world that helps us to think about and rethink our own world.

References

1. Presidência da Republica. Decreto nº 6.040, de 7 de fevereiro de (2012) http://www.planalto.gov.br/ccivil_03/_ato2007-2010/2007/decreto/d6040.htm. Accessed 14 Oct 2012
2. Scribd. Política Nacional de Desenvolvimento Sustentável dePovos e Comunidades Tradicionais—Antecedentes (2012) http://pt.scribd.com/doc/84693432/2007-Antecedentes-Da-PNPCT-Povos-e-Comunidades-Tradicionais. Accessed 10 Oct 2012
3. Ministério da Justiça. Povos Indígenas—Fundação Nacional do Índio (2012) http://www.funai.gov.br. Accessed 20 Sep 2012
4. Prof. Dr. Paulo Martins Sodero—Escola Superior de Agricultura "Luiz de Queiroz"—ESALQ/USP. Personal communication—In memoriam

Chapter 15

Participatory Mapping of the Terrestrial Landscape in Brazil: Experiences and Potentialities

Taline Cristina da Silva, Washington Soares Ferreira Jr., Patrícia Muniz de Medeiros, Thiago Antônio de Sousa Araújo, and Ulysses Paulino Albuquerque

Abstract

Maps are graphical representations of a territory. Throughout history, maps have been used by diverse groups for various purposes. Taking into account the great potential of maps for many areas of knowledge, the use of maps in ethnobotany and ethnoecology is widespread both for collecting data and for subsidizing extension actions. To accomplish this goal, this chapter identifies the main types of maps and the incentives, advantages, and disadvantages of using these tools for data collect in ethnobiology and ethnoecology, beyond others potentialities of the use of mapping tools. As an example, this chapter will use two case studies performed with local communities in the Brazilian Northeast.

Key words Ethnobiology, Ethnoecology, Participatory methods, Maps

1 Introduction

Maps are graphical representations of a territory [1]. Throughout history, maps have been used by diverse groups for various purposes, such as expanding civilizations, aiding in localization and identifying areas with resources among other goals [2].

Taking into account the great potential of maps for many areas of knowledge, the use of maps in ethnobiology and ethnoecology is widespread both for collecting data and for subsidizing extension actions. However, few efforts have been made to provide ethnobiologists and ethnoecologists with step-by-step instructions on the application of mapping tools. In this context, this chapter aims to provide a general overview of the use of mapping as a methodological tool in the collection of ethnobiological and ethnoecological data and to discuss the role of this tool in the identification of socio-environmental conflicts, biodiversity conservation and the evaluation of management programs. To accomplish this goal, this

Ulysses Paulino Albuquerque et al. (eds.), *Methods and Techniques in Ethnobiology and Ethnoecology*,
Springer Protocols Handbooks, DOI 10.1007/978-1-4614-8636-7_15, © Springer Science+Business Media New York 2014

chapter identifies the main types of maps and the incentives, advantages and disadvantages of using these tools.

As an example, this chapter will use two case studies performed with local communities in the Brazilian Northeast. The first study was performed with a community located near a Cerrado (Brazilian savanna) Protected Area, and the second study was performed with a community located near Atlantic Forest (rain forest) fragments.

2 Mapping Without the Use of Visual Stimuli

2.1 Ephemeral Cartography

Ephemeral cartography represents one of the simplest mapping techniques. In ephemeral cartography, maps representing the physical landscape and cultural ecotopes are constructed on a flat surface by the participants, who use earth, rocks, twigs, leaves and paint [3, 4]. It is believed that this technique has been in use for a long time in human history, as there are indications of ground drawings of areas that were favorable for hunting and living [5] and records of the use of ephemeral maps drawn in snow by the Inuit people of Alaska, central and eastern Canada and Greenland [6].

The map draft may be created in a totally free manner by the participants. Alternatively, the researcher may provide the contour of the area in question, thereby offering some geographical information [7]. According to [3], a map constructed with this technique is memorized by the participants and may be recalled when needed (from memory, as the original map is usually lost). This map may also be used in the creation of three-dimensional maps to raise more specific questions with the participants at different times.

Using this technique, the researcher seeks to assess, at a low cost and using few resources and materials, the participants' perceptions of the landscape [8]. However, as this technique is simple, it may be complemented by other techniques, such as the use of cartographic sketches, which are differentiated below.

2.2 Maps

Paper maps are special drawings that may be made individually or in a collaborative manner by the populations studied in ethnobiological and ethnoecological research. This tool is often used in ethnobiological and ethnoecological research to stimulate people to record on paper different aspects of the landscape and/or natural resources that they perceive and use. These maps include historical aspects of the locality and tendencies in soil occupancy, and they can complement the information obtained in assessments of natural resources, indicate the areas used by a given species at different stages of its life cycle and identify priority areas for biodiversity conservation [9] among other uses. The drawing's focus is defined by the researcher according to his/her research project (thematic map) (Box 1).

Box 1 Recommendations for the Development of a Communitarian Mapping Activity. Adapted from PETROBRAS (Petroleo Brasileiro S.A.)-SE/CONATURA [14]

1. Attach to a flat surface a white paper sheet, preferably cardboard. The sheet's length will vary according to the amount of information that the studies wishes to obtain
2. Invite the participants to approach the sheet with pencils and erasers and encourage the participants to draw the first lines on the map
3. To facilitate the participants' spatial localization, it is recommended to encourage them to start the drawing from a reference point, such as a main street or places of importance in the community
4. It is also recommended that the participants use colored pencils to differentiate between the different elements of the landscape, such as rivers, streets, and vegetation
5. Allow the people to make the map in the manner that is most convenient to them
6. Do not show impatience, as the discussions are a notably important part of the process
7. After the map is complete, encourage the participants to create legends to facilitate the interpretation of the map. It is also interesting to ask the participants to present the drawing
8. Finally, do not forget to put the participants' names on the maps

2.2.1 Property Maps

Property maps were initially created to identify the limits of properties [10]. Later, they also came to be used to show details of production (e.g., agriculture and livestock) and the social infrastructure of a property based on the owners' representations. These maps also show details that are not offered by natural resources or community maps, as the latter are usually made at a coarser resolution [11].

The same material and methodological recommendations provided in Sect. 2 in Chap. 1 may be extended to property maps. The differences between property maps and the common maps described above is that the former have a clear thematic delimitation that is linked to the socio-environmental aspects of, for instance, agricultural production and forestry, whereas the thematic delimitation of the latter is defined by the researcher based on the investigation's objectives. The construction of property maps may be individual or participatory.

Property maps have shown themselves to be an interesting tool to investigate, for example, backyards used for agroforestry, as these maps allow a detailed view of the properties' plant diversity.

2.3 Social Map

A social map seeks to gather information on the living conditions of the people in a given locality. This map may be used to determine the way that participants perceive the social strata that compose the community, the distribution of wealth, people's well-being and access to financial and other resources, recreation and agricultural production [12, 13].

This type of tool is quite interesting, as it allows the researcher to identify the socio-environmental structure of a community in addition to identifying partner organizations and possible social and environmental conflicts. In the initial stages of the research, the social map created by the researcher may also be important in establishing the first contacts necessary to complete the research.

It is necessary to form a group of community members and initiate the map with reference elements, such as the locations of services (e.g., school, church, healthcare center, recreation area, community center, grocery stores, warehouses, stores), locations of dwellings (and the number of inhabitants) and the main roads [11]. After this stage of locating the physical structures that compose the community, such questions as the following may be asked: how many members are there in each house? What do they do [12]? Social maps may also be constructed by the researchers themselves according to the socioeconomic data supplied by the community's institutions.

3 Maps Created from a Visual Stimulus

3.1 Mapping from Aerial Images

The use of aerial photography in participatory mapping may be useful to obtain diverse types of information, such as the location of resource collection zones and the occurrence of plant and animal species that are of interest for the study. A good amplified photographic image of the territory to be studied is needed along with a pencil and/or pen that would be visible on the image. In general, the printed material may be georeferenced to establish a latitude-longitude grid.

It is important to have in mind the adequate size of the printed map. Ideally, efforts should be made to (a) supply sufficient detail for a good spatial identification by the participants and a good return in terms of richness of information and (b) provide adequate space in which the mapping is performed to allow the participants easy access to the map (see [6]).

The imagery may be covered with transparent acetate sheets on which the participants can make their insertions, for example, by identifying areas with resources, the main areas where some species may be found and the changes in the size of vegetated areas (see, for example, [15]). The same images may be reused in successive stages if maps are constructed on different acetate sheets.

Building different parts of the maps on different sheets facilitates the approach, as this method does not superimpose the

Box 2 Steps to Building Participatory Maps from Aerial Images and/or Maps. Adapted from [6]

1. Attach the image to a table or flat surface with adhesive tape
2. Overlay a sheet of transparent tracing paper or acetate on the image and use adhesive tape to keep it in place. You must be able to see the satellite image through this additional paper sheet
3. Start by drawing on the paper some visible geographical features that will be used as reference points. Rivers, roads, or even the borders of the territory, when present on the image, may be used for this. This stage is notably important and must be repeated for each additional paper or acetate sheet. These are the references that will allow you to superimpose the different layers in the correct position after their creation
4. Start drawing or writing on the superimposed paper the information relevant to the involved community according to the established conventions. There are many ways of inserting these data according to the type of information that is being mapped and the material with which you are working
5. When the paper or acetate begins to contain too much information or when you change the subject being worked on (for example, from flora to fauna), it is time to remove the acetate layer and replace it with another. Repeat steps 1–4 for new information that has not yet been recorded

subjects, identifying the desired information in a more clear and direct manner. As the overlap between subjects is reduced, this approach may lead to a greater detailing of the information. Afterward, different sets of maps may be superimposed according to the researcher's plan in a similar manner to that used by certain georeferencing programs that superimpose data layers to build more detailed maps.

As for the limitations of the use of this resource, it is known that the community or a portion of its members may occasionally not be familiar with photographic resources. Additionally, certain people are incapable of interpreting a map and associating it with their spatial reality, especially when small scales (representations of large areas) are considered, due to the differences that may exist between mental maps, which are internal representations of the external reality [16], and aerial photography. The incongruence between the local representations of the landscape and the formal map structure may compromise the collection of information. Therefore, if the researcher notices that the use of aerial photography will not be successful, its use should be reconsidered.

Another point that deserves to be highlighted is the importance of obtaining high quality images without clouds and that may represent the area and time of interest to the researcher (Box 2).

Box 3 Step-by-Step Instructions for Mapping to Complement Geographic Information Systems. Adapted from [20]

1. When using a cartographic map as a stimulus, the researcher must be conscious that the maps' scales may vary. Large scales are less detailed, consequently complicating the identification of places that would have been visible at a finer scale. Conversely, small scales bring the image too close, and the researcher may have difficulties in printing the image, both for technical reasons and because the map may not be informative for the participants
2. The map must be strategically attached to a surface that favors the participants' field of view
3. Finally, the participants may be questioned with regard to different environmental aspects. The researcher may ask the informants to build, on top of the map, a second map with the main resource areas. The researcher may then ask the participants to build a third map showing whether there were modifications in the size of these areas. The result is somewhat similar to the different information layers present in GIS software

3.2 Mapping to Complement Geographic Information Systems

A geographic information system (GIS) may be defined as a system responsible for the automated treatment of georeferenced data [17]. A GIS is a tool that has been shown to be of great importance in diverse areas, as it allows the manipulation of spatial data and provides tools with which to model and analyze the interrelations among spatial data [18].

Communitarian mapping to complement GIS has been used to contribute to sustainable development [19]. In this context, GIS data projects have recently begun to take into account the perceptions of local communities in an attempt to generate more robust information that may contribute in a more efficient manner to management plans aimed at conserving natural resources; this new approach is known as public participation geographic information systems (PPGIS or PGIS) [20, 21].

This new approach (PPGIS or PGIS) is notably interesting for ethnobiological and ethnoecological studies, as these studies attempt to gain understanding of the different types of perception that people have toward the environment (Box 3).

4 Potentialities of the Use of Mapping Tools

After explaining the main types of maps used in the collection of ethnobiological and ethnoecological data, we will now exemplify how mapping has been used in a practical manner by different researchers and institutions. For example, certain researchers use mapping to select potential biodiversity conservation areas, concomitantly involving the local communities in the decision-making processes [9, 20]; to evaluate the status of a management plan to determine whether the plan's initial objectives are being reached [20, 22]; and to mediate situations involving conflicts in the use of resources [21, 23, 24]. A brief discussion of each application is presented in the following subtopics.

4.1 Mapping for Biodiversity Conservation

Participatory mapping may provide a common language between managers and local communities when the latter intend to take part in management programs for biodiversity conservation, such as when the communities depend on the biodiversity for their own subsistence [24]. For example, the results of participatory mapping with a GIS in local communities in the Shanxi province in China have been approved and supported by local authorities because information exchange between local inhabitants and managers may be facilitated by this tool during decision-making [25].

Another good example of participatory mapping for biodiversity conservation, Widayati et al. [10]. In this study, the authors aimed to assess how much of a plant resource was being collected in the Lambusango Forest region in Buton, Indonesia. The authors organized a participatory mapping with the region's collectors to verify which areas were being used to collect this resource and to direct questions regarding the patterns of this resource's collection, such as the influence of accessibility on collecting patterns [9], thus observing the extent to which this resource was being collected. This information is important in delimiting management plans for collected species, as such plans consider both the plant species' population and the needs of the local collectors of the resource.

The participatory mapping used in combination with GIS, Brown and Brabyn [21] aimed to assess the value of the landscape based on local knowledge to select areas for environmental preservation. In the same way, Bernard et al. [20] used participatory mapping combined with GIS to indicate priority areas for biodiversity protection in the Maués State Forest reserve in Amazonas, Brazil. The authors printed maps of the region obtained from satellite imagery and presented them to local residents to mark, on the map, the locations of houses, agricultural areas, hunting and fishing areas, the use of natural resources and areas used for livestock. All of these areas were indicated by the residents by means of colored markers. Complementary information was obtained for each region by means of semi-structured interviews. This information included the spontaneous and cultivated plant species used for different ends, the species of animals hunted and livestock animals. A final stage of the study located the marked areas precisely with the aid of a GPS and local specialists, who were invited to accompany the researchers to the areas indicated on the maps. Once the maps were finished, copies were provided to the communities involved and also to local authorities for use in the management and zoning of the protected area.

A somewhat different approach to reach the same goals of the previous studies, i.e., the delimitation of potential conservation areas, Fagerholm et al. [26], which was based on the landscape's environmental services. The evaluation of the area's environmental services started with a set of 19 service indicators based on the context of a rural village in Zanzibar, Tanzania. In a participatory mapping activity that used aerial photographs of the region, residents of 14 sectors of the community were invited to identify different landscapes and to attribute each of the service indicators to each identified landscape with the aid of colored markers. Studies that use this approach capture information on environmental services, as perceived by the local communities, thereby facilitating the identification of priority areas for management and conservation [26].

4.2 Mapping to Evaluate Management Programs

Participatory mapping may be used as a tool to evaluate the progress of a management plan with regard to its objectives. For example, Bourgoin [22] research presents the example of a Bouami community in Laos that was placed under a policy in the 1990s that aimed to change local farming to increase the local area covered by forest. However, the policy was implemented without presenting an alternative to the local people, who were no longer able to cultivate rice in the highlands. In this case, when performing a participatory mapping of the region followed by observations of satellite imagery, the author observed that many families use a larger area for agriculture than allowed in the local land use plan with a large difference between what the people tell the local authorities and the portion of the land that is actually used. In this case, applying mapping may propose models to solve the problems in the region by permitting a dialogue between the authorities and the community.

4.3 Mapping Conflicts in the Use of Natural Resources

Cronkleton et al. [24] research presents an application of participatory mapping in an agro-extractivist community in the Bolivian Amazon aimed at documenting property rights, as there is a conflict with regard to access rights to a species that has much use in the region: the chestnut tree. The author was invited to perform a participatory mapping of the regions with access to the species to mediate resource use conflicts. However, a strong internal conflict concerning the resource's distribution was observed during the activities, as few families had a large numbers of chestnut trees in their backyards, and many families had few or no chestnut trees [24]. Many studies have shown that participatory mapping may have a function in mediating conflicts in the use of the landscape, and a number of these studies emphasize conflicts between local communities and external authorities (see [21, 23]). However, as determined in the example above, Cronkleton et al. [24] presents an example in which mapping may provide an understanding of possible internal conflicts in the use of common resources. Under this perspective, the author indicates that it is also interesting to teach or train local residents so that the community may organize itself and produce its own maps in the future. These maps may aid the community in understanding the distribution of useful resources and of landscapes based on the community's needs, in mediating internal conflicts and in articulating requests for externally imposed policies.

Regarding the application of mapping to identify conflicts in the use of resources, many authors criticize the application of this method or even indicate some cautions that must be taken. In this manner, Brian [27] and Reyes-García et al. [28] note that maps must be made not only to delimit areas and discriminate the resources used by local or traditional communities but also to observe the history of negotiations and interference from external agents on the communities from the point of view of the people who endure the conflicts in the region. In this case, mapping would not serve to solve a land use conflict but to support future actions that will lead to the solution of the problems.

Although participatory maps may present a number of positive aspects in their applications in conflict mediation, the maps may also have negative aspects, as they may intensify internal or external conflicts [28]. For example, building maps that suggest the demarcation or protection of traditional territories may intensify the conflict between the community and external agents, such as corporations with interest in the resources located within the traditional territories. Conversely, the construction of maps may also create or intensify internal conflicts if it intensifies or creates conflicts of interest among the residents regarding the demarcation of the territories within the community [28, 29].

5 Case Studies

5.1 Mapping of Atlantic Forest Fragments in Brazil

For the reader to understand the context in which the mapping of Atlantic Forest fragments was performed in Brazil, it is necessary to explain in a succinct manner the phases preceding the use of this tool and the goals on which the project performed in the region described below was based.

This example was part of a project named *Sustainability of the remaining fragments of the Atlantic Forest in Pernambuco and their implications for conservation and local development*, performed as a partnership between Brazil and Germany, in which our team was responsible for an Ethnobotany subproject. This subproject was divided into two thematic axes: (1) local perceptions and representations and (2) use of wood and medicinal resources. The goals of the first theme, among others, were (a) to identify possible changes in the local landscape by means of the informants' representations and (b) to perform a zoning of areas and possible resource sources by means of communitarian mapping. To this end, after obtaining authorization by means of a Free and Informed Consent Form (in accordance with resolution 196/96 of the National Health Council), 260 interviews were performed (using a representative sample of the community). This sample was distributed proportionally among the ages and genders of the local community's residents. The community was located in an area close to Atlantic Forest fragments inside the territory of a sugarcane processing plant in the Igarassu-PE municipality, Northeast Brazil.

After the interviews, informants who used the forest to hunt animals and collect wood were selected, as we understood that these informants spent more time inside the fragments and travelled longer distances and different paths. A total of 60 informants were selected based on these criteria, 21 of which were women and 39 were men, with ages ranging from 18 to 70 years. However, only 30 of these informants agreed to take part in the map's construction, specifically, 17 women and 13 men aged between 18 and 68 years. Three groups were formed to facilitate the control and application of the method. The groups were asked to illustrate the local landscape starting from the locality's main street (Fig. 1). To facilitate the process of map construction, the members of each group selected two illustrators. Each group was led to name and identify each of the represented elements. The informants were

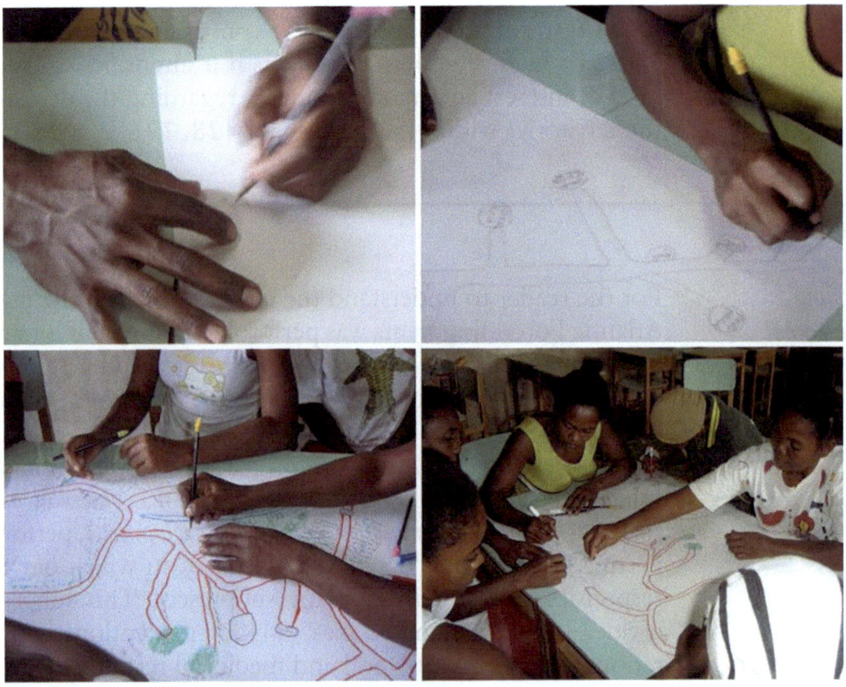

Fig. 1 The process of communitarian map construction performed by the residents of Três Ladeiras in Pernambuco, Northeast Brazil

also asked to use different colors to represent the elements, for example, rivers-blue; forest fragments-green.

During the maps' construction, the informants identified the main areas in which natural resources are collected close to the community. A total of 27 forest fragments, which may be called cultural landscapes, were illustrated. The representations of the fragments were often found in association with rivers, creeks, and springs (Fig. 2). This fact is likely to be related to the frequent incursions of the population into the forests not only to collect wood and hunt animals but also to obtain water for domestic use, as the water provision system in this community is precarious and is identified as a serious problem by the population.

There were similarities among the groups with regard to the presence of sugarcane plantations in proximity to the forest fragments. With regard to the number of streets and social service structures such as healthcare centers, schools, churches, police station, and commercial points, the representations were notably similar. However, one of the groups emphasized the community's social service structure over the forest fragments more strongly. In addition, this group's perception regarding the size of the fragments was different from the other groups, as the latter represented the forests as being smaller.

During the mapping, it was observed that the participants had difficulties in understanding the different scales and simplifying the information, as at first they drew forest fragments that were too large to the point of requiring an entire cardboard sheet for a single fragment. However, one of the participants called attention to this fact, and they started drawing a smaller map. Another aspect worth nothing was the time limit for the map construction, as certain informants wanted to draw with more detail, and such behavior required time, and the group members realized this issue themselves and had to adjust their time allocation to include all of the necessary information on the map.

Fig. 2 Rivers and creeks associated with the representation of forest fragments by residents of the Três Ladeiras District, Igarassu-Pernambuco

5.2 Mapping a Protected Area in the Brazilian Cerrado

The mapping was performed in the Cacimbas community, located in the surroundings of the National Forest of Araripe (FLONA-Araripe) in the state of Ceará, Northeast Brazil. This community depends directly and/or indirectly on forest resources. Taking this into consideration, a study was planned to take into account aspects of the relationship of the local population with their natural resources. In the first phase of this study, Rapid Participatory Mapping proved to be an important tool to gather important information concerning the community's relationship with the FLONA-Araripe, unlike a previous study in which the mapping was performed after the interviews. The participatory mapping enabled the identification of the forest products harvested by the local population and the cultural ecotopes, which, according to [30], make up the smallest component of the ecologically distinct landscape, where there is a combination of both biotic and abiotic factors identified by those responsible for the landscape's management.

To perform this activity, all of the community members who were involved in the process of collecting any resource in the FLONA-Araripe were invited. These people were identified by the president of the community's residents' association, which was our first contact. Four informants who were recognized by the community as having knowledge of the local landscape took part in the mapping.

Unlike the previous mapping, an amplified image of the FLONA-Araripe mock-up was used in the Cacimbas community to stimulate the informants to identify the locations of the main areas where resources were collected (Fig. 3).

The results of the participatory mapping permitted the identification of the main cultural ecotopes that were recognized by the Cacimbas community of the FLONA-Araripe, the main resources that were collected, the routes used by the community to collect these resources and certain elements of the landscape.

Among the cultural ecotopes, the areas that deserved emphasis were those indicated for resource collection and those indicated

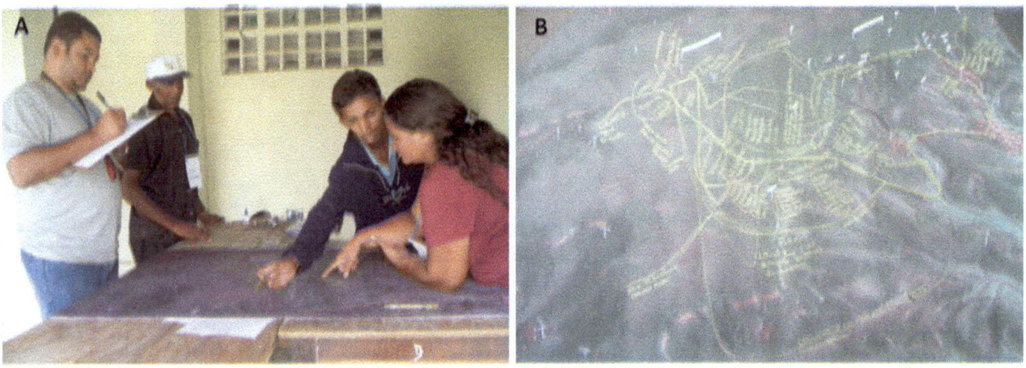

Fig. 3 Participatory mapping of the Cacimbas community, Jardim-CE municipality in Northeast Brazil

simply as routes used to reach the areas in which resources were collected but not as collection areas. This result allows the researcher, for example, to make better judgements when choosing areas of greater use pressure for future ecological studies. Regarding the names of cultural ecotopes, another interesting finding was that some of these names could be used as clues to assess the biotic and abiotic components of the landscape. For example, the cultural ecotype known by the community as Barreiro Grande (Large Mud) is a reference to the existence of an artificial lake in the forest interior, whereas the ecotopes known as Baixa do Uruçu (Uruçu Lowland) is related to the presence of large numbers of a bee, known popularly as the uruçu bee, in the forest interior. This finding demonstrates once more the importance of the tool for future management plans and studies in the FLONA-Araripe, especially when these studies intend to verify the distribution and abundance of some species.

A cultural ecotopes named "Acampamento Barreiro Novo" ("New Mud Camp") was represented as a highly important place for the community, both for the collection, sale and transportation of pequi and for social activities of the local populations, especially during the traditional "pequi feast", during which religious and secular celebrations occur. This indicates the economic and cultural relevance of the FLONA-Araripe resources, especially pequi, for the local populations.

These findings indicate that the community establishes narrow relationships with the FLONA-Araripe, with the natural resources having a remarkable role in the economic and socio-environmental dynamics of the local population. These people depend on the resources to meet the market demand and for their subsistence.

With this study, the reader may perceive the relevance of participatory mapping not only for the collection of ethnobiological and ethnoecological data but also for purely ecological studies or studies that investigate the economic, social and cultural dynamics related to the use of natural resources. However, it is important to clarify that in this case, the DRP was an exploratory study, and certain aspects that were elucidated with the aid of this method must therefore be investigated in greater detail.

Several limitations could be observed during the mapping. Among these limitations was the difficulty of spatial recognition and localization by the participants when they were given the map of the FLONA-Araripe as stimulus. The arrival of a younger (18-year-old) participant, however, allowed this difficulty to be circumvented, as this participant helped the others to interpret the map. Importantly, this episode demonstrated that the stimuli offered to the community will not always be interpreted and returned to the researchers in the way the researchers imagine. Because of this variability, it is always worthwhile to perform pilot workshops to "calibrate" the mapping method and to supply the best possible stimuli.

6 Methodological Considerations

The type of stimulus used in communitarian mapping may vary according to the desired goals of mapping, and the choice of these stimuli has implications that will influence the type of data that are collected. For example, in the two case studies presented above, the objective was to bring forth information concerning zoning and the use of natural resources. Each of these studies chose different stimuli for the mapping, and these differences were reflected in the type of data that were collected.

Because of this difference, Box 4 intends to show, based on the literature and on the experience of our research group, a number of the main aspects that the researcher must take into consideration when choosing participatory mapping for collecting ethnobiological and ethnoecological data.

Box 4 Certain Aspects That Must Be Taken into Consideration at the Moment of Choosing Mapping as a Methodological Tool for Collecting Ethnobiological and Ethnoecological Data

- There are people who have difficulties with drawing their spatial knowledge at a map scale. Because of this difficulty, it is always helpful to encourage people to feel more comfortable at such a task; usually, the most helpful individuals for this facility are the younger people
- The researcher must pay attention to the dimensions and richness of detail of the elements represented in the map, as sometimes the participants tend to place more emphasis on the landscape elements that are more important from the cultural and environmental points of view. Therefore, the researcher must always attempt to transcribe what is being discussed during the mapping
- Stimulating the map's drawing may bring forth a great richness of details. However, the excessive use of stimuli may become inductive. In addition, a large quantity of time may be demanded for this activity
- It is important that the participants identify the drawn elements by means of legends or by naming each element. Differentiation by colors is also important for the recognition of the drawn elements
- Using images as stimuli is quite interesting from a practical point of view. However, the researcher must not forget that the perceptions of, for instance, spatial scale, GIS and observation of details in the images vary among individuals and among different cultures
- Make yourself a copy of the map, as the original belongs to the people who produced it. Additionally, do not forget to include the participants' names on the map [31]

References

1. Voris HK (2001) Maps of Pleistocene sea levels in Southeast Asia: shorelines, river systems and time durations. J Biogeogr 25(5):1153–1167
2. Harley B (2009) Mapas, saber e poder. Confins 29(5):2–24
3. Acselrad H (2008) Cartografias sociais e território / Henri Acselrad. Rio de Janeiro: Universidade Federal do Rio de Janeiro, Instituto de Pesquisa e Planejamento Urbano e Regional, 168 p
4. International Fund for Agricultural Development (IFAD) (2009) Good practices in participatory mapping
5. Sztutman M (2006) Etnomapeamento: uma técnica robusta, barata e de fácil implementação para a gestão etnoambiental em terras indígenas. The Nature Conservancy, Manaus
6. Rundstrom RA (1990) A cultural interpretation of Inuit map accuracy. Geogr Rev 80(2):155–168
7. Vieira PF, Berkes F, Seixas CS (2005) Gestão integrada e participativa de recursos naturais: conceitos, métodos e experiências. APED, Florianópolis
8. Rambaldi G, Kyem PAK, McCall M et al (2006) Participatory spatial information management and communication in developing countries. Electron J Inf Syst Dev Ctries 25(1):1–9
9. Widayati A, Jones S, Carlisle B (2010) Accessibility factors and conservation forest designation affecting rattan cattan cane harvesting in Lambusango forest, Buton, Indonesia. Hum Ecol 38(1):731–746
10. Correia CS (2007) Etnozoneamento, etnomapeamento e diagnóstico etnoambiental: representações cartográficas e gestão territorial em terras indígenas do estado do acre. Tese de doutorado, Universidade de Brasília, Brasília
11. Verdejo ME (2006) Diagnóstico rural participativo (DRP Guia Prático) (Participatory rural appraisal). Emater, Rio Grande do Sul
12. Geilfus F (1997) 80 Herramientas para el desarrollo participativo: diagnóstico, planificación, monitoreo, evaluación. IICA, San Salvador
13. Young HR, Potschin M, Kienast F (2012) Indicators of ecosystem service potential at European scales: mapping marginal changes and trade-offs. Ecol Indic 21(1):39–53
14. RELATÓRIO (2004) Programa de educação ambiental com comunidades costeiras-PEAC. PETROBRAS/CONATURA, Aracaju
15. Araújo TAS, Alencar NL, Almeida ALS et al (2009) Do contato ao "retorno": relatos e desafios na pesquisa etnobotânica em uma comunidade rural. In: Araújo TAS, Albuquerque UP (eds) Encontros e desencontros na pesquisa etnobiológica e etnoecológica: os desafios do trabalho em campo. NUPEEA, Recife
16. Jones NA, Ross H, Lynam T et al (2011) Mental models: an interdisciplinary synthesis of theory and methods. Ecol Soc 16(1):46–52
17. Assad ED, Sano EE (1996) Sistema de Informações Geográficas: aplicações na agricultura. EMBRAPA, Brasília
18. Bonham-Carter GF (1994) Geographic information systems for geoscientists: modeling with GIS. Computer Methods in the Geosciences, Pergamon
19. Mcconchie J, Mckinnon JM (2002) Using GIS to produce community-based maps to promote collaborative natural resource management. Asean Biodivers 2:28–34
20. Bernard E, Barbosa L, Carvalho R (2011) Participatory GIS in a sustainable use reserve in Brazilian Amazonia: implications for management and conservation. Appl Geogr 31(1):564–572
21. Brown G, Brabyn L (2012) An analysis of the relationships between multiple values and physical landscapes at a regional scale using public participation GIS and landscape character classification. Landsc Urban Plan 107(2): 317–331
22. Bourgoin J (2012) Sharpening the understanding of socio-ecological landscapes in participatory land-use planning: a case study in Lao PDR. Appl Geogr 34(3):99–110
23. Saipothong P, Thomas D (2007) Spatial information tools for land use management networks in montane mainland Southeast Asia. Inf Dev 23(1):129–136
24. Cronkleton P, Guariguata MR, Albornoz MA (2012) Multiple use forestry planning: timber and Brazil nut management in the community forests of Northern Bolivia. For Ecol Manage 268(1):49–56
25. Wang X, Yu Z, Cinderby S et al (2008) Enhancing participation: experiences of participatory geographic information systems in Shanxi province, China. Appl Geogr 28(1):96–109
26. Fagerholm N, Käyhkö N, Ndumbaro F et al (2012) Community stakeholders' knowledge in landscape assessments—mapping indicators for landscape services. Ecol Indic 18(1): 421–433
27. Bryan J (2011) Walking the line: participatory mapping, indigenous rights, and neoliberalism. Geoforum 42(2):40–50
28. Reyes-García V, Orta-Martínez M, Gueze M et al (2012) Bolivian Study Team, TAPS. Does participatory mapping increase conflicts? A randomized evaluation in the Bolivian Amazon. Appl Geogr 34(1):650–658

29. Hodgson DL, Schroeder RA (2002) Dilemmas of counter-mapping community resources in Tanzania. Dev Change 33(2):79–100
30. Hunn ES, Meilleur BA (2009) In: Landscape Ethnoecology concepts of biotic and physical space. Leslie Main Johnson and Eugene S. Hunn (eds), Toward a theory of landscape ethnoecological classification, pp 59–82
31. Gomes MVF, Pinheiro JQ (1998) Influência do gênero em mapas cognitivos do mundo de universitários brasileiros. Estud Psicol 3(1): 139–146

Chapter 16

Chemical Tools Applied to Ethnobotanical Studies

Tadeu José da Silva Peixoto Sobrinho, Thiago Antônio de Sousa Araújo, Cecília de Fátima Castelo Branco Rangel de Almeida, and Elba Lúcia Cavalcanti de Amorim

Abstract

Medicinal plants are a main focus of research in the fields of ethnobiology and ethnoecology. Among the various techniques for studying medicinal plants, the methods of extraction and the characterisation of different classes of compounds present in these plants form the pillars of phytochemical studies. In this chapter, you will find methods detailing how to collect, store, and stabilise herbal drugs; techniques to determine which solvents and extraction methods are most appropriate; chemical enrichment methods; and protocols for characterisation of compounds by thin layer chromatography.

Key words Ethnopharmacology, Medicinal plants, Phytochemical profile, Secondary metabolism, Thin layer chromatography (TLC)

1 Introduction

The study of medicinal plants is one of the most common themes of ethnoecology and ethnobiology studies. According to Oliveira et al. [1], medicinal plants have been a predominant topic in the field of ethnobotany, as calculated by the number of presentations at the Brazilian National Congress of Botany (Congresso Nacional de Botânica—CNB) (1985–2007) and the number of ethnobotanical publications conducted in Brazil, of which 64 % of the papers published in the period from 1990 to 2007 were focused on the medicinal usage of plants. Moreover, the studies presented at CNBs and informal studies by urban populations and nontraditional individuals have also been investigated in terms of knowledge about medicinal plants; for example, the studies in the backyards of urban areas and in public markets, which add to the more formal opportunities for study, are not only restricted to the rural communities. These studies, though only a minority, address issues explicitly formulated with well-defined hypotheses or incorporate quantitative techniques for data analysis [1].

Ulysses Paulino Albuquerque et al. (eds.), *Methods and Techniques in Ethnobiology and Ethnoecology*, Springer Protocols Handbooks, DOI 10.1007/978-1-4614-8636-7_16, © Springer Science+Business Media New York 2014

If ethnoecology and ethnobiology (in particular ethnobotany) are considered interdisciplinary disciplines and potentially cover diverse themes, the study of medicinal plants can consist of an immense variety of approaches. Using medicinal plants, ethnoscientists propose, or revise formulas [2–8]; analyse the flow, transmission, and comparison of knowledge between communities and inter-communities [9–11]; test hypotheses [12–16]; relate how ecological, environmental, and biometric factors interfere with the metabolism of plants [17–20]; study conservation strategies [21–24]; and examine the dynamics of traditional markets and production chains [25–28], among several other approaches that can be undertaken.

Without a doubt, the knowledge of multiple approaches when studying medicinal plants is useful for bioprospecting that is aimed at finding new drugs. This type of approach is of great interest to the pharmaceutical industry because the sale of herbal products over the world generates billions of dollars in sales per year. Substances such as aspirin, atropine, pilocarpine, paclitaxel, vincristine, and vinblastine are examples of drugs developed from ethnobotanicals [29].

The search for new molecules and/or the development of new products using the traditional knowledge of medicinal plants involves the knowledge of various subjects, particularly phytochemistry and pharmacology, which are responsible for identifying the mechanism of action and biological properties of the substances obtained from the plants using both in vitro and in vivo models (animals and/or humans) [30].

The structural elucidation of the active ingredient of a medicinal plant is an essential step for understanding its mechanism of action [31]; in addition, it helps to improve the identification of other approaches to study medicinal plants. In seeking to collaborate with ethnobotanical studies, this chapter provides instructions on the collection, stabilisation, and storage of plant samples, as well as on the appropriate solvents, extraction methods, and methods of enrichment. In addition, this chapter provides information on how to characterise the major classes of compounds of plant metabolites by thin layer chromatography (TLC): alkaloids, terpenoids, steroids, saponins, coumarins, flavonoids, tannins, and quinones.

2 Selection of Material to be Processed

2.1 Collection of Plant Material

Sample collection is a crucial phase in the study of medicinal plant compounds. Errors in this step compromise all subsequent phases, causing lost time and money, and depending on the part of the plant, the possibility that the researcher is unable to obtain a new specimen in a short period of time.

To minimise or prevent this obstacle, the correct identification of the medicinal species indicated by popular use is the first step in the process of obtaining potential therapeutic phytochemicals. When necessary, it is important to have the consent of the owner of the studied area and the proper authority over the collection to be performed. The period of maturation or development of the species to be collected should be a controlled variable because depending on the ontogenic stage, quantitative differences can be observed [32, 33].

It is important to choose individuals and healthy organs at the time of collection; furthermore, it is imperative to obtain samples from places that do not compromise the quality of the sample, such as near farms that use pesticides or near polluted rivers and streams [33].

Another factor to be considered relates to the intraspecific genetic differences that may alter the phytochemical characteristics of the sample collected; however, the collection of a variety of samples can minimise this effect by preventing results that are random or skewed toward some plant's individual characteristics. Some studies take into account the collection of at least three individuals (*see* [8]). Moreover, because a large amount of material is needed in many cases, the collection of different individuals minimises the damage to plant populations.

2.2 Drying and Storage

An important step that precedes quantitative analysis is the process of drying the sample; i.e. after collection of the plant material, the sample should be treated to evaporate the water and to prevent enzymatic action while avoiding structural modification of substances originally present in the plant. The heat stabilisation of the material is the most suitable method for further analysis because in addition to dehydrating and blocking the hydrolysis reactions and microbial growth, this method allows for storage for long periods of time, with the constituents of the sample remaining unchanged [34]. The use of dehydrating solvents for boiling (e.g. ethanol and methanol) is effective for the immediate stabilisation of the sample; however, these chemicals might deprive the sample of some of the compounds originally present [34].

Air drying is more economical, but it is necessary to take greater precautionary measures when air drying to ensure that the dehydration is homogeneous. Leaves and flowers can be placed under ventilation and protection from the sun to prevent loss of plant metabolites due to photodecomposition. Roots and bark, after removal, may be exposed to direct sunlight [33]. The use of greenhouses is more costly but beneficial for both efficiency and process control, especially in wetter areas or periods, and it is more suitable for the temperature to not exceed 50 °C. Desiccators are also viable alternatives for reducing humidity [33].

Observing whether the material is crumbly (fragile such that it breaks easily with touch) may serve as a parameter of drying. Subsequent measurements with intervals of 1 h without observing a change of at most 1 % weight loss by dehydration should be calculated according to the equation D (%) = [(weight of fresh sample − weight of dry sample)/weight of fresh sample] × 100, where D (%) is the percentage loss by dehydration [35]. Then, the material should be stored in the appropriate packaging (amber containers or dark plastic bags) that shields the material from pests, light, and humidity and contains a label with the species name, date, place of collection, part processed, and weight [33].

When the researchers choose to use fresh materials for the preparation of extracts, they may resort to the use of refrigerators, freezers, or dry ice as alternatives to minimise enzymatic reactions and interactions with microorganisms.

3 Extraction of Compounds and Sample Preparation

3.1 Methods of Extraction: Solvents and Extraction Forms

Certain factors must be carefully considered before performing this step, such as the particle size, the standardisation sample, the part of the plant collected, the methods to be employed, and the solvents to be used for extraction. The smaller the particle diameter of the sample is, the greater the efficiency of the extraction due to the increased area of contact between the solvent and the particle. The various organs of plants have very different anatomical structures, mainly with respect to the degree of cellular organisation. In stems and roots that have lignified tissues, the extraction process is more difficult than in leaves and flowers that do not have lignin [34].

Perhaps the main dilemma of the extraction process is choosing the most appropriate solvent or system for the extraction of a particular target group involved in metabolism (Table 1). Generally, phenolic compounds are employed for the extraction of polar organic solvents (methanol, ethanol, or acetone) or a mixture of these solvents with water is utilised to optimise the process and to increase the yield [36]. However, there is great difficulty in establishing an extraction system or in obtaining a more efficient solvent extraction because many factors are involved, such as species variations, the physic-chemical properties of the substances, the time and temperature of the extraction, and the toxicity, availability, and cost of the solvent [34, 37].

According to Cechinel Filho and Yunes [38], one of the most suitable methods for chemical analysis is the pharmacological preparation of a hydroalcoholic solution (ethanol-water 50/50, v/v) to mimic the extractions performed popularly. Yet, according to these authors, the extract elicits interesting biological effects and one

Table 1
Solvents types and groups of metabolites extracted [34]

Solvent or system	Metabolites
Alkaline water	Saponins
Acidified water	Alkaloids
Alcohol mixtures	Saponins and tannins
Ethanol, methanol	Heterosides
Ethyl acetate, butanol	Flavonoids and simple coumarins
Toluene, dichloromethane, chloroform	Amines, anthraquinones, volatile oils, glycosides cardiotonic
Petroleum ether, hexane	Lipids, waxes, pigments, furanocoumarins

Table 2
Polarity indexes, densities, and boiling points of the solvents used in most plant extractions

Solvent	Polarity	Density (g/mL)	Boiling point (°C)
n-Hexane	0	0.65	69
Toluene	2.3	0.87	110.6
Dichloromethane	3.4	1.33	40
Butanol	3.9	0.81	118
Ethyl acetate	4.3	0.897	77.1
Chloroform	4.4	1.48	61.2
Ethanol	5.2	0.789	78.4
Acetone	5.4	0.79	56
Methanol	6.6	0.79	65
Water	9	1	100

should proceed using a systematic study in which the most suitable solvent to obtain the crude extract is methanol because it enables the extraction of a greater number of compounds.

When the nature of the chemical species studied is unclear, one method to identify the phytochemical samples is the separation of the crude extract in solvents of increasing polarity; thus, the sample is semi-purified in different partitions (Table 2; Fig. 1). After this process, the samples must be submitted to chromatographic procedures for the isolation and spectroscopic identification of the pure compounds [38].

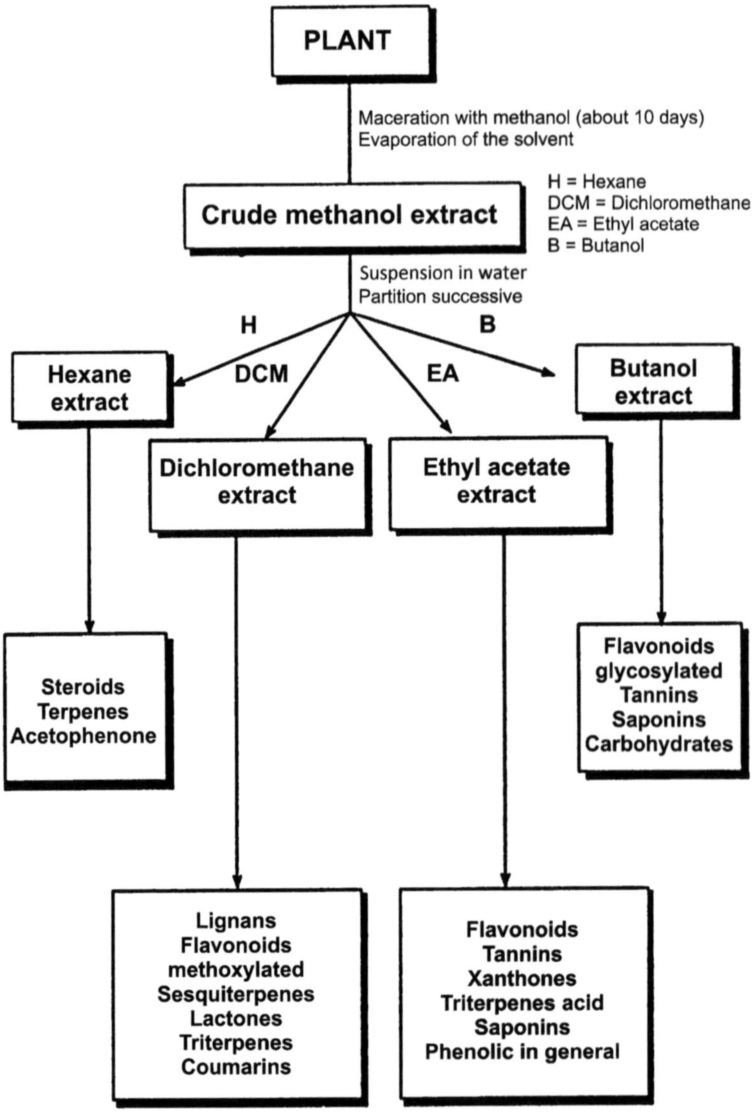

Fig. 1 General scheme partition and semi-purification through polarity of the main secondary metabolites in plants [38]

In addition to this method of separation, another alternative method for obtaining semi-purified extracts by polarity is to macerate the same plant sample for many days directly with solvents of increasing polarity, as shown in the example in Fig. 2.

Certain caveats for the extraction methods must first be considered; for example, excess spraying of the samples can interfere with the percolation process and soaking in the same way that debris with a large particle diameter requires greater extraction time. It is important to obtain standardisation of particle size, which can be achieved by sieving (sieve).

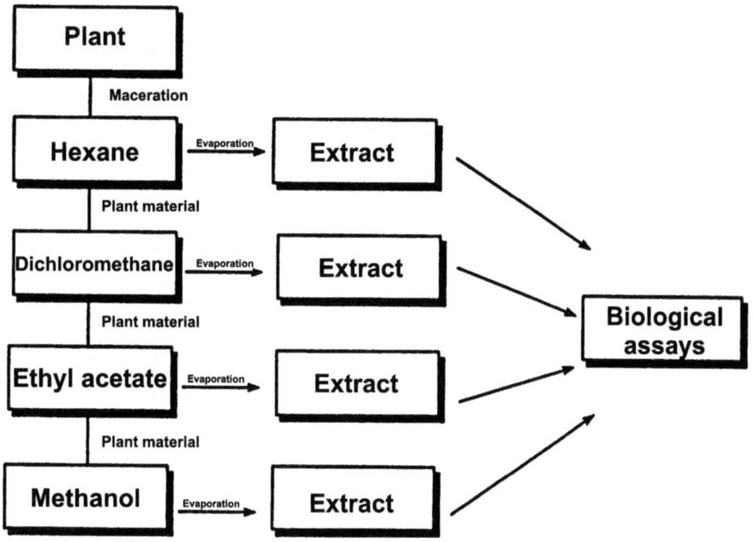

Fig. 2 General scheme for obtaining plant extracts [38]

With respect to the selection of the most suitable solvent for direct extraction according to the group of metabolites, one must also consider the solvent's inherent toxicity. With regard to temperature, one must be very careful because while heat accelerates the extraction process, it may also cause decomposition of sensitive compounds, as well as volatile compounds. We present below the main extraction processes proposed by Marques and Vigo [39].

3.1.1 Infusion

During infusion, boiling solvent is poured onto the plant material in a process similar to the home preparation of tea. The medicinal plant material is in contact with the solvent for 5–10 min before the filtration process is initiated. The weight of the plant material and solvent volume varies according to the species, with a typical ratio of 1:20 w/v. This method is suitable for the extraction of the weakest parts of the plant, such as the leaves and flowers.

3.1.2 Decoction

In this extraction process, typically the fresh plant or plant drug is placed in the solvent to heat until boiling in a closed container. Decoction is widely used to extract the compounds of the woody parts of plants, such as the bark and roots.

3.1.3 Maceration

Maceration is a longer process in which the pulverised plant material is immersed in the solvent for a suitable period of time, according to the methodology employed by the researcher. After this period, the material is filtered, the solid residue pressed, and the liquid resulting from pressing to extract the filtrate is collected. Maceration can be static, with only occasional stirring, or dynamic, with the system in constant motion.

In addition to its slow speed, this process does not lead to depletion of the plant material due to saturation of the liquid extractor and/or equilibrium between the solvent and the cell interior. All plant organs can be extracted by maceration; however, the selectivity and concentration of the solvent, the particle diameter, and cell swelling can affect the yield of the extract.

3.1.4 Turbo Extraction

This process requires a turbuliser. The plant material is extracted by the equipment after the addition of solvent, which is milled together with the plant material, by dissolving the cellular content. The yields obtained by this process are high because of the constant agitation and trituration of the system, which leads to faster extract preparation and the near exhaustion of the vegetable material.

Nevertheless, greater temperature control is required because the constant agitation may produce heat in the system, causing evaporation of volatile or degradation-sensitive compounds. Furthermore, the filtration extractive solution becomes harder, and harder parts such as stems and roots cannot be processed.

3.1.5 Percolation

This process requires a device called a percolator, which is a conical container made of steel, glass, or PVC with a cock or valve to control the output of the extract. Because the percolated plant material is in a powdered form prior to maceration, swelling and increased porosity of the cell wall will facilitate the diffusion of compounds.

The drug is deposited in a moist percolator with enough solvent to cover it. A container of solvent for the system is placed over the system to supply it continuously. When the faucet of the equipment is opened, the extractive process occurs by the force of gravity until the exhaustion of active drug from the plant. The residual extraction can be pressed, and the resulting liquid joins the leachate.

3.2 Preparation of Extracts for Characterisation

3.2.1 Rotary Evaporation

Evaporation is used to remove solvents from the extracted samples, thus resulting in more concentrated extracts. The rotary evaporator, consisting of a chamber containing a coil that condenses the solvent vapours, allows the elimination of the liquid extractant under controlled temperature at reduced pressure. Rotary evaporators are most commonly used in laboratories that perform phytochemical analysis [40].

3.2.2 Lyophilisation

Freeze drying is a dehydration process in which water is removed through direct passage from the solid state to a gas by sublimation [41]. The extremely low water content results in inhibition of the action of microorganisms and enzymes that normally degrade substances. According to Tattini Junior et al. [42], lyophilisation is divided into three steps: freezing, primary drying, and secondary drying.

Freezing. The preparation of the sample to be processed, wherein the sample is frozen at temperatures slightly below 0 °C. This step consists of immobilising the product to be lyophilised and disrupting the chemical and enzymatic reactions in the material.

Primary drying. Occurs at a melting temperature obtained by subliming ice. This process eliminates 85–90 % of the water in the sample (sample is up to 15 % humidity).

Secondary drying. Carried out at temperatures below the degradation of the product for the purpose of eliminating the last traces of residual water from the sample (up to 2 % moisture).

3.2.3 Spray Drying

This method applies the principle of increasing the specific surface of the solution, suspension, or emulsion to dry, through its spray, increasing the contact area with the fluid drying. The drying process involves spraying the product into a chamber, where it is subjected to a controlled stream of hot air, and thus, the solvents evaporate, resulting in ultra-fast separation of the solid and soluble materials with minimal degradation of the constituents of the sample. The process is completed by the recovery of the powdered product [43].

3.3 Methods of Enrichment

The goal of extraction is to remove, as completely as possible, the most active fractions or substances contained in the plant sample using solvents or mixtures of solvents [40]. However, sometimes the extractions are not selective and require partition/fractionation procedures to pinpoint the metabolites of interest and to confirm their presence.

To minimise this obstacle, Wagner and Bladt [44] described enrichment methods aimed at more selective extractions that elicit the substances of interest, as described below.

3.3.1 Amine Compounds

Alkaloids and xanthines. A pulverised sample (1 g) is thoroughly mixed with 1 mL of 10 % ammonia solution or 10 % sodium carbonate solution and subsequently extracted for 10 min with 5 mL of methanol under reflux. The filtrate is concentrated, and 50–100 μL of the total alkaloids is applied to the chromatographic plate to develop the chromatogram.

Another method to enrich the extract is to vigorously shake the pulverised sample (0.4–2 g) for 15 min in 15 mL of 0.1 N sulphuric acid, followed by filtration. The filter paper is washed with 5 mL of the same acid. Then, 1 mL of concentrated ammonia is added to the filtrate. In a separatory funnel, the mixture is extracted by stirring vigorously twice with 10 mL of ethyl ether. After separation, the two extractions are combined, and the organic phases (ethyl ether) are concentrated over anhydrous sodium sulphate, followed by filtration and evaporation of the solvent. The evaporated residue is dissolved in 0.5 mL of methanol, and 20 μL of this solution is applied to the chromatographic plate. This method is preferred for leaf samples.

3.3.2 Compounds Derived from Isoprene

Mono-, sesqui-, and diterpenes. The powdered plant sample (1 g) is extracted by constant stirring for 15 min with 10 mL of dichloromethane. The suspension is filtered and the filtrate is evaporated to dryness. The residue is dissolved in 1 mL of toluene, and 30–100 μL is used for TLC.

Triterpenes and steroids. The pulverised sample (1 g) is extracted in 10 mL of methanol for 15 min in a water bath, and 30 μL of the filtrate is used for TLC. Another method is to extract the powdered sample (1 g) with chloroform for 60 min under reflux. Then, 20 μL of the filtrate is applied to the chromatographic plate.

Saponins. Pulverise 2 g of the sample, and then extract the sample under reflux for 10 min with 10 mL of 70 % ethanol. (1) The filtrate is evaporated to approximately 5 mL, and 20–40 μL of this solution is applied to the chromatographic plate. (2) Three millilitres of the ethanol extract is agitated several times with 5 mL of butanol saturated with water (3 mL vigorously stirring distilled water and 10 mL of butanol in a separatory funnel, discarding the aqueous phase). The butanol phase is separated and concentrated to approximately 1 mL, with 20 μL used for this TLC.

The saponins may also be hydrolysed, with the pulverised sample (2 g) extracted under reflux for 60 min with 30 mL of 0.5 M sulphuric acid. The solution is filtered and stirred twice vigorously in a separatory funnel with 20 mL of chloroform. The chloroform extracts are dried and gathered under anhydrous sodium sulphate, followed by filtration and evaporation of the solvent. The residue is dissolved in 2 mL of chloroform-methanol (1:1), and 10 μL of this solution is applied to TLC.

3.3.3 Phenolic Compounds

Coumarins. The pulverised sample (1 g) is extracted in 10 mL of methanol under reflux for 30 min in a water bath. The filtrate is evaporated to approximately 1 mL, and 20 μL of this solution is used for chromatography.

Cinnamic derivatives, flavonoids and tannins. The most common procedure is to extract 1 g of the enriched sample pulverised with 10 mL of methanol for 15 min in a water bath at approximately 60 °C. The methanol extract is filtered and evaporated to 2 mL. The filtrate is mixed with 1 mL of water and 10 mL of ethyl acetate and shaken vigorously several times in a separatory funnel. The ethyl acetate phase is recovered and evaporated to 1 mL, with 10 μL applied to the plate.

Lignans. The sample (1 g) is extracted under reflux with 10 mL of 50 % methanol for 15 min. Then, 15 mL of butanol saturated with water is added to the sample (mix 3 mL distilled water and 10 mL of butanol in a separatory funnel, discarding the aqueous phase). The mixture is stirred for 5 min, followed by separation and evaporation of the butanol phase to 1 mL. Twenty to 40 μL of the samples is used for the chromatography.

Anthraquinones aglycones. To hydrolyse the carbohydrates of quinones, 0.5 g of the pulverised sample is extracted under reflux with 25 mL of 7.5 % hydrochloric acid for 15 min. After cooling, the filtrate is extracted by shaking vigorously with 20 mL of ethyl ether. The ether phase is concentrated to approximately 1 mL, and 10 μL is applied to the chromatographic plate.

Anthracenics glycosidic derivatives. In a water bath, extract the pulverised sample (0.5 g) for 5 min with 5 mL methanol. Use 20 μL of the filtrate directly on the chromatographic plate.

Naphthoquinones glycosidic. Extract 1 g of the sample for 15 min with 10 mL of methanol in a water bath. Use 30 μL of filtrate on the plate.

3.4 Pre-purification of Bioactive Compounds

This section will focus on the use of chromatography for pre-purification of active substances using a simple method of separating the components of complex mixtures, performed by distributing these closely related compounds into two phases—a stationary phase, which may be a solid or liquid supported on a solid, and a mobile phase consisting of a gas or liquid that flows continuously through the stationary phase [45].

Chromatography is a method of physic-chemical separation that is based on the differential migration of the components of a mixture during passage of the mobile phase on a stationary phase, such that the components will be selectively retained by the stationary phase due to various interactions, such as absorption, solubility, molecular weight, ionic charge, etc. [45, 46].

Because there are a wide variety of combinations of mobile and stationary phases in chromatography, it is an extremely versatile and widely applied technique [46] that is highly useful in numerous fields of science [47]. For example, chromatography can be used to identify compounds by comparing them with previously existing standards for purifying and separating the compounds from undesirable substances in a mixture.

The various forms of chromatography may be classified according to several criteria, some of which are listed by Degani et al. [46]: physical form of the chromatographic system, mobile or stationary phase employed, and the separation mode.

Figure 3 depicts the different types of chromatography according to the stationary phase used: planar or column chromatography, in which the stationary phase is arranged on a flat surface or placed on the cylindrical tube, respectively [46].

3.4.1 Planar Chromatography

In planar chromatography, the stationary phase is deposited on a support plane, which can be paper, glass, or aluminium. In any case, the mobile phase moves through the stationary phase by capillary action. This technique is widespread due to its experimental ease and low cost.

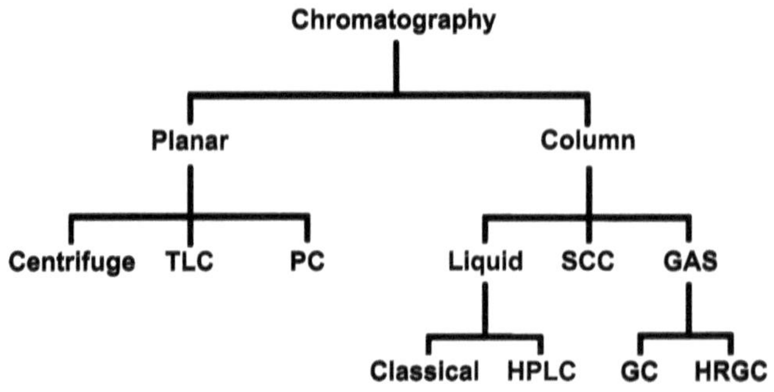

Fig. 3 Schematic representation of the different types of chromatography [46]. *TLC* thin layer chromatography, *PC* paper chromatography, *HPLC* high performance liquid chromatography, *SCC* supercritical chromatography, *GC* gas chromatography, *HRCG* high resolution gas chromatography

One important parameter of planar chromatography is the retention factor (Rf). The Rf is the ratio of the distance travelled by the components of the sample (ds) and the distance travelled by the mobile phase (dm) (Fig. 4).

In this section, we will discuss TLC in detail because the centrifugal chromatography paper has high specificity and is not useful for all classes of secondary metabolites.

3.4.2 Analytical and Preparative Thin Layer Chromatography

For this procedure, the most frequently used stationary phase is silica gel. Other chromatograph plate scan can also be produced or acquired with alumina, diatomaceous earth, or cellulose as the stationary phase. Although the technique is widely accessible, preparative TLC has some disadvantages, such as elimination of the purified substance plate, the long time required for the separation and the introduction of impurities, and residues derived from the support itself [31].

For the preparation of chromatographic plates containing silica gel, use a layer of 0.25 mm for analytical plates and 1 mm plaques for preparative plates. There are plates deposited onto aluminium foil with high efficiency that are available for purchase [46]. Samples to be analysed by TLC should be applied 1 cm from the bottom of the base plate using micro-pipettes, micro-needles, or capillary tubes with thin ends [46]. After sample application, the plate must be inserted into a chromatography tank containing the appropriate mobile phase.

There are certain items to consider when performing TLC: (1) The presence of water can significantly influence the chromatographic separations, (2) the acetone must be used with caution because it can react with some compounds containing amino groups, and (3) the presence of a constituent can mask the colouration indicative of another compound, which is a disadvantage that may lead to erroneous conclusions [48].

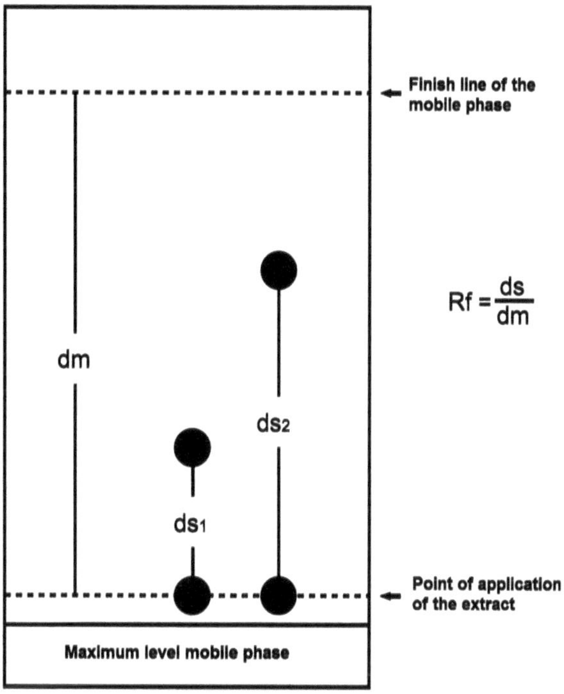

Fig. 4 Schematic representation of the calculation of retention factor (Rf) in planar chromatography [46]. *Rf* retention factor, *dm* distance travelled by the mobile phase, *ds1* distance travelled by substance 1, *ds2* distance travelled by substance 2

3.4.3 Column Chromatography

According to the internal diameter of the column tube, column chromatography is subdivided into preparative (6–50 mm), analytical (2–6 mm), and microdiameter (<2 mm) chromatography. Considering the physical state of the mobile phase, liquid, supercritical, or gas chromatography is used. Liquid chromatography is also divided into classical liquid chromatography and high performance liquid chromatography [45].

In this chapter, we chose to detail classical liquid chromatography, as other types of chromatography (HPLC, SCC, GC, and HPGC) require the use of expensive equipment and trained personnel to operate the equipment, in addition to a laboratory with the proper environment for installation.

3.4.4 Classical Liquid Chromatography

During the pre-purification step, a preparative liquid chromatography column (Fig. 5) is typically used. This technique is also called adsorption chromatography and is a technique known for being simple and low cost.

Generally, the chromatographic column is a cylindrical tube with an open upper end and a lower end that terminates in a tap that is used to control the flow of the solvent. The adsorbent (alumina, silica gel, magnesium silicate, activated carbon), which will

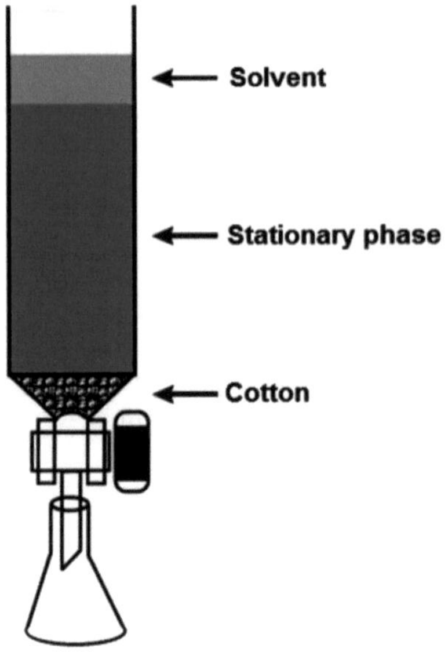

Fig. 5 Schematic representation of the liquid chromatography column for pre-purification of compounds [46]

fill the interior of the column and will become activated, must be previously determined in accordance with the chemical group of interest and characteristics of the adsorbent, respectively. The column packing material must be uniform such that air cannot become trapped between the particles and form channels that could affect the separation [49].

Monitor the TLC to determine the choice of adsorbent and solvents to be used for elution of the column. The eluents should be chosen according to their ability to elute the substances through the adsorbent. Generally, the displacement of the substances of interest becomes faster when using a range of solvents in order of increasing polarity (Table 2). As the sample travels to the lower end of the column, it should be collected in already labelled containers.

3.4.5 Open or Gravity Column Chromatography

This type of chromatography is universally used because it involves a simplified procedure, but it can only be used for separating substances based on the difference in polarity [31]. It utilises gravity to elute compounds by column chromatography. Despite being a widely used technique, it has some limitations: separation is slow, adsorption of some compounds is irreversible, and it is incompatible with small diameter particles [31].

3.4.6 Column Chromatography Under Pressure

This type of chromatography involves applying pressure to a chromatographic column and can be used to separate more difficult compounds using a high precision stationary phase with particles of small diameters [31]. This technique can vary depending on column size and pressure for separation: flash chromatography (approximately 2 bar or 30 psi), low pressure (<75 bar or 5 psi), medium pressure (5–20 bar or 75–300 psi), or high pressure (>20 bar or 300 psi).

4 Characterisation of Secondary Compounds

Among the various types of chromatography, this chapter emphasises TLC as a technique that is rapid, less expensive, and easy to standardise. Therefore, we present the method of detecting phytocompounds according to Wagner and Bladt [44]. Phytocompounds was selected because it is the most frequently investigated material.

4.1 Amine Compounds

4.1.1 Alkaloids

Elution system: toluene-ethyl acetate-diethylamine (70:20:10)

Standard: Atropine (Tropanic), Yohimbine (indole), Pilocarpine (imidazole), or quinine (quinoline)

Revelation: Developer # 1—Visible: brown or orange-brown

4.1.2 Xanthines

Elution system: ethyl acetate-methanol–water (100:13, 5:10)

Standard: Caffeine, Theobromine, or Theophylline
Revelation: Developer # 2—Visible: brown

4.2 Compounds Derived from Isoprene

4.2.1 Terpenoids

Mono-, sesqui-, and diterpenes
Elution system: toluene-ethyl acetate (97:3)

Standard: Thymol or Carvacrol

Revelation: Developer # 3—Visible: blue, violet, red, and brown

4.2.2 Triterpenos and Steroids

Elution system: toluene-chloroform-ethanol (40:40:10) for aglycones or chloroform-methanol-water (65:25:4) for glycosides

Standard: Lupeol (Triterpenes) and Sitosterol (Steroids)

Revelation: Developer # 4—UV 365: violet and blue

4.2.3 Saponins

Elution system: chloroform-glacial acetic acid-methanol-water (64:32:12:8)

Standard: Aescin or Glycyrrhizin

Revelation: Developer # 5—Visible: green, blue, violet, and red-brown, UV-365: dark green, blue, and violet

4.3 Phenolic Compounds

4.3.1 Cumarins

Elution system: ethyl acetate-glacial acetic acid-formic acid-water (100:11:11:26)

Standard: Esculetin, umbelliferone, or scopoletin

Revelation: Developer # 6—UV 365: blue and green-blue (coumarin simple) yellow and brown (coumarin derivatives)

4.3.2 Flavonoids

Elution system: ethyl acetate-glacial acetic acid-formic acid-water (100:11:11:26)

Standard: Rutin or Quercetin

Revelation: Developer # 7—365 and UV Visible: yellow-orange

4.3.3 Lignans

Elution system: chloroform-methanol-water (70:30:4)

Standard: Podophyllotoxin or Eleutheroside

Revelation: Developer # 8—UV 365: blue

4.3.4 Quinones

Anthraquinones Aglycones

Elution system: petroleum ether-ethyl acetate-formic acid (75:25:1)

Standard: Em*odin or Frangulin*

*Revelation: D*eveloper # 9—Visible: yellow; UV 365: yellow and orange-brown

Anthracenics Glycosidic Derivatives

Elution system: ethyl acetate-methanol-water (100:13, 5:10)

*Standard: A*loin or Glucofrangulin

Revelation: Developer # 6—365 and UV Visible: red (anthraquinones) and yellow (anthrone and antranol)

Naphthoquinones Glycosidic

Elution system: toluene-formic acid (99:1)

Standard: Plumbagin or Juglone

Revelation: Developer # 6—Visible: violet to brown, UV 365: yellow to brown

4.3.5 Tannins

Elution system: ethyl acetate-glacial acetic acid-formic acid-water (100:11:11:26)

*Standar*d: Catechin (condensed tannins), gallic acid, or tannic acid (hydrolysable tannins)

Revelation: Developer # 10a—(Condensed tannins)—Visible: red; Developer # 10b—(hydrolysable tannins)—UV 365: blue or violet

4.4 Developers

 # 1: Dragendorff Reagent

Solution A: Dissolve 0.85 g of nitrate (or subnitrate) bismuth in 10 mL of glacial acetic acid and 40 mL of water under reflux (filtered if necessary).

Solution B: Dissolve 8 g of potassium iodide in 30 mL of water. Stock solution: Mix solutions A + B (1:1).

Reagent: Mix 1 mL of the stock solution with 2 mL of glacial acetic acid and 10 mL of water.

Note: The colour can be intensified or stabilised by spraying an ethanolic solution of 10 % sodium nitrite in sulphuric acid or ethanol.

 # 2: Iodine-Hydrochloric Acid Reagent

Solution A: Dissolve 1 g of potassium iodide and 1 g of iodine in 100 mL of ethanol.

Solution B: Mix 25 mL of 25 % HCl in 25 mL of ethanol.

Reagent: Spray with solution A and then with solution B.

 # 3: Vanillin Sulphuric Reagent

Solution A: Prepare a solution of 1 % vanillin in ethanol.

Solution B: Prepare a solution of 10 % ethanolic sulphuric acid.

Reagent: Spray the plates with solution A and then with solution B. The plate must be heated in an oven at 110 °C for 5–10 min.

 # 4: Lieberman-Burchard Reagent

Prepare a solution with 5 mL of acetic anhydride, 5 mL of concentrated sulphuric acid, and 50 mL of ethanol. Sprinkle the plate with the solution and heat in oven at 100 °C for 5–10 min.

 # 5: Sulphuric Anisaldehyde Reagent

Mix 0.5 mL of anisaldehyde, 10 mL of glacial acetic acid, 85 mL of methanol, and 5 mL of concentrated sulphuric acid. Sprinkle the plates with 10 mL of the solution and heat in an oven at 100 °C for 5–10 min.

 # 6: Potassium Hydroxide Reagent

Prepare a solution of 5 % potassium hydroxide in ethanol. Sprinkle the board with 10 mL of the reagent.

 # 7: NEU Reagent

Solution A: Prepare a solution of 1 % 2-Aminoethyl diphenylborinate in methanol.

Solution B: Prepare a methanolic solution of 5 % polyethylene glycol in ethanol.

Reagent: Mix 10 mL of Solution A with 8 mL of Solution B.

 # 8: Vanillin Phosphoric Acid Reagent

Dissolve 1 g vanillin in 100 mL of 50 % phosphoric acid in ethanol. Spray the solution on the plate and heat for 10 min at 100 °C.

 # 9: Phosphomolybdic Acid Reagent

Prepare an ethanolic solution containing 20 % phosphomolybdic acid. Sprinkle the plates with 10 mL of the reagent and heat to 100 °C for 5 min.

 # 10a: Vanillin Hydrochloric Reagent

Prepare an ethanolic solution of 1 % vanillin. Spray the plate with 5 mL of this solution followed immediately by 3 mL of concentrated hydrochloric acid. Heat to 100 °C for 5 min.

 # 10b: Ammonium Ferrous Sulphate Reagent

Dissolve 1 g of ammonium ferrous sulphate in 100 mL of ethanol.

5 Notes

1. Mixtures of reagents must follow the order mentioned in the text.
2. The plates must be sprayed with the developers in the order mentioned.
3. All reagents containing acids must be added carefully into beakers cooled in ice baths because these reactions are exothermic.

6 Conclusions

The study of the chemistry of natural products, especially medicinal plants, has always attracted the interest of various sectors such as pharmaceutical, chemical, cosmetic and food. However, many researched believe that there is a dichotomization between obtaining data ethnodirected and their phytochemical analysis, motivated mainly by constant technical developments.

In this chapter, we presented the main tools phytochemical, accessible and easily performed, used in bioprospecting of active substances. Thus, the union of ethnobotanical methodologies and chemical analysis allow a reduction in the time of discovery of new biotechnology products.

References

1. Oliveira FC, Albuquerque UP, Fonseca-Kruel VS, Hanazaki N (2009) Avanços nas pesquisas etnobotânicas no Brasil. Acta Bot Bras 23: 590–605
2. Troter R, Logan M (1986) Informant consensus: a new approach for identifying potentially effective medicinal plants. In: Etkin NL (ed) Indigenous medicine and diet: biobehavioral approaches. Redgrave, Bedford Hills, NY, pp 91–112
3. Friedman J, Yaniv Z, Dafni A, Palewitch D (1986) A preliminary classification of the healing potential of medicinal plants, based on a rational analysis of an ethnopharmacological field survey among Bedouins in the Negev desert. Israel J Ethnopharmacol 16:275–287
4. Leaman DJ, Arnason JT, Yusuf R, Sangat-Roemantyo H, Soedjito H, Angerhofer CK, Pezzuto JM (1995) Malaria remedies of the Kenyah of the Apo Kayan, east Kalimantan, Indonesian Borneo: a quantitative assessment of local consensus as an indicator of biological efficacy. J Ethnopharmacol 49:1–16
5. Bennett BC, Prance GT (2000) Introduced plants in the indigenous pharmacopeia of northern South America. Econ Bot 54(1):90–102
6. Andrade-Cetto A, Becerra-Jiménez J, Martínez-Zurita E, Ortega-Larrocea PE, Heinrich M (2006) Disease-Consensus Index as a tool of selecting potential hypoglycemic plants in Chikindzonot, Yucatan. México J Ethnopharmacol 107:199–204
7. Leduc C, Coonishish J, Haddad P, Cuerrier A (2006) Plants used by the Cree Nation of Eeyoulstchee (Quebec, Canada) for the treatment of diabetes: a novel approach in quantitative ethnobotany. J Ethnopharmacol 105:55–63
8. Araújo TAS, Alencar NL, Amorim ELC, Albuquerque UP (2008) A new approach to study medicinal plants with tannins and flavonoids contents from the local knowledge. J Ethnopharmacol 120:72–80
9. Vandebroek I, Van Damme P, Van Puyvelde L, Arrazola S, De Kimpe N (2004) A comparison of traditional healers' medicinal plant knowledge in the Bolivian Andes and Amazon. Soc Sci Med 59(4):837–849
10. Albuquerque UP, Silva VA, Cabral MC, Alencar NL, Andrade LHC (2008) Comparisons between the use of medicinal plants in indigenous and rural Caatinga (dryland) communities in NE Brazil. Bol Latinoam Caribe Plantas Med Aromáticas 7:156–170
11. Almeida CFCBR, Ramos MA, Amorim ELC, Albuquerque UP (2010) A comparison of knowledge about medicinal plants for three rural communities in the semi-arid region of northeast of Brazil. J Ethnopharmacol 127:674–684
12. Almeida CFCBR, Silva TCL, Maia MBS, Albuquerque UP (2005) Life strategy and chemical composition as predictors of the selection of medicinal plants from the caatinga (Northeast Brazil). J Arid Environ 62(1):127–142

13. Albuquerque UP (2006) Re-examining hypotheses concerning the use and knowledge of medicinal plants: a study in the caatinga vegetation of NE Brazil. J Ethnobiol Ethnomed 2:1–10
14. Alencar NL, Araújo TAS, Amorim ELC, Albuquerque UP (2009) Can the apparency hypothesis explain the selection of medicinal plants in an area of caatinga vegetation? A chemical perspective. Acta Bot Bras 23:908–909
15. Soldati GT (2009) Produtos florestais não-madeireiros: padrão de uso e conservação de Anadenanthera colubrina (Vell.) Brenan no Agreste pernambucano. Dissertation, Universidade Federal Rural de Pernambuco, Recife
16. Alencar NL, Araújo TAS, Amorim ELC, Albuquerque UP (2010) The inclusion and selection of medicinal plants in traditional pharmacopoeias evidence in support of the diversification hypothesis. Econ Bot 64:68–79
17. Monteiro JM, Lins Neto EMF, Albuquerque UP, Araújo EL, Amorim ELC (2006) The effects of seasonal climate changes in the caatinga on tannin levels in *Myracrodruon urundeuva* and *Anadenanthera colubrina*. Rev Bras Farmacogn 16(3):338–344
18. Mccune LM, Johns T (2007) Antioxidant activity relates to plants part, life form and condition in some diabetes remedies. J Ethnopharmacol 112:461–469
19. Peixoto Sobrinho TJS, Cardoso KCM, Gomes TLB, Albuquerque UP, Amorim ELC (2009) Análise da pluviosidade e do efeito de borda sobre os teores de flavonóides em *Bauhinia cheilantha* (Bong.) Steud., Fabaceae. Rev Bras Farmacogn 19(3):740–745
20. Cabral DLV, Peixoto Sobrinho TJS, Amorim ELC, Albuquerque UP (2010) Relationship of biometric parameters on the concentration of tannins in two medicinal plants a case study. Bol Latinoam Caribe de Plantas Med Aromáticas 9(5):368–376
21. Etkin NL (1998) Indigenous patterns of conserving biodiversity: pharmacologic implications. J Ethnopharmacol 63(3):233–245
22. Albuquerque UP, Oliveira RF (2007) Is the use-impact on native caatinga species in Brazil reduced by the high species richness of medicinal plants? J Ethnopharmacol 113:156–170
23. Oliveira RLC, Lins Neto EMF, Araújo EL, Albuquerque UP (2007) Conservation priorities and population structure of woody medicinal plants in an area of caatinga vegetation (Pernambuco state, NE Brazil). Environ Monit Assess 132:189–206
24. Melo JG, Amorim ELC, Albuquerque UP (2009) Native medicinal plants commercialized in Brazil: priorities for conservation. Environ Monit Assess 156:567–580
25. Hersch-Martínez P (1997) Medicinal plants and regional traders in Mexico: physiographic differences and conservational challenge. Econ Bot 51(2):107–120
26. Olsen CS (1998) The trade in medicinal and aromatic plants from central Nepal to northern India. Econ Bot 52(3):279–292
27. Albuquerque UP, Monteiro JM, Ramos MA, Amorim ELC (2007) Medicinal and magic plants from a public market in northeastern Brazil. J Ethnopharmacol 110:76–91
28. Monteiro JM (2009) Plantas medicinais no agreste pernambucano com ênfase na valoração econômica e avaliação de processos regenerativos das cascas do caule Myracroduon urundeuva Allemão. Tesis, Universidade Federal Rural de Pernambuco, Recife
29. Balick MJ, Cox PA (1996) Plants, people and culture. Scientific American Library, New York, NY
30. Melo JG, Lins Neto EMF, Ramos MA, Soldati GT, Albuquerque UP (2010) A etnobotânica e as suas interfaces: diferentes perspectivas no estudo da interrelação entre pessoas e plantas. In: Albuquerque UP, Moura AN, Araújo EL (eds) Biodiversidade, potencial econômico e processos eco-fisiológicos em ecossistemas nordestinos, vol 6. Canal6, São Paulo, pp 99–126
31. Hostettmann K, Queiroz EF, Vieira PC (2003) Princípios ativos de plantas superiores. Edufscar, São Carlos, p 152
32. Gobbo-Neto L, Lopes NP (2007) Plantas medicinais: fatores de influência no conteúdo de metabólitos secundários. Quim Nova 30:374–381
33. Martins ER, Figueiredo LS (2009) Cultivo de plantas medicinais. In: Leite JPV (ed) Fitoterapia: bases científicas e tecnológicas. Atheneu, São Paulo, pp 143–176
34. Falkenberg MB, Santos RI, Simões CMO (2004) Introdução à análise fitoquímica. In: Simões CMO (ed) Farmacognosia: da planta ao medicamento. Editora UFRGS/UFSC, Porto Alegre/Florianópolis, pp 229–245
35. Peixoto Sobrinho TJS, Gomes TLB, Cardoso KCM, Amorim ELC, Albuquerque UP (2010) Otimização de metodologia analítica para o doseamento de flavonóides de *Bauhinia cheilantha* (Bongard) Steudel. Quim Nova 33(2):288–291
36. Oliveira MA (2005) Extração de polifenóis da semente de cacau (Theobroma cacao).

Dissertation, Universidade Federal de Santa Catarina, Florianópolis

37. Andreo D, Jorge N (2006) Antioxidantes naturais: técnicas de extração. Boletim do Centro de Pesquisa de Processamento de Alimentos 24(2):319–336
38. Cechinel Filho V, Yunes RA (1998) Estratégias para a obtenção de compostos farmacologicamente ativos a partir de plantas medicinais. Conceitos sobre modificação estrutural para otimização da atividade. Quim Nova 21(1): 99–105
39. Marques LC, Vigo CLS (2009) Preparação e padronização e extratos vegetais. In: Leite JPV (ed) Fitoterapia: bases científicas e tecnológicas. Atheneu, São Paulo, pp 170–204
40. Sonaglio D, Ortega GG, Petrovick PR, Bassani VL (2004) Desenvolvimento tecnológico e produção e fitoterápicos. In: Simões CMO (ed) Farmacognosia: da planta ao medicamento. Editora UFRGS/UFSC, Porto Alegre/ Florianópolis, pp 289–326
41. Alcarde AL, Basso LC (1997) Efeito da trealose na manutenção da viabilidade de células de leveduras desidratadas por liofilização. Sci Agric 54(3):189–194
42. Tattini Júnior V, Parra DF, Pitombo RNM (2006) Influência da taxa de congelamento no comportamento físico-químico e estrutural durante a liofilização da albumina bovina. Rev Bras Cienc Farm 42(1):127–136
43. Alvim ID (2005) Produção e caracterização de micropartículas obtidas por spray drying e coacervação complexa e seu uso para alimentação de larvas de peixes. Tesis, Universidade Estadual de Campinas, Campinas
44. Wagner H, Bladt S (1996) Plant drug analysis: a thin layer chromatography atlas. Springer, Berlin, p 384
45. Collins CH (1995) Princípios básicos da cromatografia. In: Collins CH, Braga GL, Bonato OS (eds) Introdução a métodos cromatográficos. Editora da Unicamp, Campinas, pp 11–28
46. Degani ALG, Cass QB, Vieira PC (1998) Cromatografia um breve ensaio. Quim Nova na Escola 7:1–5
47. Zanatta G, Moraes CG, Zimmer ER, Ewald L, Pettenuzzo LF, Salbego CG, Trindade VMT (2009) Prática virtual de bioquímica: cromatografia ascendente em papel de aminoácidos. Novas Tecnologias na Educação 7(3):1–10
48. Barbosa CV (2009) Avaliação do potencial antineoplásico de plantas medicinais como coadjuvantes no tratamento do câncer pelos pacientes do CACON/UFAL. Dissertation, Universidade Federal de Alagoas, Maceió
49. Vichenewski W (1995) Cromatografia por adsorção. In: Collins CH, Braga GL, Bonato OS (eds) Introdução a métodos cromatográficos. Editora da Unicamp, Campinas, pp 59–70

Chapter 17

In Vitro and In Vivo Antibacterial and Antifungal Screening of Natural Plant Products: Prospective Standardization of Basic Methods

José Vitor Lima-Filho and Rossana de Aguiar Cordeiro

Abstract

Researchers have conducted tests for antimicrobial activity of natural products by several methods, and a huge collection of results has been generated. However, the lack of standardization of basic methods of investigation has led to accumulation of non-useful data. The diversity of protocols has created divergences among specialists, and often different results are obtained with the same plant extract. Although antimicrobial tests for natural products have not been standardized by regulatory agencies, in this chapter, we recommend the use of some technical parameters for susceptibility testing reviewed by the Clinical and Laboratories Standards Institute.

Key words Agar-diffusion assay, Microdilution assay, Essential oil, Plant extract

1 Introduction

In the broad sense natural products can be defined as chemical compounds produced by living organisms with pharmacological or biological activity. In particular, the screening of natural plant products to isolate new compounds with antimicrobial activity has increased in recent years, based on three main aspects: (a) microorganisms are responsible for significant morbidity and mortality, and vaccines are available to prevent only a small percentage of infectious diseases; (b) microorganisms cause plant diseases and food spoilage; and (c) microorganisms develop resistance mechanisms to antimicrobials faster than the speed of drug discovery.

According to the World Health Organization, 7.6 million deaths occurred among children under 5 years in 2010, and of this total 2.89 million (38 %) were caused by infectious diseases [1]. Considering the youth and adult population, six diseases stand out in order of importance: acute respiratory infections (including pneumonia and influenza), HIV/AIDS, diarrheal diseases,

Ulysses Paulino Albuquerque et al. (eds.), *Methods and Techniques in Ethnobiology and Ethnoecology*,
Springer Protocols Handbooks, DOI 10.1007/978-1-4614-8636-7_17, © Springer Science+Business Media New York 2014

tuberculosis, malaria, and measles. The prevalence of each of these varies with the geographic region, but African countries and those located in southwest Asia and the eastern Mediterranean are the most affected [1]. Therefore, it is reasonable that microorganisms eligible for priority antimicrobial assays with natural products should be selected among etiologic agents whose diseases have higher incidence.

A series of emerging and reemerging pathogens has also received attention from health authorities (Table 1). The estimated economic losses from illness and quality-adjusted life caused by food-borne pathogens in the United States reach $14.0 billion a year [2]. Furthermore, the indiscriminate use of antimicrobials in human and veterinary medicine selects resistant strains, reducing the treatment options against infections [3, 4]. This problem is greatest for species of bacteria that multiply rapidly, have high mutation rates, and have developed mechanisms for transfer of genes conferring resistance to antibiotics [5, 6].

Among the sources of new antimicrobial compounds, plants have routinely been investigated because of the information accumulated over the years with their traditional use in folk medicine. A search conducted in November 2012 on the website of the US National Library of Medicine of the National Institute of Health (Pubmed), using the key words "plant" and "antimicrobial activity," produced 18,906 scientific articles on the theme. The researches have been carried out at different levels, usually through preliminary tests in silico or in vitro and preclinical studies with laboratory animals. On the other hand, only a small number are selected for clinical trials with humans, and even smaller numbers become commercial products. Particularly, the filing of patent applications relating to herbal medicines has increased proportionately to the development of new drugs. But the number of patents is still small especially in developing countries. For instance, although Brazil contains 22 % of the world's flora, only 437 documents related to patents of phytotherapics were filed by Brazilian researchers from 1995 to 2010 [7].

Many researchers have conducted tests for antimicrobial activity of natural products by several methods, and a huge collection of results has been generated. However, the lack of standardization of basic methods to assess the potential of plant species as sources of antimicrobials has led to accumulation of non-useful data. The diversity of protocols has created divergences among specialists, and often different results are obtained with the same plant extract. Although antimicrobial tests for natural products have not been standardized by regulatory agencies, we recommend following the guidelines already approved by international committees, such as the Clinical and Laboratories Standards Institute (CLSI) or the European Committee on Antimicrobial Susceptibility Testing (EUCAST).

Table 1
Emerging or reemerging bacterial and fungal pathogens of great importance for human public health

Microorganism	Diseases
Methicillin-resistant *Staphylococcus aureus* (MRSA) Vancomycin-resistant *Staphylococcus aureus* (VRSA)	Skin and soft-tissue infections; severe life-threatening infections (bacteremia, surgical site infections, pneumonia)
Penicillin-resistant *Streptococcus pneumonia* Macrolide-resistant *Streptococcus pneumonia*	Pneumonia, bacteremia, otitis media, meningitis, sinusitis
Vancomycin-resistant *Enterococcus faecalis* and *E. faecium*	Urinary tract infections, endocarditis, bacteremia, meningitis, catheter-related infections, surgical wound infection, skin and soft-tissue infection
Penicillin-resistant *Neisseria meningitidis*	Meningococcal meningitis
Fluoroquinolone-resistant Neisseria gonorrhoeae	*Gonorrhea*
Enterobacter cloacae, *E. aerogenes*, and other Enterobacteriaceae with chromosomal β-lactamases	Nosocomial infections (bacteremia, pneumonia, urinary tract infection, endocarditis, intra-abdominal infection, septic arthritis, osteomyelitis, meningitis) and community-acquired infections (skin and soft-tissue infection, urinary tract infection)
Multidrug-resistant *Pseudomonas aeruginosa*	Nosocomial infections (pneumonia, urinary tract infection, and bacteremia)
Multidrug-resistant *Stenotrophomonas maltophila*	Bacteremia, pneumonia, urinary tract infection, skin and soft-tissue infection, ocular infection, endocarditis, meningitis
Multidrug-resistant *Acinetobacter baumannii*	Bacteremia, pneumonia, meningitis, urinary tract infection, wound infection
Enterobacteriaceae with extended-spectrum β-lactamases	Community- and hospital-acquired infections (cystitis and pyelonephritis with fever, septicemia, pneumonia, peritonitis, meningitis, and device-associated infections)
Multidrug-resistant *Mycobacterium tuberculosis* Multidrug-resistant *Mycobacterium avium* complex	Tuberculosis
Candida albicans Non *albicans Candida* species (*C. tropicalis*, *C. krusei* and *C. parapsilosis* complex)	Mucocutaneous diseases, invasive diseases
Cryptococcus neoformans complex	Meningitis
Aspergillus fumigatus	Pneumonia, meningitis
Pneumocystis jirovecii	Pneumonia

In this chapter, we recommend the use of some technical parameters for susceptibility testing reviewed by the CLSI to guarantee the reproducibility and reliability of results.

2 Materials and Methods

2.1 Preparation of Stock Solutions

Typical approaches to the choice of plant species used in antimicrobial assays include the following [8, 9]: (a) selection processes based on traditional use of the plant in folk medicine or validated reports of other relevant biological activities, confirmed by different experimental models, and (b) random selection processes, based on the chemical composition or response to preliminary in vitro antimicrobial assays. Crude plant extracts are obtained through the use of water or chemical reagents used as solvents for specific groups of chemical compounds. Therefore, the preparation of stock solutions is critical before running preliminary antimicrobial assays, since the active ingredients may be lost or changed during the solvent extraction procedure.

After evaporation of solvents used in the extraction process, several studies indicate suspending the pellet with dimethyl sulfoxide (DMSO), a dispersing solvent [10, 11]. Solutions using 100 % DMSO stocks have the advantage of inhibiting proliferation of contaminating microorganisms and do not evaporate as do other organic solvents such as ethanol. However, the solvent can damage some microbial species or decrease the antimicrobial activity of essential oils [12, 13]. Thus, the final DMSO concentration in-test should not exceed 1 % [11]. In addition, due to the variety and nature of the chemical compounds present in plant extracts or essential oils, active principles responsible for the antimicrobial action can change or be lost. To reduce this probability over time, storing plant extracts in 100 % DMSO in amber glass tubes at −20 °C is recommended.

Natural plant products for use as phytotherapics should preferentially be water soluble and homogeneous in solution, with selective toxicity against microbial cells and activity at low concentrations at room or body temperature as well as having stability over time. However, even when a plant extract is highly active at low concentrations against given microbial species, it can be highly toxic to eukaryotic cell cultures or animal models [14]. Thus, minimum and maximum values for each sample type should be implemented during antimicrobial assays. These are usually based on results of toxicity tests, which allow calculation of the lethal concentration to 50 % of the cells or the laboratory animals (LC_{50}), so that a dosage about 10–100 times lower than the LC_{50} is virtually nontoxic [15, 16]. A possible scheme for preparation of dilutions from the stock solution in 100 % DMSO is given in Fig. 1.

In such example, 0.5 mL of the initial stock solution whose concentration is 100 mg/mL in 100 % DMSO is transferred

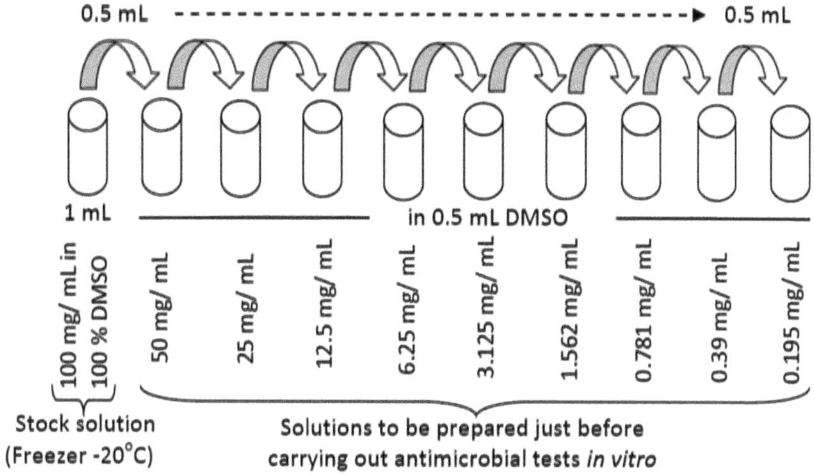

Fig. 1 Dilutions from stock solutions of plant extracts, essential oils, or pure compounds in 100 % DMSO

into 0.5 mL microtubes containing 100 % DMSO. Then 0.5 mL aliquots are transferred from tube to tube, proportionally decreasing the concentrations of test substances and DMSO. The final in-test concentration is produced after serial dilutions of each new tube in the culture medium, as described later.

2.2 Test Microorganisms

The choice of test microorganisms depends on the laboratory facilities available and the technical ability of the scientific team. Handling of microorganisms demands laboratories with appropriate biosafety level and proper training in the use of aseptic techniques. Poorly trained professionals incur serious health risks. According to the degree of pathogenicity and virulence of the microorganism, proper protection equipment must also be provided. In addition, researchers must be aware of the laws regarding biosafety (including storage and transportation) of select agents, such as *Mycobacterium tuberculosis*, *Burkholderia pseudomallei*, *Coccidioides* spp., *Histoplasma capsulatum*, and rabies virus. Further information regarding biosafety of pathogenic microorganisms can be found at the following websites: http://www.cdc.gov/biosafety/ and http://www.osha.gov/index.html.

Samples from bacterial and fungal strains can be obtained commercially from international culture collections such as the American Type Culture Collection (ATCC) or Deutsche Sammlung von Mikroorganismen und Zellkulturen (DSMZ). The collections of cultures of microorganisms recorded on all continents can be identified through a search at the website of the World Data Centre for Microorganisms (http://www.wfcc.info/ccinfo/home). Obtaining samples of collection cultures assures the researcher reliable information on the origin and identity of the microorganism at the species level.

Due to the tests involving bacteria, a panel of species should include gram positive and gram negative, but the number of isolates may vary. Preliminary assays should focus on determining whether or not the plant extract or essential oil possesses antimicrobial action to enable disregarding plants with poor prospects. Basic information about the range of antimicrobial activity of bioactive compounds can also be obtained. For example, plant extracts with broad-range antibacterial activity inhibit the growth of gram-positive and gram-negative bacteria of different species.

Due to distinct cell morphophysiological characteristics, we recommend the following species as test organisms in preliminary assays: *Staphylococcus aureus* (facultative anaerobe, gram positive, nonspore forming); *Clostridium* spp. (anaerobe, gram positive, spore forming); *Pseudomonas aeruginosa* (aerobe, gram negative, non-Enterobacteriaceae); *Escherichia coli* (facultative anaerobe, gram negative, non-encapsulated, Enterobacteriaceae); and *Klebsiella pneumoniae* (facultative anaerobe, gram negative, encapsulated, Enterobacteriaceae). Among fungal species that could be used in preliminary tests, we suggest *Candida albicans*, *Cryptococcus neoformans* species complex, dermatophytes (*Microsporum*, *Trichophyton*, *Epidermophyton*), and invasive filamentous fungi with rapid conidiogenesis, such as *Fusarium* spp., *Aspergillus* spp., *Rhizopus arrhizus*, *Pseudallescheria boydii* (*Scedosporium apiospermum*), and *Sporothrix schenckii*.

For subsequent tests, other examples of clinical isolates and/or microbial strains resistant to antibiotics are indicated in Table 1.

2.2.1 Preparation of Inocula

Preparation of bacterial inocula is usually performed following several steps. Initially the samples kept frozen or refrigerated must be activated in liquid culture medium, e.g., brain–heart infusion (BHI) broth. After 6–8 h at 35 °C, an aliquot can be transferred to a nonselective solid culture medium that is rich in nutrients, such as BHI agar or blood agar (5 % sheep blood). After 18–24 h at 35 °C, colonies of the axenic culture may be suspended in 3–5 mL of sterile saline (8.5 g/L NaCl, 0.85 % saline) and compared with the pattern 0.5 of McFarland standard (corresponding to $1–2 \times 10^8$ CFU/mL) [17]. Note that some fastidious bacteria species may require culture media with additional nutrients and better growing conditions.

The pattern 0.5 of McFarland standard, consisting of $BaSO_4$, can be prepared as follows [17] (Box 1):

The optical density of the tube containing the bacterial suspension is compared to the McFarland standard using a white card with contrasting black lines at the bottom. This procedure optimizes time and is indicated in the case of inocula of bacteria and yeasts. For antifungal susceptibility testing, the inoculum is prepared from cultures previously grown in potato dextrose agar (PDA), at the following conditions [18, 19] (Box 2):

Box 1 Preparation of the pattern 0.5 of McFarland standard

1. Add 0.5 mL aliquot of 1 % $BaCl_2$ (wt/ vol) to 99.5 mL of 1 % H_2SO_4 (vol/vol).
2. Check the turbidity with a spectrophotometer at a wavelength of 625 nm (absorbance should vary from 0.08 to 0.10).
3. Keep aliquots of 4–6 mL in tubes with screw caps of the same types used for diluting the bacterial inoculums, and store them in a dark place at room temperature.
4. Before use, shake them vigorously in a vortex agitator, making sure that it is evenly turbid.
5. Replace the tubes of 0.5 McFarland standard or check their densities every month.

Box 2 Conditions of incubation for fungal strains

Microorganism	Duration (days)	Temperature (°C)
Candida spp.	1	35
Cryptococcus spp.	2	35
Aspergillus spp., *P. boydii*, *R. arrhizus*, and *S. schenckii*	5–7	35
Fusarium spp.	2–3, followed by 4	35 25–28
Dermatophytes	Up to 15	28–30

Sterile saline solution (0.9 %) must be added to the agar slant, and the cultures are gently swabbed to dislodge the conidia or spores from the hyphal mat. Tween 20 (0.01 mL) should be added for better suspension of *Aspergillus* conidia. The resulting suspensions are allowed to settle for 5 min at 28 °C, then transferred to sterile tubes, and vortexed for 15 s, and their volumes adjusted to 4 mL with sterile saline. Optical density of yeast suspensions are compared to the 0.5 McFarland standard [18]. For filamentous fungi, the optical density of each suspension is read at 530 nm and adjusted to 95 % transmittance (dermatophytes), 80–82 % for *Aspergillus* and *S. schenckii*, and 68–70 % for *Fusarium* spp., *P. boydii*, and *R. arrhizus* [19]. These procedures produce suspensions containing approximately 1×10^6 to 5×10^6 CFU/mL for yeasts and 1×10^5 to 5×10^6 CFU/mL for invasive fungal species.

The number of cells of bacteria, yeasts, and filamentous fungi spores desirable in each test varies according to the experimental model adopted. Thus, performing dilutions in sterile saline with further counting of colony-forming units may be necessary to adjust the number of cells in the inoculum.

2.3 In Vitro Antimicrobial Activity

Dilution of extracts rich in lipid compounds and essential oils may produce nonhomogeneous and turbid solutions. This usually occurs due to decreased relative concentration of DMSO in liquid culture media, which must be kept below 1 %, as discussed above. This problem can be minimized by adding 0.5–2 % Tween 20 (v/v), a nonionic emulsifier, which has also been used to formulate microemulsion systems [20, 21]. In all cases, the introduction of positive and negative controls diminishes the chance of error in the interpretation of test results, particularly in determination of the inhibitory concentration of the extracts.

2.3.1 Agar-Diffusion Assay

The standardization for determination of bacterial susceptibility of antimicrobials was revised in the 1960s by WM Kirby and colleagues at the University of Washington School of Medicine and King County Hospital. The resulting Kirby–Bauer disk diffusion test uses filter paper disks impregnated with antimicrobial agents, which are added to plates with a solid culture medium seeded with microorganisms [22]. After an incubation period, the diameter of the inhibition halo formed around the disks is measured with a ruler or a caliper. This test is indicated for detection of sensitivity or resistance of pathogenic aerobic and facultative anaerobic bacteria to antimicrobials in order to support the physician in selecting treatment options. An update of this method by the CLSI is produced from time to time through a global consensus process. Therefore, there is an evident concern with the uniformity of the method, which is crucial for reproducibility of results.

Preliminary tests to evaluate the antimicrobial potential of natural products from plants have adapted the method of Kirby–Bauer. In this case, plant extracts are added to disks of filter paper or wells punched in the solid medium seeded with microorganisms [23] (Fig. 2). Recently the knowledge generated from the consensus indicated by CLSI has been applied by some researchers in the field of natural products. For instance, we advise the use of the Mueller–Hinton (MH) culture medium for antimicrobial assays with plant extracts, essential oils, and pure compounds. However, in spite of the guarantee of uniformity during antimicrobial assays, the following steps deserve attention on preparation of the culture medium [17]: (a) let the MH agar cool in a bath with temperature between 45 and 50 °C after autoclaving; (b) add the freshly prepared medium in Petri dishes with flat bottoms with uniform depth of approximately 4 mm (corresponding to 60–70 mL on plates with 150 mm diameter and 25–30 mL for plates with a diameter of 100 mm); (c) let the medium cool at room temperature, and store it in a refrigerator (2–8 °C), except when the plate is used on the same

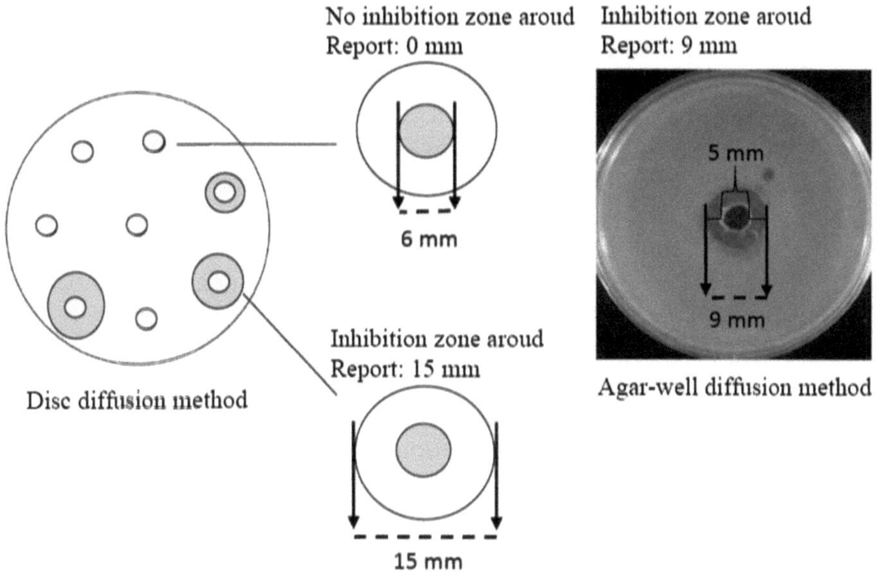

Fig. 2 Agar diffusion methods

day; (d) use the plates within 7 days after preparation; (e) confirm the sterility of the plates prior to the test by incubating them at 30–35 °C for 24 h or more; (f) check the pH of a sample of the batch of plates, which should be between 7.2 and 7.4; and (g) if necessary, avoid excessive moisture prior to the tests in plates by leaving them in the incubator at 35 °C for 10–30 min.

The inoculums of the bacterium of interest in the culture medium should be prepared from a solution whose turbidity is equivalent to the 0.5 McFarland standard. We recommend the following steps [17]: (a) 15 min after adjusting the turbidity of the inoculum suspension, dip a cotton swab in sterile microbial suspension; (b) rotate the swab and press it against the inner wall of the pipe, above the liquid, to remove excess inoculum; (c) pass the swab 2–3 times over the entire plate surface of Mueller–Hinton agar in order to produce a uniform distribution of the inoculum; and (d) leave the cover partially off for 3–5 min (maximum 15 min) to remove excess moisture. The literature reports several methodological variations for antimicrobial assays with plant extracts, essential oils, or isolated compounds [15, 24–29]. Thus, the following procedures reflect a general standard, which can be adopted.

For tests with bacteria, sterile filter paper disks 6 mm in diameter should be impregnated with a fixed volume (e.g., 20 μL) immediately before transfer to the plates. Alternatively, wells 6 mm in diameter can be punched in the agar with a bacteriological loop or sterile glass pipette (sufficient for 20–50 μL of the extracts). The disks or the wells should be spaced at least 2 cm apart (measured from the center of the disk or well). Leaving the plates at room temperature for 30–60 min is usually sufficient for diffusion of extracts around disks or wells. Essential oils can be mixed to mineral oil to facilitate diffusion. Then incubate cultured Petri dishes for 24 h at 35 °C. After this time, measure the inhibition zone around the disks or the wells with a ruler or a caliper. Use a broad-range antibiotic as positive control (e.g., oxytetracycline or nalidixic acid) and the solvent used to suspend the sample as negative control (e.g., DMSO).

For antifungal tests, a fungal suspension is inoculated onto the surface of the medium in Petri dishes containing PDA. The inoculum should be previously adjusted by turbidimetry to obtain approximately 10^6 or 10^5 CFU/mL, for *Candida* spp. or dermatophytes, respectively. Wells 6 mm in diameter are punched in the agar and the test solution is delivered to them. Griseofulvin (1 mg/mL) or amphotericin B (5 mg/L) may be used as control drugs of dermatophytes or other fungal species, respectively. After incubation for 3–5 days for *Candida* spp. and 5–8 days for dermatophytes, at 28–35 °C, the dishes are examined for zones of growth inhibition and the diameters of these zones are measured in millimeters [30]. For *Aspergillus* spp. and other fast-growing fungi, a sterile cotton swab is inoculated by dipping it into the suspension previously adjusted to approximately 10^6 CFU/mL. The swab is streaked across the surface of the agar in three directions. The plate is dried at 28 °C for 15 min before applying the paper disks (6 mm) soaked with the test solution [31].

Although some researchers conduct tests with independent replicates or implement some type of statistical analysis of the results, the interpretation of data from agar diffusion assays often only allows conclusions about the presence/absence of antimicrobial compounds in the plant extracts. Previous data have shown that comparison between results of agar diffusion and dilution tests is feasible only for some gram-positive bacteria [32]. Therefore, subsequent quantitative analyses will be necessary to determine the minimum inhibitory concentration (MIC).

2.3.2 Determination of the MIC

Prospective plant extracts have MIC values lower than 150 μg/mL (MICs higher than 150 μg/mL increase the incidence of false-positive results) [11]. The MIC of natural plant products gives quantitative results on the potency of plant extracts to control microbial growth. Before running the tests, it is advisable to prepare a series of dilutions of the stock solution. From a series of ten concentrations from 100,000 to 195.3 μg/mL in 100 % DMSO (see Fig. 1), dilute each solution 10× in liquid culture medium (Mueller–Hinton for bacteria, RPMI 1640 for yeasts) to produce working solutions from 10,000 to 19.5 μg/mL, for example (Box 3):

The volumes above can be adjusted, if necessary, depending on the total number of tests to be performed.

Box 3 Preparation of working solutions for determination of the minimum inhibitory concentration

(a) For volumes of 1 mL of the natural product diluted (sufficient for ten tests), pipette up the first volume of 0.9 mL of culture medium in each sterile test tube.

(b) Pipette 0.1 mL of the lower concentration (195.3 μg/mL) of the natural product diluted in DMSO to another tube with 0.9 mL of culture medium.

(c) Pipette 0.1 mL of a concentration 390 μg/mL and so on, throughout the range of concentrations, each time adding 0.1–0.9 mL volumes of the medium.

(d) With another pipette add 0.1 mL of 100 % DMSO in a new tube containing 0.9 mL culture medium (negative control).

Microdilution method. The microdilution technique is advantageous over the macrodilution method, considering the economy of material and the similar results. The technique is usually carried out in U-shaped plates with 96 wells and has variations from the original method described by Eloff [33]. For instance, in working solutions of plant extracts ranging from 10,000 to 19.5 μg/mL, the addition of 100 μL to wells containing 100 μL of Mueller–Hinton broth plus 5 μL of a bacterial suspension of 3.5×10^7 CFU/mL (1:10, v/v; pattern 0.5 of McFarland standard) yields final concentrations in wells of 5,000–9.75 μg/mL and approximately 5×10^5 CFU/mL. The results are usually read after an incubation period of 24–48 h/35 °C. The MIC value is the lowest concentration of the natural product that inhibited microbial growth visually.

From the tubes without visible growth, 10 μL of solution is removed and spread on Mueller–Hinton agar and incubated for another 24–48 h/35 °C to determine the minimum bactericidal concentration (MBC). The lack of colony-forming units (or growth less than 0.1 % of the initial inoculum) indicates that natural products were bactericidal. The percentage of bacterial growth inhibition by plant extracts can alternatively be calculated with the use of a spectrophotometer in comparison with positive control wells (medium plus microorganism devoid of extracts) and negative control wells (antibiotic plus microorganism) (Fig. 3).

Antifungal testing is conducted with RPMI 1640 medium buffered with 0.165 M morpholinepropanesulfunic acid (MOPS, 34.54 g/L) at pH 7.0, which is the standard medium for inoculum preparation and final dilution of test substances/control drugs. Wells are filled with 100 μL of RPMI medium and then 100 μL of a test substance. Finally, fungal suspensions are inoculated in each well, and plates are incubated at 35 °C for 48 h for *Candida* spp. or 72 h for *Cryptococcus* spp. For fast-growing fungi, the following

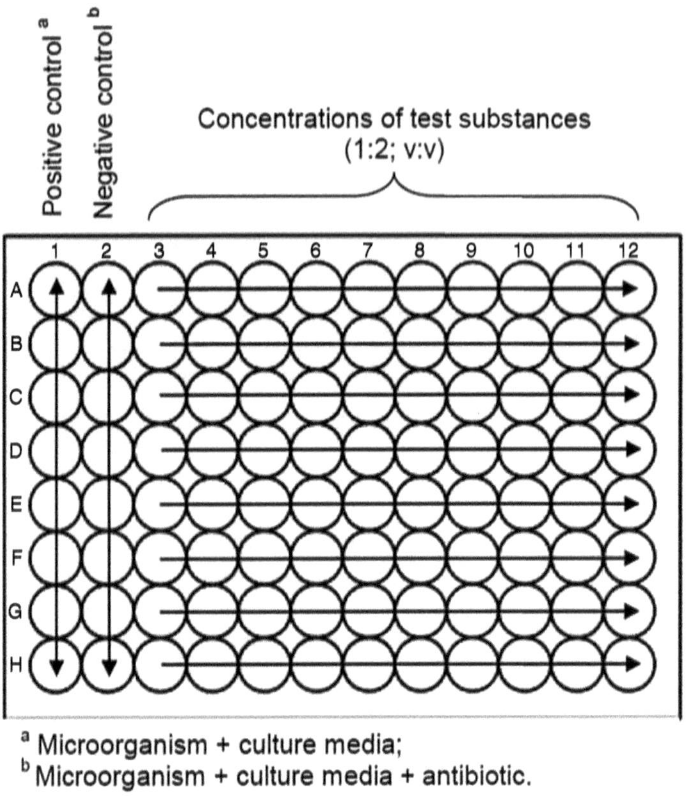

Fig. 3 Microdilution assay

incubation periods are recommended: 21–26 h for *Rhizopus* spp.; 46 hours for *Fusarium* spp., *Aspergillus* spp., and *S. schenkii*; and 70–74 h for *P. boydii* [18, 19]. Susceptibility of dermatophytes can be evaluated after incubation at 35 °C for 5 days [30]. The inoculum preparation uses stock suspensions based in the 0.5 McFarland standard diluted in RPMI medium at 1:100 and then 1:20 for yeasts and 1:50 for invasive fungal species. These procedures provide suspensions containing 5×10^2 to 2.5×10^3 CFU/mL for yeasts and 0.4×10^4 to 5×10^4 CFU/mL for filamentous fungi. Readings are performed visually, comparing the intensity of growth with control wells with RPMI medium (growth control) and wells with antifungals (negative control).

Alternatively, quantitative fungal growth may be evaluated by metabolic dyes, such as XTT and resazurin. The XTT method is based on the conversion of the dye 2,3-bis(2-methoxy-4-nitro-5-sulfophenyl)-5-[(phenylamino)carbonyl]-2H-tetrazolium hydroxide (XTT) by mitochondrial dehydrogenase into a water-soluble formazan product that is measured spectrophotometrically. The resazurin method is based on the reduction of this compound to resorufin, producing a very bright red fluorescence only by viable cells.

Researchers should choose how MICs for natural products are defined in the study. We recommend comparisons with standard antibacterials and antifungals such as nalidixic acid or amphotericin B, respectively. Besides MIC values, researchers can also calculate MIC_{50} and MIC_{90} when multiple isolates of a given species are tested (at least ten isolates). MIC_{50} is defined as the MIC value at which ≥50 % of the strains within a test population are inhibited. It is equivalent to the median MIC value. Analogously, MIC_{90} corresponds to the MIC value at which ≥90 % of the strains are inhibited [34].

2.4 In Vivo Antimicrobial Activity

Studies reporting preclinical evaluation of plant natural products on control of infections are rare [35–37]. Moreover, treatments of experimentally infected hosts with natural plant products can produce unanticipated interactions among plant compounds, microorganisms, and the host. Therefore, procedures using laboratory animals should be subject to evaluation and approval by the institutional ethics committee to ensure that animals are not subjected to unnecessary suffering. Additionally, the determination of nontoxic concentrations of the test compounds or extracts using cultured eukaryotic cells or laboratory animals is crucial. Although there are several types of infection models described in the literature, those of *Salmonella enterica* Ser. Typhimurium (gram-negative bacteria), *Listeria monocytogenes* (gram positive), and *Cryptococcus neoformans* (a yeast) can be easily reproduced in Swiss mice to cause systemic lethal or sublethal infections.

2.4.1 *Salmonella enterica Subsp. enterica* Typhimurium

Salmonella spp. are facultative anaerobes and gram-negative bacteria associated with outbreaks causing diarrheal disease in humans and various animal groups [38]. The genus *Salmonella* comprises two species that cause disease in humans: *S. enterica* and *S. bongori* [39]. The first has greater relevance for humans and warm-blooded animals and is subdivided into six subspecies, differentiated by biochemical, antigenic, and phylogenetic characteristics. The subspecies I category includes *S. enterica* serotypes, which encompasses most pathogens for humans. *S. enterica* serotype Typhimurium has been used to reproduce the symptoms of human typhoid fever (caused by serotype Typhi) in mice and thus investigate strategies to combat human infections [40].

2.4.2 *Listeria monocytogenes*

The genus *Listeria* includes six species: *L. monocytogenes*, *L. ivanovii*, *L. seeligeri*, *L. innocua*, *L. welshimeri*, and *L. grayi*. The first two are important as human pathogens. It is a gram-positive and facultative intracellular pathogen, being able to invade, survive, and proliferate in various cell types such as enterocytes, fibroblasts, hepatocytes, and professional phagocytes [41]. Although *L. monocytogenes* is infective for all human populations, more serious problems are reported in pregnant women, newborns, the elderly, and immunosuppressed individuals [42]. The bacterium is also responsible for most opportunistic infections that cause meningitis, encephalitis, and septicemia in humans and animals [43].

2.4.3 *Cryptococcus neoformans*

The encapsulated yeast *C. neoformans* is the most frequent etiologic agent of cryptococcosis, which has a remarkable ability to infect and cause disease in various hosts [44]. However, *C. albidus*, *C. laurentii*, and *C. uniguttulatus* have also been isolated as the causative agents of this mycosis [45]. *C. neoformans* var. *neoformans* harbors several virulence factors related to survival of the yeast in the host, facilitating the onset and persistence of the disease [46]. The disease is usually of pulmonary origin and tends to spread to central nervous system, causing meningitis and meningoencephalitis [46].

2.4.4 General Aspects of *In Vivo* Treatments with Natural Plant Products

The development of symptoms in susceptible laboratory animals depends on virulence of the microorganism and the number of colony-forming units in inocula. The calculation of the LD_{50} (lethal dosage of microorganisms to 50 % of animals) may be necessary so that inocula lower than the LD_{50} produce sublethal infections whereas those higher than LD_{50} produce lethal infections. However, the LD_{50} can vary with the route of inoculation (e.g., oral, endovenous, intraperitoneal), and therefore treatments can be started before or after the onset of the infection. However, curative approaches are more important considering prospective clinical trials with humans.

Although Swiss mice are susceptible to *L. monocytogenes S.* Typhimurium, before running tests with *C. neoformans* the animals should be immunosuppressed. A simple protocol can be achieved with daily administration of dexamethasone (0.4 mg/animal by i.p. route) for 3 days consecutively and then, after 4 days intervals [47]. Treatment of the experimental infections should involve at least a daily dosage of the natural product using a fixed volume (e.g., 0.2 mL), whereas the control group receives only the solvents lacking plant extracts. Pure compounds should preferably have been evaluated for their bioavailability in the blood after administration, e.g., by HPLC, before determining a dosage.

After infection, the animals should be euthanized at different time points, but the treatment time depends on the host response. Samples of the target organs of infection must be removed aseptically to count the number of colony-forming units in the organs and for histopathological analysis, for example, the spleen and the liver in *S. enterica* Ser. Typhimurium infections; spleen, liver, and the brain in *L. monocytogenes* infections; and lungs and the brain in *C. neoformans* infections. Other samples should also be obtained (e.g., blood, serum, DNA) to obtain the maximum information from the animal model. Survival curve evaluation can also be carried out after infection, but the animals should be euthanatized after the differences between the control and experimental groups become clear.

The protection exerted by administering a natural product against a systemic infection caused by *Salmonella*, *Listeria*, or *Cryptococcus* can be readily determined by assessing the levels of survival and histological damage in target organs for infection and spread in the host organism. However, in vivo antimicrobial activity of a plant extract only will be confirmed after ruling out any influence on the host immune response. Accordingly, the total and differential leukocyte cell counts in the bloodstreams (or specific T cell activation), antibody production in serum, and synthesis of inflammatory mediators (e.g., cytokines) should be investigated.

3 Perspectives

Considering the large number of infectious diseases without effective treatment, in the coming years interest should increase dramatically in finding new herbal drugs. In particular, the investigation of natural antimicrobial plants should be guided by a concern to form a global consensus on research methods and techniques, as is the case of existing tests with commercial antimicrobials produced by CLSI. Such standardization will lead to the selection of promising compounds and patents whose viability also depends on further studies using models of ex vivo or in vivo infection. The observation of these aspects is fundamental to enable testing of more herbal drugs in clinical trials with humans, particularly in developing countries.

References

1. World Health Organization (2012) World health statistics. www.who.int
2. Hoffmann S, Batz MB, Morris JG Jr (2012) Annual cost of illness and quality-adjusted life year losses in the United States due to 14 foodborne pathogens. J Food Prot 75:1292–1302
3. Hawkey PM, Jones AM (2009) The changing epidemiology of resistance. J Antimicrob Chemother 64(suppl 1):i3–i10
4. Johnson T, Jordan D, Kariyawasam S et al (2010) Sequence analysis and characterization of a transferable hybrid plasmid encoding multidrug resistance and enabling zoonotic potential for extraintestinal *Escherichia coli*. Infect Immun 78:1931–1942
5. Cercenado E (2010) Update of antimicrobial resistance in Gram-positive microorganisms. Med Clin (Barc) 135(3):10–15
6. Baquero F (2012) Metagenomic epidemiology: a public health need for the control of antimicrobial resistance. Clin Microbiol Infect 18(4): 67–73
7. Oliveira ACD, Reis SMM, Vasconcellos AG (2011) Technological prospecting for patents on herbal medicines in Brazil. Int Res J Biotechnol 2(5):78–84
8. Fabricant DS, Farnsworth NR (2001) The value of plants used in traditional medicine for drug discovery. Environ Health Perspect 109(1):69–75
9. Albuquerque UP (2010) Implications of ethnobotanical studies on bioprospecting strategies of new drugs in semi-arid regions. Open Complement Med J 2:21–23
10. Jones WP, Kinghorn AD (2012) Extraction of plant secondary metabolites. Methods Mol Biol 864:341–366
11. Cos P, Vlietinck AJ, Berghe DV (2006) Anti-infective potential of natural products: how to develop a stronger in vitro "proof-of-concept". J Ethnopharmacol 106:290–302

12. Hili P, Evans CS, Veness RG (1997) Antimicrobial action of essential oils: the effect of dimethylsulphoxide on the activity of cinnamon oil. Lett Appl Microbiol 24(4):269–275
13. Basch H, Gadebusch HH (1968) *In vitro* antimicrobial activity of dimethylsulfoxide. Appl Microbiol 16(12):1953
14. Ramos SCS, Oliveira JC, Câmara CAG et al (2009) Antibacterial and cytotoxic properties of some plant crude extracts used in Northeastern folk medicine. Braz J Pharmacog 19(2A):376–381
15. Silva AB, Silva T, Franco ES et al (2010) Antibacterial activity, chemical composition, and cytotoxicity of leaf's essential oil from Brazilian pepper tree (*Schinus terebinthifolius*, Raddi). Braz J Microbiol 41:158–163
16. Toyang NJ, Ateh EN, Keiser J et al (2012) Toxicity, antimicrobial and anthelmintic activities of *Vernonia guineensis* Benth. (Asteraceae) crude extracts. J Ethnopharmacol 144(3):700–704. doi:10.1016/j.jep.2012.10.016
17. Clinical and Laboratory Standards Institute (2003) Guideline M2-A8, standardization of sensitivity tests with antimicrobials by disc-diffusion. CLSI, Wayne, PA
18. Clinical and Laboratory Standards Institute (2008) Guideline M27-A3, Reference method for broth dilution antifungal susceptibility testing of yeast. CLSI, Wayne, PA
19. Clinical Laboratory Standards Institute (2008) Guideline M38-A2, Reference method for broth dilution antifungals susceptibility testing of conidium-forming filamentous fungi: approved standard. CLSI, Wayne, PA
20. Donsìb F, Annunziataa M, Vincensia M et al (2012) Design of nanoemulsion-based delivery systems of natural antimicrobials: effect of the emulsifier. J Biotechnol 159:342–350
21. Xu SX, Li YC, Liu X et al (2012) *In vitro* and *in vivo* antifungal activity of a water-dilutable cassia oil microemulsion against *Geotrichum citri-aurantii*. J Sci Food Agric 92(13):2668–2671
22. Bauer AW, Kirby WM, Sherris JC et al (1966) Antibiotic susceptibility testing by a standardized single disk method. Tech Bull Regist Med Technol 36(3):49–52
23. Cowan MM (1999) Plant products as antimicrobial agents. Clin Microbiol Rev 12:564–582
24. Hammer KA, Carson CF, Riley TV (1999) Antimicrobial activity of essential oils and other plant extracts. J Appl Microbiol 86:985–990
25. Dorman HJD, Deans SG (2000) Antimicrobial agents from plants: antibacterial activity of plant volatile oils. J Appl Microbiol 88:308–316
26. Rabanal RM, Arias A, Prado B (2002) Antimicrobial studies on three species of *Hypericum* from the Canary Islands. J Ethnopharmacol 81:287–292
27. Springfield EP, Amabeoku G, Weitz F et al (2003) An assessment of two *Carpobrotus* species extracts as potential antimicrobial agents. Phytomedicine 10:434–439
28. Voravuthikunchai S, Lortheeranuwat A, Jeeju W et al (2004) Effective medicinal plants against enterohaemorrhagic *Escherichia coli*. J Ethnopharmacol 94:49–54
29. Kaewpiboon C, Lirdprapamongkol K, Srisomsap C (2012) Studies of the in vitro cytotoxic, antioxidant, lipase inhibitory and antimicrobial activities of selected Thai medicinal plants. BMC Complement Altern Med 12(1):217
30. Fontenelle RO, Morais SM, Brito EH et al (2007) Chemical composition, toxicological aspects and antifungal activity of essential oil from *Lippia sidoides* Cham. J Antimicrob Chemother 59(5):934–940
31. Badiee P, Alborzi A, Moeini M et al (2012) Antifungal susceptibility of the *Aspergillus* species by Etest and CLSI reference methods. Arch Iran Med 15(7):429–432
32. Klančnik A, Piskernik S, Jeršek B et al (2010) Evaluation of diffusion and dilution methods to determine the antibacterial activity of plant extracts. J Microbiol Methods 81:121–126
33. Eloff JN (1998) A sensitive and quick microplate method to determine the minimal inhibitory concentration of plant extracts for bacteria. Planta Med 64:711–713
34. Schwarz S, Silley P, Simjee S et al (2010) Editorial: assessing the antimicrobial susceptibility of bacteria obtained from animals. J Antimicrob Chemother 65:601–604. doi:10.1093/jac/dkq037
35. Abreu A, McBainb AJ, Simões M (2012) Plants as sources of new antimicrobials and resistance-modifying agents. Nat Prod Rep 29:1007–1021
36. Lima-Filho JV, Marinho J, Silva AB et al (2010) Proteins from latex of *Calotropis procera* prevent septic shock due to lethal infection by *Salmonella enterica* serovar Typhimurium. J Ethnopharmacol 129:327–334
37. Oliveira RSB, Figueiredo IS, Freitas LB (2012) Inflammation induced by phytomodulatory proteins from the latex of *Calotropis procera* (Asclepiadaceae) protects against *Salmonella* infection in a murine model of typhoid fever. Inflamm Res 61:689–698
38. Portillo FG (2001) *Salmonella* intracellular proliferation: where, when and how? Microbes Infect 3:1305–1311

39. Dougan G, John V, Palmer S et al (2011) Immunity to salmonellosis. Immunol Rev 240:196–210
40. Libby SJ, Brehm MA, Greiner DL (2010) Humanized nonobese diabetic-scid IL2rgammanull mice are susceptible to lethal *Salmonella* Typhi infection. Proc Natl Acad Sci U S A 107:15589–15594
41. Demuth A, Goebel W, Beuscher HU (1996) Differential regulation of cytokine and cytokine receptor mRNA expression upon infection of bone marrow-derived macrophages with *Listeria monocytogenes*. Infect Immun 64:347583
42. Vázquez-Boland JA, Kuhn M, Berche P (2001) *Listeria* pathogenesis and molecular virulence determinants. Clin Microbiol Rev 14:584–640
43. Hof F, Nichterlein TE, Kretschmar M (1997) Management of listeriosis. Clin Microbiol Rev 10:345–357
44. Van Der Horst CM, Saag MS, Cloud GA et al (1997) Treatment of cryptococcal meningitis associated with the acquired immunodeficiency syndrome. N Engl J Med 337:15–21
45. Capilla J, Maffei CM, Clemons KV et al (2006) Experimental systemic infection with *Cryptococcus neoformans* var. *grubii* and *Cryptococcus gattii* in normal and immunodeficient mice. Med Mycol 44:601–610
46. Chakrabarti A (2007) Epidemiology of central nervous system mycoses. Neurol India 55:191–197
47. Medeiros CS, Pontes-Filho NT, Camara CA et al (2010) Antifungal activity of the naphthoquinone beta-lapachone against disseminated infection with *Cryptococcus neoformans* var. neoformans in dexamethasone-immunosuppressed Swiss mice. Braz J Med Biol Res 43:345–349

Chapter 18

Urban Ethnobotany: Theoretical and Methodological Contributions

Julio Alberto Hurrell and María Lelia Pochettino

Abstract

Urban Ethnobotany studies, among other issues, the botanical knowledge characteristic of those pluricultural contexts that are the urban agglomerations. The botanical knowledge and beliefs guide the strategies of selection and consumption of plants, parts thereof, and plant products. The development of a research in the conurbation Buenos Aires-La Plata (Argentina) enables the characterization of the urban botanical knowledge as a dynamic and complex corpus which includes nontraditional elements and others linked to local family traditions as well as to traditions concerning different groups of immigrants. From this starting point, this chapter includes some theoretical reflections and innovative methodological tools to understand the composition and dynamic of urban botanical knowledge, based on the evaluation of the diversity of plant elements present in the studied area and their circulation. This is performed both in the restricted context of immigrant groups (Bolivian, Chinese) and the general commercial circuit.

Key words Urban ethnobotany, Botanical knowledge, Theoretical framework, Methodological tools, Conurbation Buenos Aires-La Plata, Argentina, Immigrants

1 Introduction

This chapter includes some theoretical and methodological contributions emerging from different studies in Urban Ethnobotany, carried out at the Laboratorio de Etnobotánica y Botánica Aplicada (LEBA), Facultad de Ciencias Naturales y Museo (FCNyM), Universidad Nacional de La Plata (UNLP), Argentina. The approach focuses on the study of botanical knowledge in urban areas, which guides the selection of different vegetal elements such as plants, parts thereof, and products derived from them. The study area corresponds to the Buenos Aires-La Plata conurbation, the largest metropolitan area in Argentina both in population and size. It comprises two contiguous urban agglomerations that emerged around the cities of Buenos Aires, capital of the country, and La Plata, capital of the Province of Buenos Aires.

Ulysses Paulino Albuquerque et al. (eds.), *Methods and Techniques in Ethnobiology and Ethnoecology*, Springer Protocols Handbooks, DOI 10.1007/978-1-4614-8636-7_18, © Springer Science+Business Media New York 2014

In a broad sense, the Ethnobotany studies the complex relationships between people and plants [1–3]. Within this framework, Urban Ethnobotany is a particular chapter of recent development [4–7]. Among other issues, Urban Ethnobotany deals with the study of the peculiar characteristics of urban botanical knowledge (UBK) [8, 9]. During the course of different research in the study area, the need to consider the theoretical framework became evident, in particular to deal with the intrinsic complexity of the botanical knowledge, as well as with its particular composition and dynamic in a pluricultural context. Novel methodological tools emerged as a result of this reflection. Then, the aim of these pages is to present those contributions and to promote studies in Urban Ethnobotany from different perspectives, in order to expand their theoretical and methodological basis.

1.1 Urban Botanical Knowledge

In general terms, the botanical knowledge (BK) is a set of knowledge and beliefs that people have about the vegetal elements of their environment. The BK emerges from the relationships between people and plants, guides their criteria of selecting, and uses those resources, as well as their strategies of obtaining and processing plants, and their consumption patterns. This definition includes either the BK of the traditional societies such as the nontraditional societies. For those called *traditional societies* there are many contributions related to various aspects of the expressions of BK, as ethnoclassifications [10, 11], traditional medicinal uses that are relevant for the search of new active chemical principles [12, 13], or the rescue of traditional knowledge related to the conservation of resources, especially in areas affected by processes of environmental and cultural changes [14–18].

In the last decades, studies on the traditional BK (TBK) have proliferated because it is often close to extinction and therefore its recovery is urgent [19–22]. In contrast, studies on BK in nontraditional societies have been less frequent. The nontraditional BK is associated with Western scientific knowledge—beyond the diverse connotations of the term *Western* [23]—and it is also linked to the diffusion of that knowledge through different mass media [24], and the contents of education programs [25].

Some authors have used the expression *ethnobotanical knowledge* to refer exclusively to TBK [26–28], but Ethnobotany is a scientific discipline and, therefore, *ethnobotanical knowledge* can be referred to the scientific knowledge it produces (i.e., nontraditional), and not to TBK. This discussion arises from the meaning given to prefix *ethno-*, often referred to the knowledge system of a given culture [29]. To avoid confusion, it is preferable to talk about TBK for the heritage of traditional societies and to limit the use of the expression *ethnobotanical knowledge* to the one produced by Ethnobotany. Of course, this decision does not prevent the possibility of dialogue between the TBK and the scientific BK [3, 30, 31].

Table 1
Differences between traditional and nontraditional botanical knowledge

Traditional botanical knowledge	Nontraditional botanical knowledge
Knowledge belongs to culturally homogeneous contexts	Knowledge belongs to culturally heterogeneous contexts (pluricultural)
It is the result of long experience of the human group in its environment	It is not the result of long experience of the human group in its environment (variable residence time)
It is transmitted from generation to generation, orally and in shared practices	It is transmitted through different mass media
The relationship between production and consumption is *direct* (those who consume, produce)	The relationship between production and consumption is *indirect* (those who consume, not produce)
Production practices are known and linked to the conservation of plant resources	Production practices are little known or unknown, not linked to the conservation of plant resources

The concept of TBK is linked to that of *traditional ecological knowledge* (TEK), which is a contribution to the conservation of biocultural diversity, from an integrative perspective [20, 32–36]. The term *traditional* is not free of debate [23], however, it is usually assumed that the TBK regards to culturally homogeneous contexts, with a long experience of the human group in its environment, where knowledge and beliefs transmitted from generation to generation, orally and in shared practices. In traditional societies, the people *who consume, produce*, therefore, the relationship between production and consumption is a direct link. Also, the TBK is *adaptative*, because it allows different adjustments of the human group to changes in their surroundings, consequently TBK is neither conservative nor static but dynamic and innovative, that is, it evolves [9, 19, 37]. In contrast, the BK in large urban areas was considered nontraditional, as opposed to traditional knowledge: it is characteristic of pluricultural contexts, without a long experience of the human group in its environment, and is transmitted through the mass media. The link between production and consumption is indirect: *those who consume do not produce*. Therefore, the majority of the urban population knows little about the properties of the vegetal elements; less about their components or their origin, and the ways of obtaining and processing them are even lesser known [38]. However, this nontraditional BK is also *adaptative*, because it mediates between consumers and the supply of products of vegetal origin, guiding the choice of any of them, while others are discarded [8, 9, 39, 40]. Table 1 summarizes the main differences between TBK and nontraditional BK.

It is a common tendency to think that in urban areas nontraditional BK predominates. However, a conurbation is a pluricultural context, so its BK is heterogeneous. This is due in great part to the presence of immigrant groups of different origin and residence time, cohabiting with the local inhabitants. From the study of the BK of immigrant groups in urban areas different studies from the urban Ethnobotanical approach have been developed in many countries [5, 41–50]. In Argentina urban ethnobotanical studies on immigrants BK have also been conducted [8, 9, 51], as well as on nontraditional urban BK [39, 40, 52–60].

The segments of immigrants in metropolitan areas retain a BK relative to their areas of origin. Strictly speaking, it is not a TBK, but we consider that it is *linked to traditions* and recreated in a new cultural context. In urban areas there also coexist BK linked to other traditions, such as those for the internal migrants coming from other urban and rural areas, and those related to family traditions, such as those associated with culinary knowledge and home therapeutic practices [61–64].

From the previous statement it is concluded that there is the need to assume the complexity of BK in urban agglomerations, so as not to identify it only with the nontraditional BK. Thus, we propose to conceptualize the UBK as a complex, dynamic, and adaptative *corpus*, consisting of different types of knowledge and beliefs that guide selective actions on the use of plants, their parts, and products deriving from them. Thereby, UBK includes nontraditional components and others linked to traditions (family traditions, traditions of the origin areas in the case of immigrant groups), which coexist and interact in the same pluricultural context.

For the understanding of UBK it is important to evaluate not only its composition but also its dynamic that expressed in the movement of vegetal elements and their uses inside the pluricultural context. Many of these elements are within the general commercial circuit, and their characteristics, uses, and properties are disseminated through the mass media. For most urban inhabitants these elements are well known or *visible*. In contrast, other vegetal elements are restricted to the scope of the segments of immigrants, or family traditions, and are unknown or *invisible* to the rest of the population. For example, pears, apples, oranges, bananas, and many other fruits are common in every supermarket (*visible*), but *Syzygium samarangense* (Blume) Merr. & L.M. Perry (Myrtaceae), "Java apple," "pu tao yang" is a fruit that is sold only in certain shops in the Chinese community in Buenos Aires city, then, it is *invisible*. The distinction visible/invisible is not a definitive or invariable categorization: there are several cases between the extremes, as *Pyrus pyrifolia* (Burm. *f.*) Nakai (Rosaceae), "li shan," fruit sold in the Chinese shops and sometimes in stores of the general commercial circuit [57].

Mobility in the circulation of vegetal elements expresses the dynamic of the UBK: some of them become visible when entering the general commercial circuit. Thus, the knowledge associated

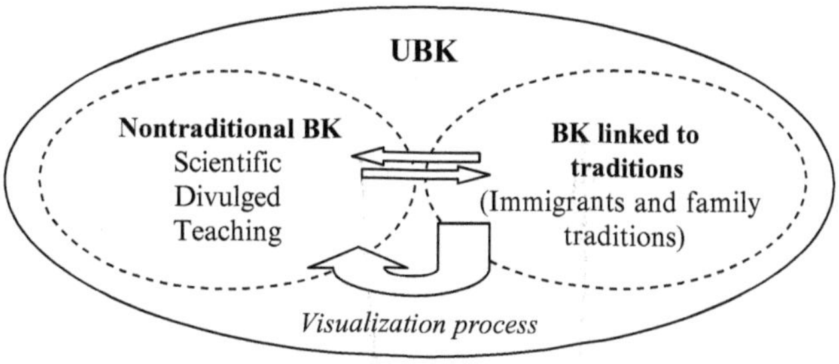

Fig. 1 UBK composition (nontraditional BK and BK linked to traditions, interacting) and UBK dynamic (visualization process)

with traditions, limited to immigrant groups, expands and becomes generalized. In this *visualization process* (Fig. 1), the mass media play a significant role to disseminate and promote the use of circulating plant elements. In this context, the example of *Pyrus pyrifolia* shows that knowledge of this species is in an expansion process (*visualization*). *Actinidia chinensis* Planch. var. *deliciosa* (A. Chev.) A. Chev (Actinidiaceae), "Kiwi," on the other hand, is already visible: it was almost unknown a few decades ago, but today their presence is common in greengrocers and supermarkets.

1.2 Study Area and Actors Involved

1.2.1 Buenos Aires-La Plata Conurbation

An *urban agglomeration* is an urbanized region extended continuously through different administrative districts, which includes a central city and various adjacent urbanizations that will be absorbed from their own growing rate. However, a *conurbation* does not emerge around a city, but that brings together different urban nuclei initially separated but contiguous. Conurbations do not require the continuity of the urbanized space: the central cities are connected by paths, roads, highways, railways that cross not urban areas and make it possible the permanent transfer of people and products.

The study area is the conurbation Buenos Aires-La Plata, including the Greater Buenos Aires, an urban agglomeration including Buenos Aires city (Federal District) and the Greater La Plata, the adjacent agglomeration comprising La Plata city (Fig. 2). Greater Buenos Aires is composed of the Federal District and 24 districts from the Province of Buenos Aires. Its total area is 3,833 km^2. Sometimes other six districts are considered as partially integrating the agglomeration [65, 66]. Buenos Aires city, with 202 km^2, has a population of 2,891,082 inhabitants, while in the 24 districts of Province of Buenos Aires overall amount is of 9,910,282 inhabitants [67]. In population, Greater Buenos Aires is the largest agglomeration in Argentina, the second in South America (after the metropolitan area of São Paulo, Brazil), the third in Latin America (after São Paulo and Mexico D.F.), the fifth of America, and the seventeenth worldwide [68]. The Greater La Plata is constituted by three districts of the Province of Buenos

Fig. 2 Satellite image corresponding to the Rio de la Plata region (NASA, April 2007), where noted the location of Greater Buenos Aires and Greater La Plata, which form a large conurbation

Aires: La Plata, Berisso, and Ensenada, its total surface is of 1,162 km^2, and its population of 793,365 inhabitants [67].

Buenos Aires-La Plata conurbation is heterogeneous, and comprises three types of spaces: (a) urbanized areas themselves, (b) nonurbanized areas with native vegetation (some of which correspond to natural protected areas), (c) *periurban areas* or transition zones between urban and nonurbanized areas (including rural zones). These latter are characterized by moving boundaries, which fluctuate according to the rhythms of urbanization [69].

In periurban sectors various horticultural undertakings are located, which are known all together as *green belt* or *horticultural belt*. It supplies fresh vegetables and fresh fruits to the urban sectors themselves and also to other provinces. The horticultural belt includes some 1,270 farms, in 8,160 ha, according to the 1998 Horticultural Census [70–72]. Local horticulture includes business ventures (which generally occupy large areas) and also different homegardens, that are low surface areas, near houses, where vegetables and fruit trees are cultivated, and that provide raw material for handmade vegetal products that are elaborated for family consumption and also for occasional sale on a reduced scale as a supplement to the domestic economy [64, 73]. Likewise, outside the horticultural belt, there are homegardens and vineyards in periurban areas, located next to native vegetation areas in the coasts of the Río de la Plata [37, 74–76].

1.2.2 Recent Immigration in the Conurbation

Argentina has received massive waves of immigration from the mid-nineteenth century and first half of the twentieth century. Most of these immigrants were of European origin: 44.9 % Italian and 31.5 % Spaniards on the total number of immigrants registered until 1940. These migration flows have helped to shape the country's cultural heritage, and many current "family traditions" have their roots in that early immigration. The waves of immigration from neighboring countries have been more or less constant in that period, but with greater presence in the borderline provinces.

In the second half of the twentieth century, a new kind of recent immigration occurred, not massive, especially from neighboring countries, whose destiny focused in Buenos Aires metropolitan area. These immigrants were oriented towards horticultural activity in periurban areas, as well as towards manufacturing industry, construction, and commerce in strictly urban areas. Most neighboring immigration comes from Paraguay and Bolivia (21.22 % and 15.24 %, respectively, of all foreigners in 2001).

The preference for the metropolitan area is expressed in some numbers: Bolivian immigrants living in Jujuy and Salta provinces represented 22 % of the total ones in the country in 2001, whereas those living in Buenos Aires city and the Province of Buenos Aires, together, accounted for 60 %. In Buenos Aires city, the total of immigrants from Bolivia, Paraguay, and Peru represented about 5 % of the total local population; for all the country, just under 2 % [77–80]. In the horticultural belt of Florencio Varela, Berazategui, and La Plata districts, 39.2 % of producers have of Bolivian origin in 2001: 75 % as renters and 25 % of owners who used only work force proceeding from Bolivia [64, 70, 81, 82].

Bolivian immigrants are an example of the pattern that connects the periurban areas (production) with the strictly urban areas (consumption). Immigrants dedicated to horticultural production provide food for the urban sector, where other Bolivian immigrants commercialize those products: e.g., at Bolivian traditional market of neighborhood of Liniers, placed in the west of the Buenos Aires city [9, 83, 84].

Another recent immigration in the conurbation corresponds to Far Eastern countries such as China, Korea, and Japan. Asian immigration in 2001 represented almost 2 % of all foreigners in the country, meager value compared to 67.96 % coming from American countries and 28.22 % from European countries [77]. However, immigrants concentrated in Buenos Aires city have a conspicuous presence and are engaged in the manufacture of clothing and commerce (in the categories of clothing and food). Chinese immigration in the first half of the twentieth century was low and settled in periurban areas, dedicated to horticultural practices. In the late twentieth century, Chinese immigrants exceeded in number of the Japanese and Koreans, who previously dominated [85–88].

At present, the Chinese community in Buenos Aires city has several restaurants, shops, and supermarkets concentrated in one area of the neighborhood of Belgrano, called *Chinatown*. Supermarkets offer products that satisfy the demands of the community itself, other communities (Japanese, Korean, Arabic), and local residents looking for new products. In many establishments the Chinese owners give employment to Bolivian immigrants. It is estimated that Chinatown receives some 15,000 visitors every weekend. For the Chinese community it embodies a space for local tourism: restaurants offering Chinese food oriented to the "Argentine taste." Restaurants that offer dishes preferred by Chinese immigrants, however, are not located in Chinatown. Nevertheless, its markets introduce a wide variety of novel plant products and are a center for disseminating those products to the local urban residents [89].

2 Methodological Inputs

2.1 General Considerations

The studies on the UBK *composition* were oriented to evaluate the components linked to traditions in two reference groups of immigrants: the Bolivians (at traditional market of Liniers neighborhood), and the Chinese (at supermarkets in Chinatown of Belgrano neighborhood). For their traditions of origin, they represent centers of conservation of knowledge and beliefs about plants and their uses. Thus, the UBK linked to its traditions can be extrapolated from the analysis of the vegetal products in circulation. To assess the UBK *dynamic*, the vegetal elements restricted to the groups of immigrants were analyzed that are in process of visualization. For this, the routes of entering the general commercial circuit were inquired.

For Bolivian immigrants segment, the traditional market of Liniers, like other traditional markets, is considered as a dynamic system whose components (actors and social networks, exchange and distribution of products, their origin and destination), beyond the geographical and anthropological approaches, need to be explored from an ethnobotanical perspective [90]. Latin American traditional markets are valuable sources of information for the study of the BK and also germplasm banks that help to preserve plant diversity, through the use of different species [91–98]. However, in Argentina there are not many precedents about researches in traditional markets placed in a strictly urban area as the one here studied. In that sense, this work is a contribution from Urban Ethnobotany. In contrast, Chinatown is not a traditional market, because of the generalist commercial feature of their premises, although they also sell products linked to their own traditions.

The ethnobotanical data collection followed the usual qualitative techniques and methods [99–102], such as participant observation, free listings, open-ended, and semi-structured interviews, focused according to specific criteria for traditional markets [90]. In all the cases samples of plants and their products were collected, which were deposited in the LEBA. The available literature relevant to the observed species and their uses was revised. The information from labels, leaflets, and advertisements of many products, both printed and available in electronic media, was evaluated because this information guides the general public in the selection of products to be consumed.

Salesmen have been considered as *qualified informants*, people of both sexes and different ages that demonstrate their knowledge about the properties of the vegetal products they sell. In Chinatown, the study is still under development in three major supermarkets, with 10 qualified informants interviewed. In traditional market of Liniers, 30 outlets (street stalls and premises) have been relieved, and 50 informants have been interviewed (from a total of 95 salesmen).

2.2 Contributions to the Study of the UBK Composition

Plants, parts thereof, and products of plant origin circulating within the restricted scope of the immigrant groups provide information about the UBK linked to their traditions of origin. The work focused on those vegetal elements that are marketed both for food and therapeutic purposes. This decision is based on the fact that these elements are the core of attention of both the Bolivian traditional market of Liniers and Chinatown of Belgrano, but also it has been assumed that the boundary line between these categories is not always clear and precise and many plants *serve to eat* while they *are used to heal* [62, 103, 104]. This argument is in line with the broad concept of health as a state of complete physical, mental, and social welfare, not just the absence of disease [105]. Among immigrants, the integrative notion of *plants to eat and heal* is related to their own traditions. For the rest of the inhabitants of the conurbation, that notion is mediated by the communication systems that help to spread the idea of *healthy foods*.

The analysis of the plant elements useful to eat and heal at Liniers Bolivian market shows that from a total of 160 recorded plant species that are sold as food, 54 are valued as food plants and at the same time as medicinal plants. Of these, vegetal products for 20 species are marketed that are found only in this market, that is, they are *exclusive* elements (Table 2). It is noteworthy that exclusivity refers to products and not to species. For example, for *Phaseolus vulgaris*, cultivars called "panamito" and "canario" are exclusive to Liniers market, but other cultivars of beans are sold as well ("alubia," "colorado," "negro," "regina," among others),

Table 2
Parts of plants and plant products belonging to 20 species from ten botanical families, exclusively commercialized in the Bolivian market of Liniers, Buenos Aires city

Families	Scientific names	Locales names	Part/products
Asteraceae	*Porophyllum ruderale* (Jacq.) Cass.	Quirquina	Fresh aerial parts
	Smallanthus sonchifolius (Poepp. & Endl.) H.Rob.	Yacón	Fresh roots and jam
	Tagetes minuta L.	Huacatay	Fresh aerial parts
Basellaceae	*Ullucus tuberosus* Caldas	Ulluco/papa lisa	Fresh tubers
Brassicaceae	*Lepidium meyenii* Walp.	Maca	Roots in powder or as flour
Cucurbitaceae	*Cucurbita ficifolia* Bouché	Cayote/alcayota	Fresh fruits
	Cyclanthera pedata (L.) Schrader	Caiwa/achojcha	Fresh fruits
	Sechium edule (Jacq.) Sw	Papa del aire/chayote	Fresh fruits
Euphorbiaceae	*Plukenetia volubilis* L.	Sacha inchi	Seeds in snacks, liquid, ointment, and powder
Leguminosae	*Arachis hypogaea* L. cvs.	Maní boliviano	Dry seeds
	Lablab purpureus (L.) Sweet	Poroto japonés	Fresh fruits
	Lupinus mutabilis Sweet	Tauri/tarwi	Dry seeds
	Pachyrhizus ahipa (Wedd.) Parodi	Ajipa	Fresh roots
	Phaseolus vulgaris L. cvs.	Porotos *canario* y *panamito*	Dry seeds
Oxalidaceae	*Oxalis tuberosa* Molina	Oca	Fresh tubers
Poaceae	*Zea mays* L. cvs.	Maíz *morado*	Whole dry spikes or powder
		Maíz *pelado* o *mote*	Dry or cooked grains
		Maíces *blanco, colorado,*	Dry or cooked grains
		chuspillo y *huillcaparu*	Dry grains
Rubiaceae	*Coffea arabica* L.	Sultana	Seeds (seed coat)
Solanaceae	*Capsicum annuum* L. cvs.	Ajíes *picante, escabeche, amarillo* y *campanita*	Fresh and dry fruits
	Capsicum pubescens Ruiz & Pav.	Locoto/rocoto	Fresh fruits and powder
	Solanum tuberosum L. cvs.	*Chuño* y papines	Dried and fresh tubers

which are also usually expended in the general commercial circuit shops. These plant elements, as opposed to exclusive, have been named *generalized* [8].

The same situation occurs with cultivars of *Capsicum annuum* and *Zea mays* cited as exclusive, while other products (peppers, corn) are generalized. The cultivar of *Arachis hypogaea* called "maní boliviano" ("Bolivian peanut") is exclusive of the Liniers Bolivian market, while others peanuts are widespread. The fresh fruits of *Cucurbita ficifolia* are also exclusive, but alcayota jam is sold in several supermarkets (generalized). These cases express that one species can be represented by exclusive and generalized elements in the same area of study, even in the same market [8, 9].

The distinction *exclusive/generalized* is a methodological tool, because that categorization of plant elements enables to express the underlying BK: generalized elements are *visible* to the population and linked to nontraditional component of the UBK, the exclusive ones are *invisible* and refers to the component linked to traditions. Regarding Liniers Bolivian market, the species listed in Table 2 correspond to Andean crops, excluding the cases of *Sechium edule*, of Mesoamerican origin, and *Lablab purpureus* and *Coffea arabica*, both from Africa; however, they are plants and products commonly used in Andean countries. Bolivian immigrants enter their Andean elements that satisfy the demands of their own community and also open up possibilities for local inhabitants to increase urban plant diversity with plants with therapeutic and alimentary value, such as *Lupinus mutabilis*, relevant source of protein, and some microthermic tubers with antioxidant properties [106–108]. In addition, biocultural diversity is incremented with the incorporation of knowledge and beliefs linked to traditions in the local UBK domain.

2.3 Contributions to the Study of the UBK Dynamics

The study of the UBK dynamics involves a look on the movement of vegetal elements, especially the passage of such elements from a restricted context (such as immigrants segments) into the broader general commercial circuit. From the distinction exclusive/generalized or, if it is preferred, *invisible/visible*, it is possible to identify many different plant elements located between the extreme cases, more or less common in immigrant outlets and also in the general commercial circuit, therefore, with varying degrees of *visualization*. In evaluating the visualization process it has been particularly useful to consider the vegetal elements that serve to eat and heal. In the metropolitan area of Buenos Aires, as elsewhere in the world, the general commercial circuit shops called *dietéticas* or *health-food stores* [109] are the focus that concentrates the interest on plants that contribute to health, and are privileged sites chosen by the people for the supply of healthy foods, dietary supplements, and materials for infusions—in general for therapeutic purposes. Urban residents value these shops as a place where they can obtain safe products, and in the discourse of sellers and consumers there are no defined limits in its dual character of food sale and therapeutic elements dispensing [9, 39].

Decades ago, the sale of plant material for medicinal purposes was the patrimony of *herbal shops* or *herboristerías*, but at present, these places are almost gone, and herbal products are sold routinely in the health-food stores or dietéticas [58, 109].

These shops are furthermore true dispersal centers for plant products, enhanced by the mass media. It has been confirmed in the health-food stores the presence of different products coming from the restricted contexts of immigrant groups, both the Bolivian and Chinese ones. Once in the health-food stores, those invisible products gain the general commercial circuit and become visible. In this sense, those stores are *visualization agents*. As a contribution to the study of the UBK dynamics, the exploration of these shops is an adequate methodological tool.

Table 2 includes two species whose products are exclusive in Liniers Bolivian market, *Lepidium meyenii*, "maca" and *Plukenetia volubilis*, "sacha inchi." Both of them originate from Peru, where their properties are known from pre-hispanic times [54] and enter directly to Buenos Aires from Bolivia. Recently, health-food stores expend other products of these species (usually tablets); therefore, they are immersed in the visualization process. This is amplified by the diffusion of its therapeutic properties, especially on the Internet. A similar case, even more widespread, is *Morinda citrifolia* L. (Rubiaceae), "noni," a Polynesian plant, which is sold in health-food stores, in powder or liquid; sometimes it is sold in the Liniers Bolivian market as well, where it enters from Peru through Bolivia. In a short time this species has become popular worldwide due to the wide spectrum of therapeutic effects that are attributed to it [54, 110]. Another example, but in this case related to the segment of Chinese immigrants, is *Lycium barbarum* L. (Solanaceae), "goji," whose dry fruits are sold for its therapeutic effects at Chinatown, in Belgrano neighborhood [57]; it is now sold in certain health-food stores, so its visualization is recent in Argentina (in other countries has already achieved wide diffusion).

3 Final Remarks

Relationships between people and plants are intrinsically complex and Ethnobotany must reflect this complexity. In the same sense, Urban Ethnobotany must reflect the complexity of the UBK, in order to understand how this knowledge is embodied in specific human actions such as strategies of selection and consumption of plant origin products.

As a result of researches on UBK complexity in vogue in the pluricultural context of a conurbation, there are two methodological tools considered relevant for the comprehension of the UBK composition and dynamic. Its relevance lies in assuming the complexity of the phenomena and that our explanation result consequent.

Thereby, this contribution gives a theoretical reflection and some methodological inputs to Urban Ethnobotany, a recently installed field, especially in Argentina, that becomes a fertile ground to develop new perspectives.

Acknowledgments

The authors acknowledge Dr. Ulysses P. de Albuquerque, Prof. Emilio Ulibarri, and Dr. Patricia M. Arenas, the different contributors to the field work carried out, the Facultad de Ciencias Naturales y Museo de la Universidad Nacional de La Plata, and the Consejo Nacional de Investigaciones Científicas y Técnicas (CONICET) of Argentina for their diverse supports, both direct and indirect, to the implementation of this work.

References

1. Hurrell JA (1987) Las posibilidades de la etnobotánica y un nuevo enfoque a partir de la ecología y su propuesta cibernética. Rev Esp Antropol Am 17:235–258
2. Hurrell JA (1990) Interpretación de relaciones en ecología, a partir de la noción de sistema en Santa Victoria e Iruya (Salta, Argentina). Tesis Doctoral 548, Fac Ci Nat y Museo, Universidad Nacional de La Plata
3. Albuquerque UP, Hurrell JA (2010) Ethnobotany: one concept and many interpretations. In: Albuquerque UP, Hanazaki N (eds) Recent developments and case studies in ethnobotany. SBEE-NUPEEA, Recife, pp 87–99
4. Minter S (1999) The resurgence of ethnobotany in Europe. Roots 1(19):36–38
5. Balick MJ, Kronenberg F, Ososki A, Reiff M, Fugh-Berman A, O'Connor B (2000) Medicinal plants used by Latino healers for women's health conditions in New York. Econ Bot 54:344–457
6. Balick MJ, Lee R (2001) Looking within: urban ethnomedicine and ethnobotany. Altern Ther Health Med 7(4):114–115
7. Etkin NL (2004) The future of ethnobotany: moving fast, going where? Delpinoa 45:245–251
8. Hurrell JA, Ulibarri EA, Puentes JP, Buet Costantino F, Arenas PM, Pochettino ML (2011) Leguminosas medicinales y alimenticias utilizadas en la conurbación Buenos Aires-La Plata, Argentina. Bol Latinoam Caribe Plant Med Aromat 10(5):443–455
9. Pochettino ML, Puentes JP, Buet Costantino F, Arenas PM, Ulibarri EA, Hurrell JA (2012) Functional foods and nutraceuticals in a market of Bolivian inmigrants in Buenos Aires (Argentina). Evid Based Complement Alternat Med 2012:1–4. doi:10.1155/2012/320193
10. Roca Alcázar F (2008) Las palmeras en el conocimiento tradicional del grupo indígena amazónico Aguaruna–Huambisa. Rev Peru Biol 15:147–149
11. Leyew Z (2011) Wild plant nomenclature and traditional botanical knowledge among three ethnolinguistic groups in northwestern Ethiopia. African Books, Addis Ababa
12. Yusuf M, Wahab MA, Chowdhury JU, Begum J (2006) Ethno-medico-botanical knowledge from Kaukhali proper and Betbunia of Rangamati district. Bangladesh J Plant Taxon 13(1):55–61
13. McClatchey WC, Mahady GB, Bennett BC, Shiels L, Savo V (2009) Ethnobotany as a pharmacological research tool and recent developments in CNS-active natural products from ethnobotanical sources. Pharmacol Ther 123(2):239–254
14. Albuquerque UP, Andrade LH (2002) Conhecimento botánico tradicional e conservação em uma área de caatinga no estado de Pernambuco, Nordeste do Brasil. Acta Bot Brasilica 16:273–285
15. Zarger RK, Stepp JR (2004) Persistence of botanical knowledge among Tzeltal Maya children. Curr Anthropol 45(3):413–418
16. Hilgert NI, Gil GE (2005) Traditional Andean agriculture and changing processes in the Zenta river basin, Salta, northwestern Argentina. Darwiniana 43:30–43
17. Hilgert NI, Gil GE (2007) Reproductive medicine in northwest Argentina: traditional and institutional systems. J Ethnobiol Ethnomed 3:19. doi:10.1186/1746-4269-3-19
18. Pesek TJ, Helton L, Reminick R, Kannan D, Nair M (2008) Healing traditions of southern India and conservation of culture and biodiversity: a preliminary study. Ethnobot Res Appl 6:471–479
19. Castellano MB (2000) Updating aboriginal traditions of knowledge. In: Sefa Dei G, Hall BL, Rosenberg BL (eds) Indigenous

knowledges in global contexts. Multiple readings of our world. University of Toronto Press, Toronto, pp 21–36
20. Maffi L (2001) On the interdependence of biological and cultural diversity. In: Maffi L (ed) On biocultural diversity. Linking language, knowledge and the environment. Smithsonian Institution Press, Washington, pp 1–50
21. De Britto J, Mahesh R (2007) Exploration of Kani tribal botanical knowledge in Agasthyamalai biosphere reserve, south India. Ethnobot Leafl 11:258–265
22. Ramírez CR (2007) Etnobotánica y la pérdida de conocimiento tradicional en el siglo 21. Ethnobot Res Appl 5:241–244
23. McClatchey WC (2005) Exorcising misleading terms in ethnobotany. Ethnobot Res Appl 3:1–4
24. Arenas PM (2006) Etnobotánica en áreas urbanas. Bol Latinoam Caribe Plant Med Aromat 5(5):95
25. Wagner G (2008) Botanical knowledge of a group of college students in South Carolina, U.S.A. Ethnobot Res Appl 6:443–458
26. Azaizeh H, Fulder S, Khalil K, Said O (2003) Ethnobotanical knowledge of local Arab practitioners in the Middle Eastern region. Fitoter 74(1–2):98–108
27. Reyes-García V, Marti N, McDade TW, Tanner S, Vadez V (2007) Concepts and methods in studies measuring individual ethnobotanical knowledge. J Ethnobiol 27(2):108–203
28. Shivanna MB, Rajakumar N (2010) Ethno-medico-botanical knowledge of rural folk in Bhadravathi taluk of Shimoga district, Karnataka. Indian J Tradit Knowl 9(1): 158–162
29. Chaves Alves AG, Albuqerque UP (2010) "Ethno what?" terminological problems in ethnocience with a special emphasis on the Brazilian context. In: Albuquerque UP, Hanazaki N (eds) Recent developments and case studies in ethnobotany. SBEE-NUPEEA, Recife, pp 67–79
30. Mazzocchi F (2006) Western science and traditional knowledge. Despite their variations, different forms of knowledge can learn from each other. EMBO Rep 7(5):463–466
31. Hurrell JA, Albuquerque UP (2012) Is ethnobotany an ecological science? Steps towards a complex ethnobotany. Ethnobiol Conserv 1(4):1–16
32. Berkes F (1993) Traditional ecological knowledge in perspective. In: Inglis JT (ed) Traditional ecological knowledge: concepts and cases. IDRC, Ottawa, pp 1–9
33. Gadgil M, Berkes F, Folke C (1993) Indigenous knowledge for biodiversity conservation. AMBIO 22(2–3):151–156
34. Petch V (2000) Traditional ecological knowledge: an anthropological perspective. In: Oakes J, Riewe R, Koolage W, Simpsony L, Schuster N (eds) Aboriginal health, identity and resources. Native Studies Press, Winnipeg, pp 150–164
35. Luna Morales C (2002) Ciencia, conocimiento tradicional y etnobotánica. Etnobiología 2:120–135
36. Reyes-García V, Vadez V, Tanner S, McDade T, Huanca T, Leona W (2006) Evaluating indices of traditional ecological knowledge. J Ethnobiol Ethnomed 2:21. doi:10.1186/1746-4269-2-21
37. Pochettino ML, Hurrell JA, Lema VS (2012) Local botanical knowledge and agrobiodiversity: homegardens at rural and periurban contexts in Argentina. In: Luna Maldonado AI (ed) Horticulture. InTech, Rijeka, pp 105–132
38. Balick MJ, Cox P (1996) Plants, people and culture. The science of ethnobotany. W. H. Freeman, New York
39. Pochettino ML, Arenas PM, Sánchez D, Correa R (2008) Conocimiento botánico tradicional, circulación comercial y consumo de plantas medicinales en un área urbana de Argentina. Bol Latinoam Caribe Plant Med Aromat 7(3):141–148
40. Hurrell JA, Ulibarri EA, Delucchi G, Pérez ML, Anglese N (2010) Plantas condimenticias utilizadas en la ciudad de Buenos Aires y en el noreste de la provincia de Buenos Aires (Argentina). In: Pochettino ML, Ladio AH, Arenas PM (eds) Tradiciones y transformaciones en etnobotánica. CYTED-RISAPRET, San Salvador de Jujuy, pp 344–349
41. Reiff M, O'Connor B, Kronenberg F, Balick MJ, Lohr P, Fugh-Berman A, Johnson KD (2003) Ethnomedicine in urban environment: Dominican healers in New York city. Hum Organ 62(1):12–26
42. Fuck SB, Athanazio JC, Lima CB, Ming LC (2005) Plantas medicinais utilizadas na medicina popular por moradores da área urbana de Bandeirantes, Paraná, Brasil. Semin Cienc Agr 26:291–296
43. Pieroni A, Muenz H, Akbulut M, Başer K, Durmuşkahya C (2005) Traditional phytotherapy and trans-cultural pharmacy among Turkish migrants living in Cologne, Germany. J Ethnopharmacol 102(1):69–88
44. Sandhu DS, Heinrich M (2005) Use of health foods, spices and other botanicals in the Sikh

community in London. Phytother Res 19(7): 633–642
45. Ceuterick M, Vandebroek I, Torrya B, Pieroni A (2008) Cross-cultural adaptation in urban ethnobotany. The Colombian folk pharmacopoeia in London. J Ethnopharmacol 120(3): 342–359
46. Pieroni A, Gray C (2008) Herbal and food folk medicines of the Russlanddeutschen living in Künzelsau/Taläcker, south-western Germany. Phytother Res 12(7):889–901
47. Pieroni A, Vandebroek I (eds) (2009) Traveling cultures and plants. The ethnobiology and ethnopharmacy of migrations. Berghahn Books, Oxford
48. Volpato G, Godínez D, Beyra A (2009) Migration and ethnobotanical practices. The case of Tifey among Haitian immigrants in Cuba. Hum Ecol 37(1):43–53
49. Yöney A, Prieto J, Lardos A, Heinrich M (2009) Ethnopharmacy of Turkish-speaking Cypriots in greater London. Phytother Res 24(5):731–740
50. Petkeviciute Z, Savickiene N, Savickas A, Bernatoniene J, Simaitiene Z, Kalveniene Z (2010) Urban ethnobotany in Samogitia region, Lithuania. J Med Plant Res 4(1): 64–71
51. Richeri M, Beeskow AM, Ladio AH (2010) Las plantas y la salud en la comunidad boliviana de Puerto Madryn (Argentina). In: Pochettino ML, Ladio AH, Arenas PM (eds) Tradiciones y transformaciones en etnobotánica. CYTED-RISAPRET, San Salvador de Jujuy, pp 297–302
52. Pochettino ML, Martínez MR, Itten B, Zucaro M (1997) El uso de plantas medicinales en la atención primaria de la salud: estudio etnobotánico en una población urbana (La Plata, Buenos Aires, Argentina). Parodiana 10:141–152
53. Arenas PM (2007) Suplementos dietéticos: estudio etnobotánico en zonas urbanas. Kurtziana 33(1):193–202
54. Arenas PM, Cristina I, Puentes JP, Buet Costantino F, Hurrell JA, Pochettino ML (2011) Adaptógenos: plantas medicinales tradicionales comercializadas como suplementos dietéticos en la conurbación Buenos Aires-La Plata (Argentina). Bonplandia (Corrientes) 20(2):251–264
55. Hurrell JA, Ulibarri EA, Delucchi G, Pochettino ML (2008) Biota rioplatense XIII. Plantas aromáticas condimenticias. Ed. Lola, Buenos Aires
56. Hurrell JA, Ulibarri EA, Delucchi G, Pochettino ML (2009) Biota rioplatense XIV. Hortalizas. Verduras y legumbres. Ed. Lola, Buenos Aires
57. Hurrell JA, Ulibarri EA, Delucchi G, Pochettino ML (2010) Biota rioplatense XV. Frutas frescas, secas y preservadas. Ed. Lola, Buenos Aires
58. Hurrell JA, Ulibarri EA, Arenas PM, Pochettino ML (2011) Plantas de herboristería. Plantas medicinales comercializadas en herboristerías de la Ciudad de Buenos Aires. Ed. Lola, Buenos Aires
59. Hilgert NI, Higuera ML, Kristensen MJ (2010) La medicina herbolaria en el contexto urbano. Estudio de caso en un barrio de la ciudad de Tandil, Argentina. Bol Latinoam Caribe Plant Med Aromat 9(3):177–190
60. Ladio AH, Molares S (2010) Aspectos do estudo da dinâmica do uso de produtos etnobiológicos não tradicionais. In: Albuquerque UP, Lucena RF, Cunha LV (eds) Métodos e técnicas na pesquisa etnobiológica e etnoecológica. NUPEEA, Recife, pp 267–278
61. Volpato G, Godínez D (2004) Ethnobotany of pru, a traditional Cuban refreshment. Econ Bot 58:381–395
62. Pieroni A, Price L (2006) Eating and healing: traditional food as medicine. Hawthorn, New York
63. Eichemberg MT, Amorozo MC, de Moura LC (2009) Species composition and plant use in old urban homegardens in Rio Claro, southeast of Brazil. Acta Bot Brasilica 23(4):1057–1075
64. Pochettino ML (2010) Huertos periurbanos como aporte a la diversidad agrícola, provincia de Buenos Aires, Argentina. In: Pochettino ML, Ladio AH, Arenas PM (eds) Tradiciones y transformaciones en etnobotánica. CYTED-RISAPRET, San Salvador de Jujuy, pp 186–192
65. Atlas Ambiental Buenos Aires (2010) http://www.atlasdebuenosaires.gov.ar. Accessed 21 Oct 2012
66. Gemini M (2003) ¿Qué es el gran Buenos Aires? INDEC, Buenos Aires
67. Instituto Nacional de Estadística y Censos (2012) INDEC. Censo 2010. http://www.censo2010.indec.gov.ar. Accessed 21 Oct 2012
68. Forstall RL, Greene R, Pick JB (2004) Which are the largest? City Futures Conference, Chicago
69. Barsky A (2005) El periurbano productivo: un espacio en constante transformación. Introducción al estado del debate y referencias al caso de Buenos Aires. Scripta Nova (Barcelona) 9: 194(36). http://www.ub.

edu/geocrit/sn/sn-194-36.htm. Accessed 21 Oct 2012
70. Barsky A (2010) La agricultura de "cercanías" a la ciudad y los ciclos del territorio periurbano. Reflexiones sobre el caso de la región metropolitana de Buenos Aires. In: Svetlitza A (ed) Agricultura periurbana en Argentina y globalización. Escenarios, recorridos y problemas. FLACSO, Buenos Aires, pp 15–29
71. García M (2010) Inicios, consolidación y diferenciación de la horticultura platense. In: Svetlitza A (ed) Agricultura periurbana en Argentina y globalización. Escenarios, recorridos y problemas. FLACSO, Buenos Aires, pp 69–85
72. Feito M (2007) Modalidades de intervención social sobre los horticultores bonaerenses: una mirada antropológica. Avá (Posadas) 10:78–96
73. Nazarea V (1998) Cultural memory and biodiversity. University of Arizona Press, Tucson
74. Marasas M, Velarde I (2000) Rescate del saber tradicional como estrategia de desarrollo: los viñateros de la costa. Bol ILEIA 16(2):23–24
75. Hurrell JA, Buet Costantino F, Puentes JP, Ulibarri EA, Pochettino ML (2011) Huertos familiares periurbanos de las costas de Ensenada-Berisso y de la isla Martín García (Buenos Aires, Argentina). Bonplandia (Corrientes) 20(2):213–229
76. Buet Costantino F, Ulibarri EA, Hurrell JA (2010) Las huertas familiares en la isla Paulino (Buenos Aires, Argentina). In: Pochettino ML, Ladio AH, Arenas PM (eds) Tradiciones y transformaciones en etnobotánica. CYTED-RISAPRET, San Salvador de Jujuy, pp 479–484
77. Instituto Nacional de Estadística y Censos (2012) INDEC. Censo 2001. http://www.indec.mecon.ar/webcenso/index.asp. Accessed 21 Oct 2012
78. Benencia R (2008) Migrantes bolivianos en la periferia de ciudades argentinas: procesos y mecanismos tendientes a la conformación de territorios productivos y mercados de trabajo. In: Novick S (ed) Las migraciones en América Latina. CLACSO, Buenos Aires, pp 13–30
79. Cerrutti M (2009) Diagnóstico de las poblaciones de inmigrantes en la Argentina. Dirección Nacional de Población, Buenos Aires
80. Sassone S (2009) Breve geografía histórica de la migración boliviana en la Argentina. In: Maronese L (ed) Temas de patrimonio cultural 24: Buenos Aires Boliviana. Comisión para la Preservación del Patrimonio Histórico Cultural, Buenos Aires, pp 389–402
81. Benencia R (1997) De peones a patrones quinteros: movilidad social de familias bolivianas en la periferia bonaerense. Estud Migr Latinoam 35:63–102
82. Benencia R, Karasik G (1994) Bolivianos en Buenos Aires: su integración laboral y cultural. Estud Migr Latinoam 27:261–299
83. Gavazzo N (2009) Identidad boliviana en Buenos Aires: políticas de integración cultural. In: Maronese L (ed) Temas de patrimonio cultural 24: Buenos Aires Boliviana. Comisión para la Preservación del Patrimonio Histórico Cultural, Buenos Aires, pp 403–430
84. Pizarro C (2009) Espacios socioculturales "bolivianos" trans-urbanos en el Área Metropolitana de Buenos Aires. In: Maronese L (ed) Temas de patrimonio cultural 24: Buenos Aires Boliviana. Comisión para la Preservación del Patrimonio Histórico Cultural, Buenos Aires, pp 37–52
85. Gómez S, Onaha C (2008) Asociaciones voluntarias e identidad étnica de inmigrantes japoneses y sus descendientes en Argentina. Migraciones (Madrid) 23:207–235
86. Higa J, Laumonier I (1990) La otra inmigración. Asoc Universitaria Nikkei, Buenos Aires
87. Mera C (1998) La inmigración coreana en Buenos Aires. EUDEBA, Buenos Aires
88. Sassone S, Mera C (2007) Barrios de migrantes en Buenos Aires. Identidad, cultura, cohesión socioterritorial. Brussels: V Congr Europ CEISAL Latinoamer. http://www.reseau-amerique-latine.fr/ceisal-bruxelles/MS-MIG/MS-MIG-1-Sassone_Mera.pdf. Accessed 21 Oct 2012
89. Bogado Bordazar L (2003) Migraciones internacionales: influencia de la migración china en Argentina y Uruguay. Tesis Doctoral 12, Fac. Ci. Jurid Soc., Universidad Nacional de La Plata
90. Cunningham AB (2001) Etnobotánica aplicada. Pueblos, uso de plantas silvestres y conservación. Nordan-Comunidad, Montevideo
91. Macía MJ, García E, Vidaurre PJ (2005) An ethnobotanical survey of medicinal plants commercialized in the markets of La Paz and El Alto, Bolivia. J Ethnopharmacol 97: 337–350
92. Azevedo SK, Silva I (2006) Plantas medicinais e de uso religioso comercializadas em mercados e feiras livres no Rio de Janeiro, RJ, Brasil. Acta Bot Brasilica 20(1):185–194
93. Albuquerque UP, Monteiro J, Ramos M, Amorim E (2007) Medicinal and magic plants from a public market in northeastern Brazil. J Ethnopharmacol 110:76–91

94. Maioli V, Fonseca V (2007) Plantas medicinais e ritualísticas vendidas em feiras livres no municipio do Rio de Janeiro, Brasil: estudo de caso nas zonas norte e sul. Acta Bot Brasilica 21:263–275
95. Godinez D, Volpato G (2008) Plantas medicinales que se venden en el mercado El Río, Camagüey, Cuba. Rev Mex Biodivers 79(1):243–259
96. Giraldo D, Baquero E, Bermúdez A, Oliveira-Miranda M (2009) Caracterización del comercio de plantas medicinales en los mercados populares de Caracas, Venezuela. Acta Bot Venez 32:267–301
97. Bussmann R, Sharon D (2010) From collection to market and cure. Traditional medicinal use in northern Peru. In: Albuquerque UP, Hanazaki N (eds) Recent developments and case studies in ethnobotany. SBEE-NUPEEA, Recife, pp 189–212
98. Katz E (2010) Cruzeiro do sul market (Acre, Brazilian Amazon): reflection of the regional culture and agricultural diversity. In: Pochettino ML, Ladio AM, Arenas PM (eds) Tradiciones y transformaciones en etnobotánica. CYTED-RISAPRET, San Salvador de Jujuy, pp 531–539
99. Martin GJ (1995) Ethnobotany. A methods manual. Chapman & Hall, London
100. Alexíades MN, Sheldon JW (1996) Selected guidelines for ethnobotanical research: a field manual. The New York Botanical Garden, New York
101. Albuquerque UP, Lucena RF (2004) Métodos e técnicas na pesquisa etnobotánica. Livro Rápido/NUPEEA, Recife
102. Etkin NL, Ticktin T (2010) Advancing an ethno-ecological perspective that integrates theory and method in ethnobotany. In: Albuquerque UP, Hanazaki N (eds) Recent developments and case studies in ethnobotany. SBEE-NUPEEA, Recife, pp 33–57
103. Etkin NL, Ross PJ (1982) Food as medicine and medicine as food: an adaptive framework for the interpretation of plant utilization among the Hausa of northern Nigeria. Soc Sci Med 16:1559–1573
104. Chen N (2009) Food, medicine, and the quest for good health: nutrition, medicine and culture. Columbia University Press, New York
105. World Health Organization (2012) WHO-UN. http://www.who.int/en. Accessed 21 Oct 2012
106. King SR, Gershoff S (1987) Nutritional evaluation of three underexploited Andean tubers: *Oxalis tuberosa* (Oxalidaceae), *Ullucus tuberosus* (Basellaceae), and *Tropaeolum tuberosum* (Tropaeolaceae). Econ Bot 41:503–511
107. Gladstones J, Atkins C, Hamblin J (eds) (1998) Lupins as crop plants. CABI International, London
108. Tapia M (ed) (2000) Cultivos andinos subexplotados. FAO, Santiago de Chile
109. Pochettino ML (2003) Comer o curarse: qué son las dietéticas de centros urbanos de la argentina. In: II International symposium of ethnobotany disciplines proceedings, La Paz (CD-ROM)
110. Dixon AR, McMillen H, Etkin NL (1999) Ferment this: the transformation of noni, a traditional Polynesian medicine (*Morinda citrifolia*, Rubiaceae). Econ Bot 53:51–68

Chapter 19

The Dynamics of Use of Nontraditional Ethnobiological Products: Some Aspects of Study

Ana H. Ladio and Soledad Molares

Abstract

Up to the present time, the use of ethnobiological resources that do not belong to a certain culture or environment has been little studied. This situation is found especially in urban and semi-urban areas, where dwellers are greatly influenced by globalization processes. Moreover, due to the phenomenon of immigration, cities are often transformed into multicultural societies where interaction occurs between different traditions relating to the use of plants and animals. As well as this, the ecological process of species invasion of new territories, or "species contamination," brings new, useful resources into contact with cultures that are not familiar with them. In this chapter, theoretical and methodological aspects of the study of the use of nontraditional resources will be analyzed. The complexity involved requires a multidisciplinary approach that integrates cultural, historical, and social factors, as well as particular components related to biology and psychology.

Key words Nontraditional resources, Traditions, Cities, Change, Invasions, Immigration

1 Introduction

Nontraditional ethnobiological resources can be defined as useful resources of plant origin that do not belong to a certain culture or environment at the time of study, although in the future they may become part of the cultural background of the community [1]. These are distinguished from traditional resources in that their relation with members of a culture is minimal and/or of short duration. The knowledge used in their recognition, naming and use has not been generated locally; it has not been passed on by cultural transmission from generation to generation, nor from shared experience; in its construction, it is possible that no age-old adaptive processes have been involved. These resources make their way to the heart of local cultures through introduction, intentional or otherwise, from other cultures or cultural tendencies, or are promoted by recent ecological-environmental changes in the landscape, whether of anthropomorphic origin or otherwise.

Ulysses Paulino Albuquerque et al. (eds.), *Methods and Techniques in Ethnobiology and Ethnoecology*,
Springer Protocols Handbooks, DOI 10.1007/978-1-4614-8636-7_19, © Springer Science+Business Media New York 2014

Human beings, distinguished by their capacity to learn and innovate, have, throughout history, incorporated new ethnobiological elements into their daily lives as they widen their distribution ranges and come into contact with new cultures and environments [2]. For this reason these nontraditional resources cannot be included in a static definition, but rather form part of a process, since some of them may, over a period of time, be incorporated into other cultural contexts. Gradually, with continuous use, they may receive names in local dialects, be managed in a traditional way or be included in spiritual or symbolic practices.

Many examples can be taken from the European colonization in America to demonstrate the acquisition of new resources by different cultures. Such is the case of the introduction (or rather, imposition) of the use of wheat, the sheep, and the cow, elements that were rapidly assimilated throughout the entire American continent, but whose presence brought about drastic changes in local habits and landscapes [3]. This process has been interpreted by Rapoport [4] as real "gastronomic imperialism," which led to the substitution of American foods such as the viscacha (*Lagostormus maximus*), huauzontle (*Chenopodium nuttalliae*), la alegría (*Amaranthus leucocarpus*), quinoa (*Chenopodium quinoa*), and Chilean tarweed (*Madia sativa*) for foods of Eurasian origin. Many of these cultivars were even sought out by colonizers, who catalogued them as Devil's food, and their cultivation was prohibited. In fact, America offered the world over 88 species of edible plants [5], including maize and potatoes, with significant social and economic consequences that led to the advent of the European Industrial Era, following almost two centuries during which resources such as the potato had not been accepted by Europeans because they were thought to be poisonous [4].

Up to the present time, the use made by human populations of ethnobiological resources which are not part of their culture and/or environment has been little studied (e.g., [6, 7]). Nevertheless, to ethnobiologists these topics can be of great interest, not only in scientific terms, since they enable us to understand the dynamics of novel processes but also in practical terms, since these cases of new plants or animals that take on importance generally have direct implications for the receiving society from a sanitary, bromatological, and/or environmental perspective.

Questions on the use of resources that lack real historical and cultural continuity of use in a community could begin with the aspects detailed in Table 1. Assuming that the approach can vary in depth, a gradient for complexity of study can also be established, as can be seen in the same table.

Table 1
Guidelines and study topics relating to nontraditional ethnobiological products

Complexity	Guidelines and topics
	• The identification of a recently arrived ethnobiological resource and the possible ways and dates it was introduced into the new socio-environmental context
	• History of usage in place of origin. Sociocultural and environmental aspects of place of origin
	• Evaluation of local methods of usage, forms of cultural transmission, and new cultural interpretations
	• If dealing with a new wild resource, determine local abundance, level of invasibility, and ecological impact
	• Follow up over time of changes in cultural use consensus and study of social, economic, cultural, and ecological factors involved

2 Cultural Contexts of Interest in the Study of Nontraditional Ethnobiological Resources

It is widely known that the majority of ethnobiological studies have focused on rural and/or indigenous communities [2, 8, 9]. Little attention has been paid to urban contexts, despite the fact that many ethnobiological resources play a significant role in the lives of their inhabitants [1, 10–12].

However, it is in urban regions that the incorporation of the use of new items of plant or animal origin coming from totally different cultural contexts arises. It has been widely suggested that people in urban centers are greatly influenced by processes of globalization and market economy, which has a direct influence on the use of ethnobiological products [13, 14].

For example, this is the case of the phenomenon of the resurgence and diffusion of alternative and natural medicines and food items that use both animal and plant resources, [e.g., algae (*Fucus vesiculosus*, Spirulina, etc.), Andean cereals (*Chenopodium quinoa*, quinoa; *Salvia hispanica*, chia, etc.), infusions and oriental spices (*Centella asiática*, centella; *Ginkgo biloba*, ginko, *Panax ginseng*, ginseng etc.)] in combination with philosophies related to greater holistic care [1, 15, 16]. Ethnobotanical studies carried out by Arenas and collaborators [7, 17, 18] on the use of algae as a slimming aid in Argentina show that a lack of tradition on the part of the population in the use of these resources leads to usage that is lacking in shared experience, and is principally based on the nonspecialized instructions given by the sales company or person. According to this author, this means they are called by the all-encompassing name "algae" without reference to any taxonomical characteristics or even to whether they are of marine or freshwater origin.

It is also the cities that, due to immigration, are often transformed into multiethnic communities where different traditions interact in terms of the use of plants and animals. Different ethnobotanical studies involving populations that have immigrated or have been forced to live in new environments have revealed, for example, that dwellers try to maintain their medical traditions, and so they obtain their medicinal plants from their home countries by importation or by cultivation in their gardens. Slowly, however, they incorporate phytotherapeutic elements from the receiving society [19–22]. In this sense, Medeiros et al. [22] have suggested that depending on the specific context and conditions of migrations, different processes, such as adaptation, maintenance, and/or replacement, that determine the selection of resources may become predominant in each case.

It is also worthy of note that, due to the social transformation our societies are going through at the moment, rural populations are also exposed to new ethnobiological resources, and could therefore be contexts suitable for the study of these topics. At the present time, it is not uncommon for country people to live in an urban center while also maintaining a dwelling in the countryside, and the use of plants and animals no longer depends solely, or principally, on local conditions. This can be attributed to the concept of post-communality or trans-community [23]. This concept asserts that rural populations are experiencing a widening of their social arena, promoted mainly by greater mobility due to improved access to means of transport and communication, as well as migration (temporary or permanent) to cities. As a result, rural and urban zones interact continuously, and bidirectionally, generating the incorporation of nontraditional resources in a new context. Sears et al. [24] showed how Amazonian farmers who work temporarily in the lumber industry have enriched their management practices, incorporating useful new species and technologies, as well as new conservation strategies and commercialization.

Another scenario to be studied here is when, due to the ecological invasion processes of species moving into new territories or "species contamination," useful new resources come into contact with cultures that are not familiar with them [25]. Such is the case when a foreign plant or animal seeds naturally in natural or anthropic environments, producing an important change in the composition, structure, and processes of the ecosystem it enters [26]. By way of illustration we can mention the case of the sweet briar (*Rosa rubiginosa*), a species that has invaded much of the northwest of Argentinean-Chilean Patagonia. Introduced by

European colonists at the beginning of the twentieth century, this species spread rapidly, occupying roadsides, steppe, and artificial clearings in the forests [27]. Ethnobotanical studies in areas with a high abundance of this species show that in less than 100 years it has become one of the most important ethnobotanical resources for the population, who use its fruit as a food and a medicinal and cosmetic resource. In addition, the harvesting of this fruit is not only relevant for domestic consumption but it also constitutes a significant source of earnings for low income families, both urban and rural. Despite its cultural importance, the harvesting of the sweet briar is little documented, but it is known that the fruit is picked principally by women and children, in autumn, and that approximately 50 g/day of fruit can be collected for commercial purposes [28–30].

3 Some Theoretical Concepts of Interest in the Study of Nontraditional Ethnobiological Resources

The complexity of the study of the use of nontraditional resources requires a multidisciplinary approach that integrates cultural, historical, and social aspects, and also particular characteristics related to biological and psychological factors. Models or theoretical approximations are needed which can describe the nonlinear, highly unpredictable trajectory of a new resource that begins to be used by more and more individuals within a population.

In Table 2 below, some brief, non-exhaustive concepts are presented which can be used as starting points in these studies: cultural adaptation, cultural transmission, resilience, syncretism, hybridization, acculturation, and transculturation. Although the theoretical development of these topics is under permanent discussion and also, strictly speaking, lies outside the scope of this book, they have been chosen because up to the present time they have dominated most of the reference literature.

4 Methods and Approaches

These studies can be carried out using the different methodological tools presented in other chapters of this book, particularly those related to research done in markets and fairs (Chap. 2), and the micrographic studies applied in ethnobotany (Chap. 17), since plant or animal material is generally found in powder form, sometimes with some larger fragments present. Basically, what is necessary is identification and ethnographic work with the consumers, suppliers, intermediaries or, if they exist, those responsible for selling the nontraditional resource (see Chap. 2). If, on the other hand, new wild resources are being dealt with, then plant surveys

Table 2
Some theoretical concepts applicable to the study of the use of ethnobiological resources

Concept	Definition	Some references
Cultural adaptation	The process of adjustment of a population to new conditions in which individuals' genetic and ontogenetic information does have a role to play, but most important is the socially transmitted information, which forms part of the group's cultural inheritance. Thus, learning and social transmission are fundamental elements that enable cultural change	[32–35]
Cultural transmission	The process that allows information to be acquired and shared socially. It depends on different means of transmission: (a) vertical (parents to children), (b) horizontal (between peers), (c) oblique (unrelated adult individuals), (d) indirect (from a key individual), and (e) is dependent on the frequency of opinions (majority opinion)	[32, 33, 36, 37]
Resilience Cultural	The ability or capacity of a system to absorb disturbance, assimilate it and reorganize itself after the change, maintaining essential functions, structure, and identity. Depends on the level of self-organization and the ability to learn from new experiences and adapt	[38–43]
Cultural synchretism	A process that unites two doctrines or different elements. This assimilation is generally spontaneous, a consequence of exchanges between different cultures	[44, 45]
Cultural hybridization	The ability of local populations to effectively integrate traditional or local knowledge with scientific or modern knowledge	[24, 46, 47]
Acculturation	A process that leads to a group acquiring aspects of a new culture, in general at the expense of their own culture	[42, 48]
Transculturation	The gradual process by which a group gradually adopts cultural traits from another cultural system	[48]

should be carried out in order to estimate how the abundance of these resources influences their level of use (see Chaps. 8 and 9).

For example, Díaz-Betancourt and collaborators [31] have estimated the edible biomass (kg/ha) of weeds growing in natural, anthropic and kitchen garden environments, so as to estimate the amount of food resources which are little used by the population due to lack of knowledge about their usefulness.

It should be pointed out that for a great many studies on the use of nontraditional resources, it is necessary to work with the sales points that are common in market societies: pharmacies, herbalists, whole food stores and even supermarkets. These sales points have distinctive characteristics that lead to a different rapport (sometimes difficult) between researcher and informants. Faced with the tremendous whirlwind of commercial activity at these sites, it is necessary to arrange interviews outside commercial hours and obtain signed authorization to go round the shops and interview their staff [16]. It is essential to pay attention here to the enormous number of identifying factors that refer to the culture of origin, and also to the advertising material referring to the products [1]. As well as this, it is possible that these stores are not attended to by specialists, and often the sales people know very little about the resources they offer [15, 16].

Since most of these studies are carried out in cities, sampling will involve a lot of work if the data are to be representative and reflect the heterogeneity of the population being studied, and so random selection of informants is recommended, following the guidelines in Chap. 1. In the case of studies in immigrant populations, these families, stores, markets or street vendors are often found in neighborhoods that are very clearly delimited, and in general are more homogenous, with a higher level of cultural cohesion, so there is a greater likelihood of finding specialists and more willingness to speak of traditions.

Finally, the ethnographic information obtained and the plant survey must be complemented with an exhaustive bibliographical review that will provide information on the usage of the resources in question in their place of origin.

Acknowledgements

We would like to thank the Consejo Nacional de Investigaciones Científicas y Técnicas de la Argentina and the Universidad Nacional del Comahue. This chapter has been financed by the ANPCyT of Argentina (PICT-07-02289) and CONICET (PIP 0337).

References

1. Pochettino ML, Arenas P, Sánchez D, Correa R (2008) Conocimiento botánico tradicional, circulación comercial y consumo de plantas medicinales en áreas urbanas de Argentina. Bol Latinoam Caribe Plant Med Aromat 7(3):141–148
2. Anderson EN, Pearsall DM, Hunn ES, Turner NJ (2011) Ethnobiology. Wiley-Blackwell, Hoboken
3. Torrejón F, Cisternas M (2002) Alteraciones del paisaje ecológico araucano por la asimilación mapuche de la agroganadería hispano-mediterránea (siglos XVI-XVII). Rev Chil Hist Nat 75:729–736
4. Rapoport EH (1987) Lo bueno y lo malo tras el descubrimiento de América. El punto de vista ecológico y biogeográfico. Alter Natura, Bariloche
5. Le Comte E (2007) Sabiduría Tradicional de Plantas Americanas. Le Comte. Bs. As
6. Pemberton RW, Lee NS (1996) Wild food plants in South Korea; market presence, new crops, and exports to the United States. Econ Bot 50:57–70
7. Arenas PM (2004) Relevamiento etnofarmacológico, análisis micrográfico y potenciales efectos fisiológicos de sumplementos dietéticos conteniendo algas en su composición. Dissertation, Facultutad Ciencias Naturales y Museo, Universidad Nacional de La Plata
8. Balick MJ, Kronenberg F, Ososki AL, Reiff M, Fugh-Berman A, O'Connor B, Roble M, Lohr P et al (2000) Medicinal plants used by Latino healers for women's health conditions in New York city. Econ Bot 54(3):344–357
9. Cunningham AB (2001) Etnobotánica aplicada. Pueblos, usos de plantas silvestres y conservación. Plantas y pueblos. Nordan Comunidad, Uruguay
10. Macía MJ, García E, Vidaurre PJ (2005) An ethnobotanical survey of medicinal plants

commercialized in the markets of La Paz and El Alto, Bolivia. J Ethnopharmacol 97: 337–350

11. Batista de Lima CB, Takaki Bellettini NM, Sardinha da Silva A, Cheirubim AP, Kassen Janani J et al (2007) Uso de Plantas Medicinais pela População da Zona Urbana de Bandeirantes-PR. Revista Brasileira de Biociências 5:600–602
12. Ferreira FS, Albuquerque UP, Coutinho HDM, Almeida WO, Alves RRN (2012) The trade in medicinal animals in Northeastern Brazil. Evid Based Complement Altern Med 2012:20. doi:10.1155/2012/126938 (Article ID 126938)
13. Reyes-García V (2001) Indigenous people, ethnobotanical knowledge, and market economy: a case study of the Tsimane' Ameridians of lowlands Bolivia. Dissertation, University of Florida
14. Rocha JM (2005) Measuring traditional agro-ecological knowledge: an example from peasants in the Peruvian Andes. Field Methods 17(4):356–372
15. Arenas PM (2007) Suplementos dietéticos: estudio etnobotánico en zonas urbanas. Volumen especial de etnobotánica. Kurtziana 33(1):193–202
16. Cuassolo F (2009) Estudio etnobotánico de las plantas medicinales nativas y exóticas comercializadas en la ciudad de Bariloche (Patagonia, Argentina). Dissertation, Universidad Nacional del Comahue
17. Arenas PM, Correa RF, Cortella AR (1997) Algas (Phaeophyta) presentes en productos comerciales utilizados para adelgazar. Acta Farmacéutica Bonaerense 16(3):151–160
18. Arenas PM, Cortella AR (1996) Análisis micrográfico de muestras comerciales de Spirulina. Acta Farmacéutica Bonaerense 15(1):11–19
19. Ceuterick M, Vandebroek I, Torry B, Pieroni A (2007) Cross-cultural adaptation in urban ethnobotany: the Colombian folk pharmacopoeia in London. J Ethnopharmacol 120:342–359
20. Pieroni A, Muenzc H, Akbulut M, Husnu Can Baser K, Durmuskahya C (2005) Traditional phytotherapy and trans-cultural pharmacy among Turkish migrants living in Cologne, Germany. J Ethnopharmacol 102:69–88
21. Richeri M, Beeskow AM, Ladio AH (2010) Las plantas y la salud en la comunidad boliviana de Pto. Madryn (Argentina). In: Pochettino ML, Ladio AH (eds) Traditions and transformations in ethnobotany. CYTED, La Plata
22. Medeiros PM, Soldati GT, Alencar NL et al (2012) The use of medicinal plants by migrant people: adaptation, maintenance, and replacement. Evid Based Complement Altern Med 2012:807452. doi:10.1155/2012/807452 (Article ID 807452)
23. Gundermann H, González H, Ruyt L (2009) Migración y movilidad mapuche a la Patagonia Argentina. Magallana 37(1):21–35
24. Sears RR, Padoch C, Pinedo-Vasquez M (2007) Amazon forestry tranformed: integrating knowledge for smallholder timber management in Eastern Brazil. Hum Ecol 35: 697–707
25. Nuñez MA, Simberloff D (2005) Invasive species and the cultural keystone species concept. Ecol Soc 10(1):4, http://www.ecologyandsociety.org/vol10/iss1/resp4/
26. Cronk QCB, Fuller JL (1995) Plant invaders. Chapman and Hall, Cambridge
27. Damascos MA (2008) La rosa mosqueta y otras rosas. Centro Regional Universitario Bariloche, Bariloche
28. Ladio AH, Rapoport EH (1999) El uso de plantas silvestres comestibles en una población suburbana del noroeste de la Patagonia. Parodiana 11(1–2):49–62
29. Ladio AH, Lozada M (2001) Non-timber forest product use in two human populations from NW Patagonia: a quantitative approach. Hum Ecol 29(4):367–380
30. Oxfam (2003) Recolectoras de frutos silvestres. Taller de Acción Cultural, Santiago
31. Díaz-Betancourt ME, Ghermandi L, Ladio AH, López Moreno IR, Raffaele E, Rapoport EH (1999) Weeds as a source for human consumption. A comparison between tropical and temperate Latin America. Rev Biol Trop 47(3):329–338
32. Cavalli-Sforza LL, Feldman MW (1981) Cultural transmission and evolution: a quantitative approach. Princeton University Press, Princeton
33. Boyd R, Richerson PJ (1985) Culture and the evolutionary process. University of Chicago Press, Chicago
34. Laland KN, Odling-Smee J, Feldman MW (2000) Niche construction, biological evolution and cultural change. Behav Brain Sci 23:131–175
35. Folke C, Colding C, Berkes F (2003) Synthesis: building resilience and adaptive capacity in social-ecological systems. In: Berkes F, Colding J, Folke C (eds) Navigating social-ecological systems: building resilience for complexity and

change. Cambridge University Press, Cambridge, pp 352–387
36. Boesch C, Tomasello M (1998) Chimpanzee and human cultures. Curr Anthropol 39(5): 591–614
37. Lozada M, Ladio AH, Weigandt M (2006) Cultural transmission of ethnobotanical knowledge in a rural community of NW Patagonia. Hum Ecol 60(4):374–386
38. Berkes F, Colding J, Folke C (2000) Rediscovery of traditional ecological knowledge as adaptive management. Ecol Appl 10:1251–1262
39. Berkes F, Folke C (2002) Back to the future: ecosystem dynamics and local knowledge. In: Gunderson LH, Holling S (eds) Panarchy, understanding transformations in human and natural systems. Island Press, Washington, pp 121–146
40. Davidson-Hunt IJ, Berkes F (2003) Nature and society through the lens of resilience: toward a human-in-ecosystem perspective. In: Berkes F, Colding J, Folke C (eds) Navigating social-ecological systems: building resilience for complexity and change. Cambridge University Press, Cambridge
41. Tengo M, Belfrage K (2004) Local management practices for dealing with change and uncertainty: a cross-scale comparison of cases in Sweden and Tanzania. Ecol Soc 9(3):4
42. Ladio AH, Lozada M (2008) Medicinal plant knowledge in rural communities of North-western Patagonia, Argentina. A resilient practice beyond acculturation. In: Albuquerque UP, Ramos AM (eds) Current topics in ethnobotany. Research Signpost, India, p 40
43. Ladio AH (2011) Traditional knowledge of edible wild native and exotic plants in the context of cultural change in human populations of arid Patagonia. Bioremediat Biodivers BioAvailab 5(1):60–64
44. Citarella L, Conejeros AM, Espinossa B, Jelves I, Oyarce AM, Vidal A (1995) Medicinas y culturas en La Araucanía. Programa de Atención Primaria en Salud. Cooperación Italiana, Santiago
45. Steward C (1999) Syncretism and its synonyms: reflections on cultural mixture. Diacritics 29(3):40–62
46. Vessuri H (2004) La hibridización del conocimiento. La tecnociencia y los conocimientos locales a la búsqueda del desarrollo sustentable. Convergencia 11(35):171–191
47. Agrawal A (2005) Community, intimate government, and the making of environmental subjects in Kumaon, India. Curr Anthropol 46:161–190
48. Berry JW, Sam DL (1997) Acculturation and adaptation. In: Berry JW, Segall MH, Kagitzibasi C (eds) Handbook of cross-cultural psychology: social behavior and applications, vol 3, 2nd edn. Allyn & Bacon, Needham Heights

Chapter 20

Plant Anatomy in Ethnobotanical Research: Micrographic Techniques and Applications

Soledad Molares and Ana H. Ladio

Abstract

In this chapter you will find a general discussion on the significance of plant anatomy in ethnobotanical research: its scope and limitations. Some detailed examples have been given of the usefulness of micrographic methodology for clarifying doubts on the use of plants in different cultural and temporal contexts, and in different use categories. Finally, the most commonly used applied plant anatomy techniques will be described.

Key words Plant anatomy • Ethnobotany • Taxonomic identification • Quality control • Selection and use patterns • Paleoethnobotany

1 Introduction

Ethnobotany, a science nourished by contributions from several different disciplines and methods, involves the collection and analysis of data related to interrelationships between plants and human beings. Plant micrographics in particular is one of the disciplinary areas of considerable importance in ethnobotanical studies. It brings together a series of laboratory techniques, which, by means of microscope observations, facilitate the finding and description of elements and structures that characterize plant samples. Using this method, along with histochemical tests, a great variety of components of the primary and secondary metabolism of plants can also be detected.

The detection of these substances and structures is important because they contain elements of high value to human populations in terms of usefulness and symbolism, and which have been taken advantage of since ancient times for different uses. Through both individual and social learning processes over many generations, humans have, directly or indirectly, explored the plant resources in their natural surroundings. The search for essential nutrients (carbohydrates, proteins, lipids, vitamins, etc.), metabolites with

Ulysses Paulino Albuquerque et al. (eds.), *Methods and Techniques in Ethnobiology and Ethnoecology*,
Springer Protocols Handbooks, DOI 10.1007/978-1-4614-8636-7_20, © Springer Science+Business Media New York 2014

effective medicinal effects (e.g., essential oils, alkaloids, tannins) and plant fibers for the working of handcrafts and dyes (tannins, chlorophyll, etc.), among other plant components, would therefore constitute a complex evolutionary strategy influenced as much by sociocultural and ecological factors as by botanical and anatomical ones, and would have resolved problems in people's everyday lives and so improved their quality of life [1, 2].

The information obtained from micrographic study can be of considerable relevance to diagnostic work, such as the taxonomic identification of species and plant complexes (mixtures of two or more species which share the same common name and/or organoleptic qualities) when this cannot be determined botanically with conventional macroscopic techniques.

From the point of view of economic botany, finding and characterizing structures in the organs where phytochemicals of interest to us are located would enable us to determine the best method for their extraction, avoiding the inclusion of unwanted material and thus reducing the time and cost of production [3].

At this point the feasibility of applying basic micrographic techniques to ethnobotanical studies should be highlighted. Although certain materials and equipment are required, as well as dedication and training in their use, micrographics can be valuable in the resolution of certain hypotheses and questions that arise in the field and laboratory, while at the same time providing a wider and more integrated look at human and his environment.

Below, you will find a brief description of some uses of micrographics in everyday ethnobotany, along with the most general and specific micrographic techniques for the study of ethnobotanical material.

1.1 The Use of Micrographics in the Taxonomic Identification of Plant Samples

The collection and identification of voucher specimens is a routine, necessary practice in ethnobotanical work. On the one hand, correct taxonomical identification of plant material is essential for the qualitative and quantitative analysis of data obtained in fieldwork, for the drawing up of the species list, and for the estimation of plant richness and diversity. On the other hand, it is a necessary requirement for the placing of ethnobotanical collections in official herbaria, and for the publication of results in prestigious scientific journals [4]. On many occasions, the ethnobotanical material collected consists of complete samples of plant species, that is, it includes leaves, flowers, and fruits that facilitate easy identification using keys and herbaria for reference. However, on some occasions samples consist of the incomplete organs of a species, which may form part of play items, baskets, musical instruments, etc., or they may be plant products of medicinal and/or nutritional value that come from markets, herbalists, etc., which include only small pieces of wood, parts of the root, leaves or petals, etc. Many of these samples tend to be in powder form or partially decomposed, making botanical identification extremely difficult by mere observation of external morphological characteristics (Sects. 2.3–2.6). In these cases, the detection of diagnostic microscopic elements becomes indispensable for the taxonomic identification of the samples.

Of these diagnostic characteristics the most important are the epidermal cells of the leaves and primary stems, the medullary rays of stems, the elements of the vascular system, the secondary cortex, the crystals, the secretory structures (cavities, ducts, glandular trichomes, nectaries, and idioblasts), the stomata, and the hairs, among others [2, 5–9]. The study of these characteristics includes analysis of their position, frequency, shape, and size [10, 11]. At the same time as the anatomy is being described, histochemical techniques are frequently used for the detection of cell wall components (Sect. 2.2) that can be associated with structures and taxa. An example of this is shown in Table 1, where some of the principal diagnostic features of various economically important families are described.

Recognition of these elements can lead to identification of the unknown species through comparison with published data in reference catalogues, scientific journals, and official pharmacopeia, or at least, an approximation can be made to identifying the pertinent genus or family [12].

1.2 Micrographics Applied to the Quality and Safety Control of Medicinal and Edible Species

Control of the quality and safety of commercial food and medicinal plants is particularly useful in large urban centers where products of plant origin can commonly be found on sale in small pieces or processed, with no official quality control [13–15].

Micrographics, together with macroscopic, organoleptic inspection, and in some cases physicochemical testing, makes it possible to identify the plant species used in the elaboration of these products, and to detect contaminants, substitutes, and adulteration which may have taken place during the production process [14, 16, 17].

The adulteration of a plant drug or natural food refers to the partial or total elimination of the valuable component of the substance, replacing it with another of inferior quality or different origin. This adulteration may be accidental, that is, a consequence of carelessness during processing or storage, or it may be intentional. In most cases replacement is partial, and involves a substance of inferior quality, but in some cases the original product may be completely replaced by a substance of no medicinal or nutritional value, or which may even be toxic [14].

Microscopic methods include observation and characterization of diagnostic features through the use of an optical microscope (Sect. 2). Macroscopic methods are carried out with the naked eye or a magnifying glass, and consist of observation and

Table 1
Microscopic diagnostic features of leaves and shoots of some of the main economically important families

Diagnostic structure	Botanical family
Leaves	
Laticiferous cells	Apocynaceae, Papaveraceae, Convolvulaceae
Mucilaginous epidermis	Malvaceae, Salicaceae
Scales	Bromeliaceae
Sclereids	Oleaceae, Theaceae
Anisocytic stomata	Brassicaceae, Plumbaginaceae
Anomocytic stomata	Berberidaceae, Liliaceae, Ranunculaceae
Diacytic stomata	Caryophyllaceae
Paracytic stomata	Poaceae, Juncaceae, Rubiaceae, Magnoliaceae
Salt glands	Frankeniaceae
Hypodermis	Lauraceae, Piperaceae
Capitate trichomes	Lamiaceae, Sapindaceae
Dendritic trichomes	Melastomataceae, Solanaceae
Silicified trichomes	Poaceae
Urticating trichomes	Urticaceae, Loasaceae, Euphorbiaceae
Stems	
Cystoliths	Moraceae, Urticaceae, Cannabinaceae
Druses	Cactaceae, Chenopodiaceae, Malvaceae, Rutaceae, Tiliaceae, Urticaceae
Multiple cambia	Amaranthaceae, Chenopodiaceae, Nyctaginaceae
Intraxylary phloem	Apocynaceae, Convolvulaceae, Cucurbitaceae, Lythraceae
Medullary bundles	Begoniaceae, Asteraceae, Piperaceae, Saxifragaceae, Brassicaceae
Raphides	Liliaceae, Rubiaceae
Deep cork layer	Bignoniaceae, Hypericaceae, Rosaceae, Casuarinaceae
Surface cork layer	Asteraceae, Lamiaceae, Brassicaceae

Taken in part from Cutler [7]

description of the shape, size, and color of the samples, in addition to an organoleptic analysis of consistency, texture, and in some cases, flavor and odor. Physicochemical tests, on the other hand, are based on the extraction of chemical components and the carrying out of a series of experiments to identify the main substances of interest, which are then analyzed further to obtain values for specific physical indices, such as optical rotation, refractive index, and relative density. [16]. In addition to this, the moisture content of plant samples is also measured, since moisture content values higher than those registered in the official pharmacopeias may favor enzyme activity and the proliferation of decomposing microorganisms, which would have a negative effect on the quality and safety of the product [13, 16].

For example, the organoleptic and chemical analysis of a powdered food item may indicate a high level of starch, suggesting the presence of flour; the micrographic study of the starch grains would then complement this, determining whether the flour is from wheat, rice, potato, or maize [18] (Sects. 2.3 and 2.6).

These techniques can also identify and quantify non-plant foreign matter, such as insect eggs or larva, sand, soil, rodent excrement, or live insects [19].

In the case of products that are more or less mass-produced, there are quality standards set to mark the tolerance limits for the presence of foreign matter not declared on the labels, as well as lists of the main contaminants that may be found. These values are useful as a guide to their identification, control, and safety.

1.3 Micrographics in the Analysis of Plant Selection and Use Patterns

Micrographics can contribute information that is useful in resolving questions regarding the selection of species, plant organs, methods of preparation and associated cultural practices, both in present and past contexts [20–25]. For example, *Baccharis obovata* (Asteraceae) is an aromatic species belonging to the traditional pharmacopeia of the Mapuche of Patagonia, Argentina. According to Molares et al. [26] the anatomy of this species reveals that the leaves and the stem present cavities that secrete essential oils. By means of an ethnobotanical study, these authors noted that a decoction of the branches is for topical use, for the treatment of alopecia, dandruff, diverse forms of dermatitis and headaches. It was also observed that during the process of preparing the medicine, the intensity of the aroma emanating from the plant as it boiled, along with the color taken on by the water, are important indicators as to the state of progress of the recipe. A penetrating aroma and a yellow-green color mark the point when the preparation is ready. These results allowed the authors to infer that the interviewees can detect and evaluate different aromatic and volatile components during the boiling process, through sensory mechanisms. The gradual liberation of aromatic vapors is a consequence of the particular anatomy of the plant, which "encloses" essential oils in cavities, isolating them from the external environment and freeing them progressively as these secretory structures are broken up by the boiling process [26].

1.4 Micrographics in Archaeobotanical and Paleobotanical Research

Worthy of particular note is the importance of micrographics in archaeobotanical and paleobotanical research for the identification of artifacts and products made from plants (baskets, weapons, musical instruments, foods, etc.), and also in the interpretation of their use in the past. The anatomic study of mummified coprolites has also allowed the identification of many components of the diet of prehistoric man.

These remains have often suffered destructive processes, whether mechanical, chemical, or biological, through direct human action (cooking, soaking, carbonization, etc.), and also due to the effects of the long years themselves, and are frequently found broken into pieces, dehydrated, oxidized, salinized, or even reduced to powder [27]. In this state, the remains tend to lack most of the diagnostic morphological features, and their recognition by direct comparison with modern herbarium samples is extremely complicated. In these cases, modern plant material such as seeds and pieces of wood are often submitted to processes similar to those which the unknown material may have gone through, and the structural changes produced in them are documented, so as to compare them with their archaeological counterparts, thus improving their interpretation and identification [22]. This technique makes possible the revision of certain resistant microscopic characteristics present in the epidermis, starch grains, phytoliths (formed by the mineral precipitation of substances from plant metabolism, in general silicious or calcareous), pollen grains, etc. [27].

The epidermis is a structure of great importance in this kind of study, since it is a tissue with high structural resistance over time, due to the presence of a continuous, waxy, protective film called the cuticle, which is inseparable from the epidermal cells and enables them to stay together [28]. The study of the epidermal cells themselves, the cuticular designs, cell contents, and associated structures such as the stomata and trichomes, has been pointed out by many authors as having important diagnostic value (Sects. 2.4 and 2.5) [24, 28–30].

In many cases, starch grains are also preserved in archaeobotanical samples and their identification has been of particular importance in elucidating paleohorticultural patterns [22] (Sect. 2.6). Their micrographical importance lies in the specificity of their location at species level, their morphological diversity, size, and birefringence, i.e., their capacity to split a ray of incident light into two perpendicularly polarized rays [31–33].

Other plant structures of high diagnostic value for this kind of study are spores and pollen grains. They present characteristic structure and ornamentation at species level that allow their identification by comparison with pollen keys [34]. On occasions these studies are complemented with the analysis of phytoliths, especially the silicaphytoliths, which are hydrated silica crystals made by plants from the silicates absorbed from groundwater. Deposited in areas of major evapotranspiration in the plants, they take on distinctive shapes which are preserved in the soil when the organic material has disappeared. When the phytoliths, spores, and the pollen grains are recovered from the soil, or rather, from soil attached to instruments, containers, etc., we can make an approximate identification of the species [35].

2 Materials and Methods

2.1 Methodology for the Observation of Tissues in Longitudinal and Transverse Sections

2.1.1 The Selection and Fixation of Organs

Plant organ samples should be selected whose development is representative of the species, that is, of average size, healthy, of characteristic color, etc.

The plant material collected that is not for immediate use can be preserved in 70 % ethyl alcohol. Although it is important to take into account that ethanol tends to dehydrate and shrink the tissue, this can be controlled by adding a small amount of glacial acetic acid (approximately 5 % of the total volume). Percentages of ethanol above 75 % make the tissue rubbery, and it becomes difficult to cut.

Another option for fixation is the FAA fixative (formaldehyde 10 %: alcohol 50 %: acetic acid 5 %: water 35 %). The advantages of this include rapid, complete penetration of the tissue by the formaldehyde, the possibility of more prolonged storage of the material and a balance in the contraction–expansion produced by the alcohol and acetic acid. Disadvantages include the toxicity of formaldehyde [36, 37].

2.1.2 Sections

Fresh material is cut directly, while that preserved in 70 % ethyl alcohol is washed in water several times before being cut. In the case of dried samples, the material must be hydrated in hot water to make it easier to cut, and when it is very hard and woody, it should be boiled for a few minutes or up to some hours, until soft. Depending on the consistency of the material, the sections may be cut freehand with a razor blade, or with a sliding or freezing microtome. The stem and root can be cut without the support necessary for cutting leaves and petioles. For these we suggest placing the material between two semicylinders of polystyrene or high-density Styrofoam, then cutting them together. The cut material is picked up with a brush and preserved in water to prevent drying out and damage. It is very important that the cutting tool has a sharp enough blade, otherwise excessive pressure may be put on the segment, or a saw-like movement may be generated, which would deform or crush it. Following this, the thinnest and most uniform sections are selected using a magnifying glass. These should always be kept in water until they are stained.

2.1.3 Clarification and Direct, Fast, and Simple Staining for the Observation of Plant Structures

The sections are placed in a beaker with 50 % sodium hypochlorite for clarification. When they turn white, wash them well several times with water. Put them in 50° ethanol, then 70° ethanol. Pick the sections up with a brush and transfer them to a watch glass or beaker with safranin diluted in 80 % alcohol and leave them for 2–5 min, or the time needed for them to take up the color. Wash with water.

2.1.4 Mounting

Temporary preparations (to be observed immediately) can be mounted in water-glycerin 50:50 %. Put one or more drops of this liquid on a slide and with a brush pick the selected section up and place it on the slide, then cover it with the coverslip. Permanent mounting media include Canada balsam and synthetic resins [36].

2.1.5 Observation with Optical Microscope

Observations and descriptions of the stained material should be made, as well as measurements of the diagnostic structures. The lignified walls and the cuticle will be seen in an intense pink color and the parenchyma cells a soft pink color.

2.1.6 Labeling

If the specimen is to be preserved, a label is placed on one side of the slide, with the name of the species, the technique applied, and the date of preparation.

2.2 Methodology for the Identification of Cell Contents [36]

For the detection of the principal cell contents and components of the cell wall, the untreated section is placed on the slide, and according to the test to be carried out, one or more drops is added of:

1. Sudan III in 90° ethanol to detect lipids and essential oils. A positive reaction gives a strong orange color.
2. Lugol's reagent to detect starch. The starch will take on an intense blue color.
3. 1 % Phloroglucinol in 96° ethanol-HCL to detect lignin. The lignin will stain a violet-red color.
4. 5 % Ferric trichloride to identify tannins. A positive reaction gives a blue-green color.
5. Concentrated sulphuric acid for saponins. If the tissue sample contains saponins, it will first take on a yellow color, which will change to red after 30 min and finally, it will turn violet or blue-green.
6. A 1 % solution of picric acid for aleurones. The aleurones will stain yellow.
7. Cupric acetate for calcium oxalate. If there are calcium oxalate crystals present, they will dissolve and the oxalic acid will spread throughout the intercellular spaces, where they will form crystals of cupric oxalate.

2.3 Methodology for the Observation of Powdered Samples

The powder is taken on the tip of a spatula and placed between the slide and coverslip with a drop or two of water. Instead of water, water-glycerine 50:50 % can be used, which will keep the section hydrated for longer [38].

In this case, the previously described histochemical techniques for detection of cell and cell wall components can be used.

2.4 Methodology for the Study of the Epidermis

Important diagnostic features are found in the epidermis [39]. Techniques for surface observation of this tissue are varied, the simplest consisting in separating the epidermis from the rest of the organ with forceps. Since in many cases this is particularly complicated, other techniques are also employed, based on cutting a section of the plant material and rendering the mesophyll transparent, or diaphanous.

Two of the most frequently used techniques for diaphanization are:

2.4.1 The Dizeo de Strittmater Technique [40]

Boil the material in 96° ethanol for 10 min, then place it in a solution of equal parts of 96° ethanol and 5 % sodium hydroxide, and boil for 5 min. Rinse with water several times. Next, put the material in a solution of 50 % sodium hypochlorite and leave it there until it becomes transparent. Rinse with water. Place pieces of the material in a solution prepared with 5 g chloral hydrate in 20 mL of water so that it loses its opacity. Finally, rinse in water and stain with diluted safranin, mount in water-glycerin 50:50 % and observe.

2.4.2 Determination of Quantitative Parameters in Leaves

Clarify pieces of the material in 50% Sodium Hypochlorite, rinse in water, and observe without further treatment. The diaphanization of the material will allow the determination of quantitative parameters in leaves, such as the palisade index, the stomatal index, and vein-islet index. These indices are often used in the comparison of similar species belonging to the same genus, ecotypes, and specimens in a fragmented state [41]. The following formulae are used for their calculation:

- Palisade index: the number of palisade parenchyma cells covered by one single epidermal cell.

$$PI = no.\ of\ palisade\ parenchyma\ cells\ /\ 4\ epidermal\ cells$$

- The stomatal index: the number of stomata per 100 epidermal cells measured in an area of 1 mm^2, marked out with a micrometer lens.

$$EI = no.\ of\ epidermal\ cells + no.\ of\ stomata\ /\ 100 \times no.\ of\ stomata$$

- The vein-islet index: the area of photosynthetic tissue surrounded by the ultimate terminations of the leaf vein-islet in 1 mm^2.

$$VII = no.\ of\ vein\ islets\ /\ mm^2$$

In all cases the data should be taken from the mid part of the adult leaf surface, and at least 30 repetitions should be made of the estimation of the average values and their standard deviations.

2.5 Maceration and Dissociation

In order to study complete, individual cells, methods of maceration or dissociation techniques are used to dissolve the middle laminas that keep the cells together [42].

There are two main types of dissociation, a gentle one and a strong one:

2.5.1 Gentle Dissociation Is Used for Leaves, Stems, and the Cortex

This technique involves boiling a piece of material in a 5 % NAOH solution for 5 min. It is then cooled and passed through filter paper. A portion of the retained material is picked up with a spatula and placed on a slide with a drop of water. The coverslip is put in place and the sample gently flattened. This technique preserves the crystals, but not the starches.

2.5.2 Strong Dissociation Is Used to Separate Wood or Hard Seed Cells

This technique involves boiling a piece of material in a 10 % KOH solution for 10–15 min. It is then cooled and passed through filter paper. Following this, it is rinsed with water and put in another glass beaker with a 25 % chromic acid solution for 30 min. It is again rinsed with water and passed through filter paper. A portion of the retained material is picked up with a spatula and placed on a slide with a drop of water. The coverslip is put in place and the sample flattened gently. This technique does not preserve starch or crystals.

2.6 Methodology for the Analysis of Starch Grains

A technique used for the identification of cereal flours (wheat, rice, maize, etc.) fruit and underground storage organs (roots, tubers, bulbs, etc.) [31, 43].

It is worthy of note that starch grains may be of two types: transitory or storage. The former are produced in green leaves and stems and used quickly in plant catabolism. The latter are found completely developed and fulfill the function of long-term storage. The storage starch grains are therefore the most reliable as diagnostic features [44].

The storage organs are scraped with a razor blade or scalpel. With the material thus obtained, semipermanent preparations are assembled in water-glycerin, which allows rotation of the grains. They can be stained with Lugol for differential coloration. Observations are made with an optical microscope at maximum magnification, with both normal and polarized light.

The flours are treated as specified in the item "samples in powder form" (Sect. 2.3).

The qualitative and quantitative attributes commonly observed in starch grains are: length and width (μm); relationship length/width; multiplicity: simple or compound grain (formed by two, three, or more granules); relationship between number of simple grains/number of multiple grains; shape (oval, triangular, reniform, irregular, pyriform, etc.); the hilum of the grain (the point where the layers of starch begin to be deposited): presence or absence, visibility and shape; growth rings: presence or absence; and extinction cross (presence indicates a higher level of organization of amylose and amylopectin chains): shape and number of arms.

In all cases repetitions of the observations should be carried out in order to calculate the averages and standard deviations.

3 Final Remarks

Other micrographic techniques not considered in this chapter may also be used in the study of ethnobotanical material. For example, environmental scanning electron microscopes are useful in situations where taxonomic definition is difficult, for the analysis of adaptations and particular shapes, etc. It should be noted, however, that its use requires sophisticated equipment and the costs are high.

Optical micrographics, on the other hand, is a fast, low-cost method, relatively easy to carry out, which allows the proportion of a component in a mixture to be asserted (a common situation when studying medicinal uses of plants), the origin of the material to be identified, etc.

However, a problem frequently faced by an ethnobotanist when trying to identify a modern, fragmented specimen using the micrographic method, is a lack of reference catalogues with anatomical descriptions. Because of this situation in many parts of the world, it is necessary to promote and develop keys of micrographic characteristics which will make possible the identification of species even when samples are incomplete or in bad condition. It is also necessary to develop human resources in this field and encourage interdisciplinary work between botanists who are plant anatomy experts.

Acknowledgements

We would particularly like to thank the *Consejo Nacional de Investigaciones Científricas y Técnicas de la Argentina* and the *Universidad Nacional del Comahue*. This chapter has been financed by the ANPCyT of Argentina (PICT-07-02289) and CONICET.

References

1. Molares S, Ladio A (2009) Chemosensory perception and medicinal plants for digestive ailments in a Mapuche community in NW Patagonia, Argentina. J Ethnopharmacol 123: 397–406
2. Molares S, Ladio A (2012) Plantas aromáticas con órganos subterráneos de importancia cultural en la Patagonia Argentina: una aproximación a sus usos desde la etnobotánica, la percepción sensorial y la anatomía. Darwiniana 50(1):7–24
3. Duke SO (1994) Glandular trichomes. A focal point of chemical and structural interactions. Int J Plant Sci 155(6):617–620
4. Benitez G, González-Tejero MR, Baena L (2009) Primera exsiccata de flora etnobotánica en el ámbito ibero-macaronésico. Bol AHIM 11:34–36
5. Cutter EG (1969) Plant anatomy: experiment and interpretation. I: Cells and tissues. E. Arnold, London
6. Cutter E (1969) Plant anatomy: experiment and interpretation. II: Organs. E. Arnold, London
7. Cutler D (1987) Anatomía vegetal aplicada. Librería Agropecuaria, Buenos Aires
8. Fahn A (1990) Plant anatomy. Pergamon Press, Oxford
9. Esau K (1993) Anatomía de las Plantas con Semillas. Hemisferio Sur, Buenos Aires
10. Molares S, Ladio AH, Vobis G (2007) Etnobotánica y anatomía foliar de las especies

de *Chiliotrichum* Cass. (Asteraceae: Astereae): *Ch. rosmarinifolium* Less. y *Ch. diffusum* (Forst.) O. Kuntze. Kurtziana 33(2):5–16
11. Arambarri AM, Freire SE, Bayón ND et al (2009) Morfoanatomía foliar de árboles medicinales de la Provincia Biogeográfica de las Yungas (Argentina). BLACPMA 8(5):342–379
12. González-Tejero MR, Casares-Porcel M (1996) La anatomía vegetal como método de identificación en etnobotánica. Monograf Jard Bot Córdoba 3:33–37
13. Nascimento JE, Lacerda EU, Nascimento VT et al (2005) Produtos à base de plantas medicinais comercializados em Pernambuco—Nordeste do Brasil. Acta Farmacéutica Bonaerense 24:113–122
14. Arenas PM (2009) Algas empleadas en la elaboración de suplementos dietéticos: abordaje etnobotánico en algunas áreas urbanas de Argentina. In: Arenas PM (ed) Etnoficología aplicada: estudio de casos en relación a la salud y la alimentación en ambientes rurales y urbanos. RISAPRET-CYTED, Jujuy, Argentina
15. Molares S, Arenas P, Aguilar A (2012) Etnobotánica urbana de los productos vegetales adelgazantes comercializados en México DF. BLACPMA 11(5):400–412
16. WHO (World Health Organization) (1998) Quality control methods for medicinal plant materials. WHO Library, Geneva
17. Rivera-Arce E, Gattuso M, Alvarado R et al (2007) Pharmacognostical studies of the plant drug *Mimosae tenuiflorae* cortex. J Ethnopharmacol 113(3):400–408
18. Rivas MA, Bassols G, Schimpf J et al (2009) Identificadores micrográficos de interés para el control de calidad botánico de alimentos elaborados con harinas provenientes de cultivos andinos. In: Borneo R, Nassetta M, Passalacqua N (eds) Ciencia y Tecnología de los Alimentos, Avances en Inocuidad y Microbiología. Trabajos completos presentados al III Congreso Internacional de Ciencia y Tecnología de los Alimentos. Libro II, Córdoba, Argentina, pp 170–176
19. Cuassolo F, Ladio AH, Ezcurra C (2009) Aspectos de la comercialización y control de calidad de las plantas medicinales más vendidas en una comunidad urbana del NO de la Patagonia Argentina. BLACPMA 9(3):166–176
20. Rodríguez MF (2000) Woody plant species used during the archaic period in the Southern Argentine Puna. Archaeobotany of Quebrada Seca 3. J Archaeol Sci 27:341–361
21. Herron S (2003) Catnip, *Nepeta cataria*, a morphological comparison of mutant and wild type specimens to gain an ethnobotanical perspective. Econ Bot 57(1):135–142
22. Lema V, Capparelli A, Pochettino ML (2008) Taxonomic identification of dry and carbonized archaeobotanical remains of *Cucurbita* species through seed coat micromorphology. Veg Hist Archaeobot 17(suppl 1):277–286
23. Marconetto MB (2008) Linnaeus en el ambato. El uso de la clasificación taxonómica en arqueobotánica. In: Archila S, Giovannetty M, Lema V (eds) Arqueobotánica y teoría arqueológica. Discusiones desde Suramérica. Universidad de Los Andes Facultad de Ciencias Sociales–Ceso, Bogotá, pp 143–166
24. Molares S (2010) Flora medicinal aromática de la Patagonia: características anatómicas y propiedades organolépticas utilizadas en el reconocimiento por parte de la terapéutica popular. Tesis Doctoral, Universidad Nacional del Comahue, Bariloche
25. Simpson BB, Conner-Ogorzaly M (1986) Economic botany: plants in our word. McGraw-Hill, New York
26. Molares S, González S, Ladio A et al (2009) Etnobotánica, anatomía y caracterización físico–química del aceite esencial de *Baccharis obovata* Hook. et Arn. (Asteraceae: Astereae). Acta Bot Brasilica 23(2):578–589
27. Bertone G, Bellido E, Jing Na L (2008) La arqueobotánica peruana: del objeto de estudio al objeto de conocimiento. In: Archila S, Giovannetti M, Lema V (eds) Arqueobotánica y teoría arqueológica. Discusiones desde Suramérica. Universidad de los Andes Facultad de Ciencias Sociales—Ceso, Bogotá, pp 127–140
28. Pelliza A (1993) Acerca de la microhistología. Comunicación técnica 32, INTA-EEA, Bariloche
29. Bonzani NE, Barboza GE, Bugatti MA et al (2003) Morpho-histological studies in the aromatic especies of *Chenopodium* from Argentina. Fitoterapia 74:207–225
30. Bonzani NE, Costaguta M, Barboza GE (2007) Anatomical studies in *Mentha* (Fam. Lamiaceae) from Argentina. Arnaldoa 14(1):77–96
31. Cortella AR, Pochettino ML (1994) Starch grain analisis as a microscopic diagnostic feature in the identification of plant material. Econ Bot 48(2):171–181
32. Torrence R, Barton H (2006) Ancient starch research. Left Coast Press, Walnut Creek
33. Giovannetti MA, Lema VS, Bartoli CG et al (2008) Starch grain characterization of *Prosopis chilensis* (Mol.) Stuntz and *P. flexuosa* DC, and the analysis of their archaeological remains in Andean South America. J Archaeol Sci 35: 2973–2985
34. Consuegra EM (1996) Arqueopalinología y etnobotánica. Monograf Jard Bot Córdoba 3:23–31

35. Piperno DR (2006) Phytolith. A comprehensive guide for archaeologist and paleoecologist. Altamira Press, San Diego
36. D'Ambrogio A (1986) Manual de técnicas en histología vegetal. Hemisferio Sur, Buenos Aires
37. Ruzin SE (1999) Plant microtechnique and microscopy. Oxford University Press, New York
38. Eschrich W (1988) Pulver-Atlas der Drogen. Gustav Fischer Verlag, Stuttgart
39. Barboza GE, Bonzani N, Filippa EM et al (2001) Atlas histomorfológico de plantas de interés medicinal de uso corriente en Argentina. Museo Botánico de Córdoba. Serie especiel I, Córdoba
40. Dizeo de Strittmater C (1973) Nueva técnica de diafanización. Bol Soc Argent Bot 15(1):126–129
41. Albrecht RA, Bassols GB, Gurni AA (2005) Aplicación de parámetros cuantitativos en el estudio de *Aloysia citrodora* Palau (Verbenaceae). Acta Farmacéutica Bonaerense 24(2):239–241
42. Gurni AA, Wagner ML (2007) Guía de Trabajos Prácticos de Farmacobotánica. Centro de Estudiantes. Facultad de Farmacia y Bioquímica, UBA Buenos Aires
43. Babot MP, Oliszewski N, Grau A (2007) Análisis de caracteres macroscópicos y microscópicos de *Phaseolus vulgaris* (Fabaceae, Faboideae) silvestres y cultivados del noroeste Argentino: una aplicación en arqueobotánica. Darwiniana 45(2):149–162
44. Torrence R, Wright R, Conway R (2004) Identification of starch granules using image analysis and multivariate techniques. J Archaeol Sci 31:519–532

Chapter 21

Methods and Techniques for Research on the Supply Chains of Biodiversity Products

Rafael Ricardo Vasconcelos da Silva, Laura Jane Gomes, and Ulysses Paulino Albuquerque

Abstract

A "supply chain" comprises the set of transactions needed to produce a good or service, including the path from production to the end user. In ethnoecological and ethnobiological studies, the focus on supply chains allows a wider contextualization of the socioeconomic importance of biodiversity products. This wider context includes the agents outside the local communities and allows a more thorough analysis of the factors that influence the commercial use of native species by local populations. In this chapter, we discuss the primary methods and techniques used to study the supply chains of biodiversity products.

Key words Supply chain, Extractivism, Biodiversity products, Socio-biodiversity, Socioeconomics

1 Introduction

The notion of a supply chain, as formulated by Goldberg in 1967 [1], represents an approach to studying the flow of products based on the analysis of the agents that constitute the links of the chain, which ends with the end user. In this sense, we consider a "supply chain" to be a set of transactions leading to the production of a good or service, including the path from production to end user, as well as the institutional and organizational influences along the production steps[1] (Fig. 1).

A large fraction of the biodiversity products are obtained in an extractivist manner [2, 3]. An analysis using the supply chain perspective indicates that under these conditions, the extractivist products of biodiversity might be associated with remarkable environmental, economic, and social fragility that is even greater when strategies and public policies to promote the sustainable use

[1] We refer to the different stages involved in the transformation of a raw material into the final product as the "production steps," i.e., the various phases of production from harvest to end user.

Ulysses Paulino Albuquerque et al. (eds.), *Methods and Techniques in Ethnobiology and Ethnoecology*,
Springer Protocols Handbooks, DOI 10.1007/978-1-4614-8636-7_21, © Springer Science+Business Media New York 2014

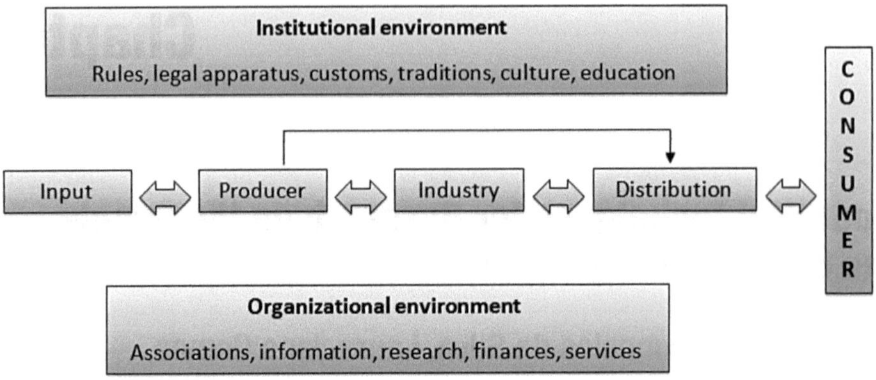

Fig. 1 Overall model of the supply chain. Adapted from [14]

of these products are not available [4]. In Brazil, for instance, the federal government began to approach this topic as a matter of public policy only in 2009, when the National Socio-Biodiversity Plan was instituted to promote an environment favorable to the development of specific work plans and to formulate a strategic view that supports and stimulates the local and regional productive arrangements based on the idea of socio-biodiversity product chains. Only through these means could the extractivists gain strength [5].

The number of studies focusing on supply chains has increased, resulting in attempts to approach this subject from an interdisciplinary perspective, particularly as it concerns the interface between economics, government, and the social sciences. In the fields of ethnoecology and ethnobiology, for instance, the analysis of the factors influencing the extractivism of native species for commercial purposes [6] might be placed into context by focusing on the supply chains. These types of studies might eventually also contribute to the elaboration of strategies for the management of these species [7].

The current chapter discusses the primary methods and techniques used to study the supply chains of biodiversity products and how these chains might contribute to a better understanding of the complex relationship between nature and culture.

2 Methods and Techniques

2.1 Sampling

The supply chains related to biodiversity usually exhibit wide variability in terms of their socioeconomic profile and the number of the agents involved in the various steps from harvest to commercialization. Therefore, a systemic view of the supply chain of interest must be developed to define the type of approach to use for the study of each of these steps.

An indispensable step in the preliminary characterization of the investigated supply chain is a survey of the information available in the literature, reports, and databases of official agencies among other sources of secondary data. In addition, and whenever possible, initial interviews must be conducted with the representatives of the extractivist communities and the institutions involved in the exploitation of the species under study. At this step, priority must be given to the informants who are identified as being well-acquainted with the chain.

In this preliminary characterization, a collection of data regarding the approximate number of people or families involved in the various production steps; the times and sites of harvesting; the methods for storage, improvement, and commercialization; the existence of extractivist associations or cooperatives; and the identity of the buyer companies is important among other data that are considered relevant for the definition of the sample, the selection of informants, and the study planning. One must keep in mind that this information is used to define a sample that must include the various agents involved in each link of the chain, e.g., extractivists, middlemen, transporters, processors, members of associations or cooperatives, retailers, end users, entrepreneurs, landowners, public managers, and representatives of the involved institutions (e.g., non-governmental institutions—NGOs, research agencies).

Based on the collected information, one might choose to perform a "probabilistic" or "non-probabilistic" sampling at each production step [8]; the latter is commonly used in studies concerning the supply chains of biodiversity products.

Probabilistic sampling should be applied when the number of agents involved in the investigated production step is quite large and exhibits wide variability, as is often the case with extractivist communities. This type of sampling is advantageous because it grants each element the same probability of participating in the sample, and thus the results represent the entire population.

However, when the subject matter is extractivism, a probabilistic sampling does not always yield satisfactory results. In the cases where a given species or the extractivist practice is subjected to legal constraints, the informants tend to feel inhibited and unwilling to provide information, even when they are aware that their anonymity will be protected. In these situations, it is advisable to perform non-probabilistic intentional sampling because it allows the informants who are involved in the extractivism and commercialization of the species and who are able to convey information more easily to be selected.

"Non-probabilistic" sampling must also be used for the production steps that involve a relatively low number of informants and little variability in regard to the desired information (variables of interest). That type of sampling allows the informants to be selected in an intentional manner when probabilistic sampling is impossible or inappropriate. In these cases, the "snowball technique" is recommended, whereby an informant is first interviewed, and later, at the end of interview, he or she is asked to indicate other possible informant(s) involved in the various production steps (e.g., extractivism, transportation, processing, or commercialization).

2.2 Data Collection

The main tools for data collection in the field are structured and semi-structured interviews [9]. The forms used in the interviews must be adjusted and applied to the various agents involved in the chain from the extractivist to the end user and must include representatives from the different institutions that influence the chain.

Structured interviews are conducted using forms that contain basic exploratory questions (e.g., the interviewees' socioeconomic data). This type of interview is recommended for the initial contact with the informants because it will provide information on their profile and help draft the forms for the semi-structured interviews. The latter, in turn, include questions that seek to dig deeper into the most relevant subjects. In both cases, the questions must be drafted based on the information expected from each link of the chain. In addition, each interview should be planned based on previous interviews, observations, statements, the reports of events, the analysis of documents, minutes from meetings, reports, and contracts [10]. Box 1 illustrates several examples of the basic questions that could be answered by the various links in a supply chain.

Box 1 Examples of Interview Questions for Studies on the Supply Chains of Biodiversity Products

GENERAL FORM FOR INFORMANT IDENTIFICATION

No.° of questionnaire:	Municipality:	District:	Community:
Informant's name: Surname:	Age:		Birthplace:
Marital status:			
Educational level:			
Occupation:			
Address:			
Complementary information:			
Telephone or other contact means:			
Residents at the same household: () Father () Mother () Children (how many?): () Other (how many?):			Age of children:
Name of spouse:			
Spouse's age:			
Spouse's educational level:			
Spouse's occupation:			
Complementary information:			

(continued)

Box 1 (continued)

GENERAL SCRIPT FOR INTERVIEWS WITH EXTRACTIVISTS

1. **Extractivism history:** When did the harvesting start (month/year)? What products do you harvest? Reason (e.g., market value)? How long have you been familiar with the extracted species? How and from whom did you learn this activity?
2. **Access, handling, and knowledge:** Where do you harvest? Do you harvest from private properties (identify them)? Do you pay a fee for the right to harvest (how much)? What do you do with the harvested products? What is the use of each product? What is the harvesting season of each product? During what month of the year does each species exhibit the greatest yield? What is the harvesting frequency (how many days per week)? How many hours do you work per harvesting day? What is the ideal site from which to harvest the product? Do you select or choose the trees for harvesting? What criteria do you apply? How is harvesting performed (tools used)? How much do you harvest per day? How do you measure the harvested product (measure used)? Does harvesting differ when performed closer or farther from the original harvesting site (how)? What is the destination of the harvested product? How do you transport the harvested product? Do you store the harvested products? How and where? Do you treat the products before selling them? What do you treat them with? Is production increasing or decreasing? Why? Can something be done to increase or maintain the level of production? On a regular harvesting day, can something be done to ensure a satisfactory return?
3. **Income:** What are your family's income sources? What is your family's main income source? How much is your family's monthly income? How many members contribute to your family's income? What is the price of the product per kilo (or other measure)? How is the product paid for? How much does the sale of each harvested product contribute to the family's income?
4. **Interactions among extractivists:** Is harvesting performed in groups or individually? Who participates in harvesting (relatives, friends, and/or neighbors)? What can and cannot be done while harvesting? Is there conflict among extractivists (which ones?) What do you do when conflicts arise while harvesting? Is there conflict between the extractivists and other people? Do people help each other while harvesting? What do you do when you are not harvesting (other activities)? At what time of the year (months)? Do you participate in groups or associations (which ones)? Are there other activities performed in groups in addition to harvesting?
5. **Social organization of extractivism:** Who organizes the harvesting (leader)? What is done to begin harvesting (how do you organize yourselves)? What does one need to do to be an extractivist (who does one contact?) Are there people who cannot be extractivists (why)? Are there specific functions in harvesting (things only some people may do)? Do extractivists divide into groups? Do longer-term and more recent extractivists hold the same rights? Who are the people who buy the harvested products? Who pays for the activity of harvesting? How is payment calculated (per kg, per day)? Who earns more from harvesting and why (name them)? Are other people from the community aside from the extractivists involved in extractivist activities? What do you know about the person or company that buys the harvested product? Does someone hinder the harvesting of products at ... (who and why)? What are the extractivists' opinions on harvesting? What is your opinion regarding the requirement of a permit for harvesting?
6. **Conflict:** Are there hindrances at the time of harvesting (causes, which products, how, where)? Have you ever been detained due to harvesting? By whom (where, when, cause)? Have you ever faced a difficult situation while harvesting? Are there problems due to the money paid to the extractivists?
7. **Community:** What does the community do to make a living? Who are the oldest members in the community (name them)? Who are the large landowners in the community or in the surrounding area? What are the productive activities performed by the community (agriculture, trade, extractivism, fishing, apiculture)? Are extractivists landowners? In your opinion, has the community changed since extractivism began? What do you like in your community? What you do not like in your community? What are the community's problems? What is the future that you imagine and plan for your community? In your opinion, what does your community need? Which people do you consider to be community leaders?
8. **Environment:** What does the environment mean to you? What is damaging for the environment? How is the environment in your community? Tell what the harvesting environment was like in the past and now? What changes occurred (damages)? What is the effect of extractivism on the environment? What is done to keep from destroying the environment? Who cares for the environment in your community?

(continued)

Box 1 (continued)

GENERAL SCRIPT FOR INTERVIEWS WITH MIDDLEMEN, PROCESSORS, AND TRADERS (MARKETS OR COMPANIES)

How did you come to work in this job? How long ago? What do you buy? From whom do you buy it (names, institutions, sites)? How much do you pay? To whom do you sell (names, institutions, sites)? For how much do you sell? What are the products' uses (medicinal, food, seasoning, oil, etc.)? Where are the processing plants located? How was the potential of this species discovered? Where are the products processed and traded? What criteria define the selection of the buying sites? From how many communities do you buy? How much of the processed product is exported? Who are the main consumers? Are there competitors in other countries? Which ones? What proportion of the supply comes from cropping and what from extractivism? Are there quality standards for the products (size, color, format)? What tests are performed to assess the quality or standards of the products? Are there differences in the active principles content per area of occurrence? What tests demonstrate these differences? What are the supplier states and what amounts do they produce? Is the processing waste used for some purpose (animal feed, fertilizers, fuel)? What is the proportion of product lost and/or that does not meet the buyers' standards? What amount must the company buy to meet the demand? How much do you pay for the product occurrence sites to be permitted to harvest in the states? Do you have difficulties in harvesting in the states? Which states? Has the company ever been penalized by environmental agencies in the states where it buys its products? Which agencies?

GENERAL SCRIPT FOR INTERVIEW WITH PUBLIC MANAGERS

Are you acquainted with extractivism? Are there policy guidelines for that activity? What products are authorized for harvesting and trading? Are permits needed to use the existing species? Are there cession agreements? What is the environmental situation, and what is the policy for the area? Which companies perform harvesting? What is the procedure adopted by the agency that authorizes harvesting? What are the standards (technical, inspection, period, amount, licensing, and/or permit fees)? How much product comes from extractivism (kg/ton)? What are the regional exports? Do you know the product's destination (state/country)? Is there a register/records of the people who practice extractivism? What is the economic market value of extractivist products? How much (in BRL) does extractivism contribute in taxes to the area? What is the agency's position in regard to extractivism?

Adapted from [20] and [21]

The data collected might be subjected to qualitative and/or quantitative analysis. The aim of a qualitative approach would be to collect subjective and detailed information on the socio-environmental relationships of the extractivist activity, and it is widely used in research on the supply chain for biodiversity products. The quantitative approach allows the data collected to be quantified for subsequent statistical treatment, and it is equally important for the data analysis in studies on the supply chain. It is worth noting that there is complementarity, not contradiction, between the quantitative and qualitative approaches [11].

Depending on the study, data collection might be complemented by other research tools (Box 2), such as the so-called participatory methods [12, 13], which have been applied to studies on supply chains. These tools allow detailed information to be collected in a shorter period of time and permit a more informal relationship between the investigator and the informant.

Box 2: The Primary Instruments for Data Collection Used in Studies on the Supply Chains of Biodiversity Products

Methods	**Aim**	**Description**
Analysis of documents	To develop a general view of a given supply chain, identifying its potential and its weak points. To supplement the information collected through interviews and other sources whenever it is necessary	Survey of the data is usually performed before the first visit to the community. All of the existing municipal, state, and federal projects and programs must be analyzed in addition to the projects of the nongovernmental organizations that are active in the area, which might contribute to the strategic and operational planning of the study. During the development of the study, the document analysis might complement the information collected from other sources (oral, iconographic), thus increasing the robustness of the results
Direct observation	To allow investigators to perform an "*in loco*" analysis of the various activities related to the chain	Ideally, the observation should cover all of the production steps, from harvest to storage, processing, and trade with the companies, fair markets, and markets. The data thus generated might serve to complement the information collected from interviews and other sources as well as to elicit further questions or to identify contradictions between the interviewees' *corpus* (knowledge) and *praxis* (behaviors)
Direct measurement	To obtain data on the amounts (in kilograms or other units) commercialized of a given product, distances travelled, and work time, among other data	Measurement allows for the collection of quantitative data and for the calculation of estimates that are representative of the chain
Questionnaires	To collect specific information relevant to the study such as the community's socioeconomic profile	A questionnaire might contribute to all of the study stages, particularly to planning, by providing basic information on the agents involved in the chain
Semi-structured interviews	To collect more thorough information on different topics that are relevant to the study	Semi-structured interviews are based on a form containing a series of questions that are relevant to the study, which allows for a more thorough analysis of the responses and the inclusion of further questions. The questions must be applied to the various social actors who are directly or indirectly participating in the chain
History of the activity	To collect objective and subjective information on the history of commercial extractivism from the various agents involved	The informants are invited to answer questions on the origin of extractivism in the community, the first residents to practice it, the creation of associations or cooperatives, the most important events that occurred in the community, the changes occurring after the onset of the activity (in regard to income, the environment, habits, etc.), among others

(continued)

Box 2 (continued)

Mapping	To identify harvesting sites (harvesting areas), sales points, processing units, existing infrastructure (highways), etc.	In a participatory workshop, the participants are guided by a facilitator to draft a map (or use a pre-existing map) where the sites related to the activity are highlighted. Once the map is drafted, it is advisable to perform a verification "*in loco*" together with the participants to observe the features that are most relevant to the study
Seasonal calendar	To identify the main production-related activities performed by the community during 1 year	In a participatory workshop, the informants are requested to list all of the income-producing activities, as well as those directly related to the local traditions and customs. The data thus produced serve as indicators of the importance of the extractivist activity for the participants' lifestyles and production modes. In addition, the calendar allows for the appropriate organization of the chronogram of the study tasks
Daily routine forms	To verify the distribution of the daily harvesting, processing, and trading activities among men, women, and youths	The daily routine forms require a consolidated trust-based relationship between the investigator and the informants. The latter must understand clearly the procedures for filling out the forms, which must be written in a clear and understandable style. The forms might produce detailed information on the amounts harvested, number of hours worked, the distance travelled, and the resulting income, among other information
Guided tour	To accompany (e.g., on foot, by car, or by boat) the informant up to a site of interest (e.g., harvesting or sales sites)	The investigator must observe and record in a field notebook, using a tape-recorder, photo camera, and GPS, the features that are relevant for the characterization of the visited site (e.g., landscape, state of the vegetation, and infrastructure). During the tour, the informant must be asked about the subjects of interest

Another important step in the studies on supply chains is visits to the companies and processing units involved in the investigated activity and employee interviews. Observations and interviews *in loco* offer the investigator a detailed image of the important production steps, which, in many cases, summarize and consolidate the process of transforming a component of biodiversity into an economic resource.

It is worth noting that the data collection in studies on the biodiversity chain must also include information on the product supply and demand, production costs, selling prices, profit margin, the localization and socioeconomic characteristics of the extractivist communities, the product's overall characteristics, its importance for its industrial uses (including cosmetics, pharmacological, and nutrition), and its importance in the lifestyles of the extractivist communities [14].

In addition, the ecological information related to the populations of native species used in an extractivist manner ought to be sought whenever possible. The data on the density, abundance, frequency, and reproductive processes of these species, among other characteristics, will allow inferences to be made on the possible impact of the supply chain on the conservation of the investigated native species [15].

2.3 Analysis and Interpretation of Data

The analysis and interpretation of data must be performed according to the study aims and hypotheses. In ethnobiology and ethnoecology, special attention must be paid to the relationship between the extractivist human population and biodiversity. Accumulating theories aimed at the identification and understanding of human behavioral patterns in situations that are characterized by interaction in the search for resources can create a background for hypotheses that might reveal interesting information (see Box 3).

In addition, a focus on the supply chain allows for a more thorough analysis of the socioeconomic importance of the extractivist products for the agents who are outside of the local communities [16]. In this regard, the "progressive contextualization" technique [17] might provide important contributions to the data analysis. According to [11], that technique allows for a broader scope beyond the local level, including the regional, national, and even the global levels. These authors emphasize the importance of different scales of analysis because a large part of the human local population is subjected to external influences that might play a defining role in their approach to the handling and use of native species.

It is worth noting that the product's trajectory from harvest to production and market distribution, including the various phases of processing, must be described and analyzed in a clear and objective manner. In this regard, quantitative data analysis procedures are widely used in the studies examining biodiversity product supply chains.

Box 3 Theories with Possible Applications to Ethnoecological and Ethnobiological Studies Involving the Supply Chains of Biodiversity Products

The data originating from the studies on supply chains are usually interpreted to elucidate the positive or negative influence that the supply chains exert on the life of the native species and on the people involved in the various production steps. Based on these diagnostics, the aim of these studies is to indicate possible paths to strengthen the chains by reducing the asymmetries found along them and by implementing management and trade techniques that contribute to the maintenance of the investigated activity. Little attention has been paid to the behaviors and decisions made by the agents involved in the investigated activity.

Within that context, we highlight the following theories because they might contribute to the formulation and testing of hypotheses on the influence that a given supply chain might exert on the relationship between a human extractivist population and biodiversity.

Game theory: provides models that are useful for the description and analysis of situations characterized by interactions among people (or organizations) who recognize the mutual interdependence of their individual decisions and act in a rational and strategic manner, always aiming at achieving the best results (profit maximization) [22, 23]. This theory might contribute to the analysis of situations where the rational decision of each individual ends in a negative (irrational) result for the group [24], which can establish the dilemma presented by the "tragedy of the commons" hypothesis. According to that hypothesis, formulated by [25], when a natural resource is shared by many, competition and the resulting pressure tend to exhaust it.

Optimal foraging theory: suggests that the problem posed by the distribution and optimization of energy is analogous to a cost-benefit analysis performed in the field of economics. In studies on supply chains performed with extractivist communities, the benefit corresponds to the profit generated by the extractivist activity and cost corresponds to the time (or money) required to ensure returns. An application of this theory might allow for an evaluation of the assumption that several features of human behavior are selected by evolution and adaptation in regard to plants [26]. In this regard, Begossi [27] observes that optimal foraging theory might be a useful tool to understand the decisions made regarding the interaction between individuals and resources in an area under collective management.

Box 4 An Example of a Flowchart for Studies on the Supply Chains of Biodiversity Products

Flowcharts are important tools to organize and present data. This technique generates graphical representations that allow for the schematization and visualization of all of the processes included in a supply chain in an objective, clear, and summarized manner, and thus facilitate understanding.

Below is an example (Fig. 2) of a flowchart that was drafted by [28] in a study of the supply chain of Brazilian pepper, i.e., the fruit of *Schinus terebenthifolius* Raddi, a tree species native to Brazil, the extractivist production of which meets the market demand.

In qualitative studies, data collection and analysis usually occur in an integrated manner. For the final presentation of the data, summaries, network diagrams, and flowcharts are drafted (Box 4), as well as matrices with accompanying text [11]. After the systematization and organization of the data set, inferences and appropriate discussions of the data become possible.

It is advisable that the data analysis be based on at least three different sources of information whenever possible, as recommended by the so-called triangulation technique [18]. The triangulation technique consists of using multiple methodological and theoretical perspectives, data, and investigators to attain the widest

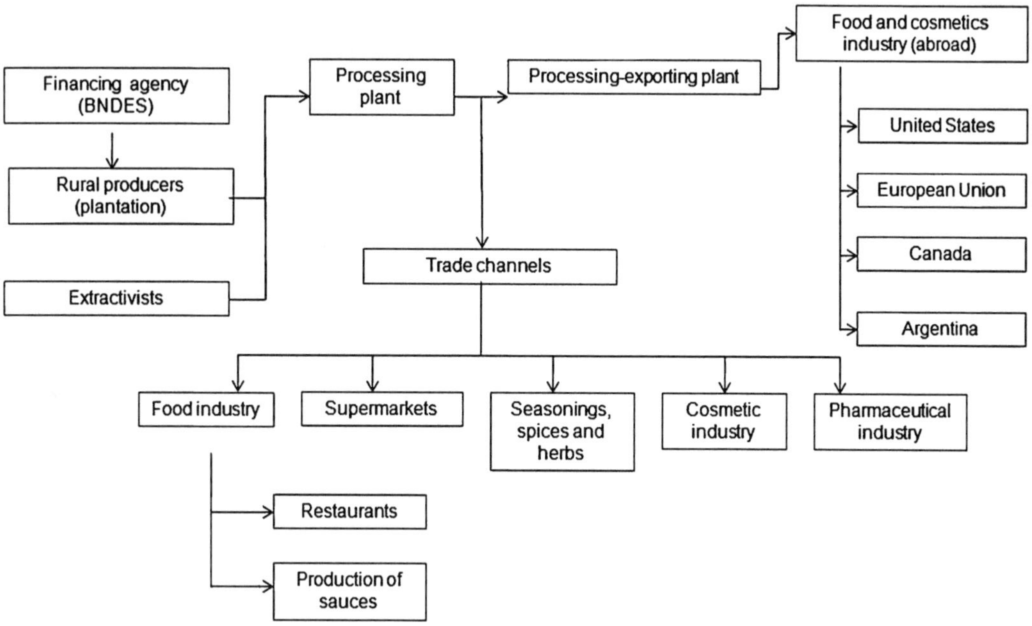

Fig. 2 Flowchart of the Brazilian pepper supply chain (production, partners, exportation, commercialization). *Source*: [28]

possible amplitude in the description, explanation, and understanding of the topic under study [16, 19]. Thus, the collected data can be analyzed more thoroughly and with greater precision by cross-referencing the information obtained from secondary sources, semi-structured interviews, and the various tools for quick participatory diagnosis, for example.

3 Final Considerations

The studies related to the supply chains of biodiversity products in ethnobiological and ethnoecological research allow a wider contextualization of the factors that influence the relationships between the human extractivist population and biodiversity. In addition, these studies provide a context for the socioeconomic importance of these products for the agents who participate in the supply chain and are outside the extractivist community. This analytic perspective transcends the boundaries of the production environment and the local human population as isolated elements, while seeking to establish the influence that the market exerts on the behavior of a human extractivist population.

Thus, these studies might make important contributions to the management strategies for the supply chains of biodiversity products. As a possible offshoot, not only is the conservation of the involved native species expected but also an enhancement of the extractivist populations' lifestyle and, consequently, also an enhancement of the processes through which those human populations acquire and perpetuate their knowledge regarding the environment.

References

1. Santana AC (2002) Caracterização do mercado de produtos florestais madeireiros e não-madeireiros da região de Mamuru-Arapiuns. Universidade Federal Rural Da Amazônia, Instituto Socioambiental e Recursos Hídricos, Instituto de Desenvolvimento Florestal do Pará (Relatório) [Characterization of the timber and non-timber forest product market in the Mamuru-Arapiuns area. Federal Rural University of Amazonia, Socio-Environmental and Water Resources Institute, Institute of Forest Development of Pará (Report)], Belém, Pará
2. Neumann RP, Hirsch E (2000) Commercialisation of non-timber forest products: review and analysis of research. STM Grafika, Putera, Indonédia
3. Alexiades MN, Shanley P (2004) Productos forestales, medios de subsistencia y conservación: estúdios de caso sobre sistemas de manejo de productos forestales no maderables [Forest products, means of subsistence and conservation: case studies on non-timber forest product management systems]. In: Alexiades MN, Shanley P (ed) Productos forestales, medios de subsistencia y conservación: estúdios de caso sobre sistemas de manejo de productos forestales no maderables [Forest products, means of subsistence and conservation: case studies on non-timber forest product management systems]. CIFOR, America Latina
4. Gomes LJ, Gomes MAO, Jesus NB (2010) Aspectos socioambientais da atividade extrativista de produtos florestais não-madeireiros: os casos da Fava-D'Anta (*Dimorphandra*sp.) e da Aroeira-da-praia (*Schinusterebinthifolius* Raddi) [Socio-environmental features of the extractivist activity of non-timber forest products: the cases of *Dimorphandra* sp. and the Brazilian pepper tree (*Schinus terebinthifolius* Raddi)]. In: Albuquerque UP, Hanazaki N (eds) Árvores de valor e o valor das árvores: pontos de conexão [Trees of value and the value of trees: connection points]. NUPEEA, Recife
5. Brasil (2009) Plano nacional de promoção das cadeias de produtos da sociobiodiversidade. [The National Plan for the promotion of sociodiversity product chains]. MDA, MMA, MDS, Brasília
6. Ruiz-Pérez M, Belcher B, Achdiawan R, Alexiades M, Aubertin C, Caballero J, Campbell B, Clement C, Cunningham T, Fantini A, De Foresta H, García Fernández C, Gautam KH, Hersch Martínez P, De Jong W, Kusters K, Kutty MG, López C, Fu M, Martínez Alfaro MA, Nair TR, Ndoye O, Ocampo R, Rai N, Ricker M, Schreckenberg K, Shackleton S, Shanley P, Sunderland T, Youn Y (2004) Markets drive the specialization strategies of forest peoples. Ecol Soc 9(2):4 [online]
7. Varghese A, Ticktin T (2008) Regional variation in non-timber forest product harvest strategies, trade, and ecological impacts: the case of Black Dammar (*Canariumstrictum*Roxb.) use and conservation in the Nilgiri Biosphere Reserve, India. Ecol Soc 13(2):11 [online]
8. Albuquerque UP, Lucena RFP, Neto EMFL (2010) Seleção dos participantes da pesquisa [Selection of the study participants]. In: Albuquerque UP, Lucena RFP, Cunha LVFC (eds) Métodos e técnicas na pesquisa etnobotânica [Methods and techniques in ethnobotanic research]. NUPEEA, Recife
9. Albuquerque UP, Lucena RFP, Alencar NL (2010) Métodos e técnicas para coleta de dados etnobiológicos [Methods and techniques for the collection of ethnobiological data]. In: Albuquerque UP, Lucena RFP, Cunha LVFC (eds) Métodos e técnicas na pesquisa etnobotânica [Methods and techniques in ethnobotanic research]. NUPEEA, Recife
10. Alencar E (1999) Introdução à metodologia de pesquisa social [Introduction to the methods for social research]. UFLA, Lavras
11. Amorozo MCM, Viertler RB (2010) A abordagem qualitativa na coleta e análise de dados em etnobiologia e etnoecologia [The qualitative approach to the collection and analysis of data in ethnobiology and ethnoecology]. In: Albuquerque UP, Lucena RFP, Cunha LVFC (eds) Métodos e técnicas na pesquisa etnobotânica [Methods and techniques in ethnobotanic research]. NUPEEA, Recife
12. Kummer L (2007) Metodologia participativa no meio rural: uma visão interdisciplinary. Conceitos, ferramentas e vivências [Participatory methods in the rural setting: an interdisciplinary view. Notions, tools and experiences]. GTZ, Salvador
13. Sieber SS, Albuquerque UP (2010) Métodos participativos na pesquisa etnobiológica

[Participatory methods in ethnobiological research]. In: Albuquerque UP, Lucena RFP, Cunha LVFC (eds) Métodos e técnicas na pesquisa etnobotânica [Methods and techniques in ethnobotanic research]. NUPEEA, Recife
14. Enríquez GEV (2008) Desafios da sustentabilidade da amazônia: biodiversidade, cadeias produtivas e comunidades extrativistas integradas. [Challenges to Amazonian sustainability: biodiversity, supply chains, and integrated extractivist communities]. Doctoral dissertation, Universidade de Brasília (Tese), Brasília
15. Fernandez ASF, Antunes PC, Leandro M, Zucco CA (2012) How sustainable is the use of natural resources in Brazil? Natureza & Conservação 10(1):77–82
16. Jensen A (2009) Valuation of non-timber forest products value chains. Forest Policy Econ 11:34–41
17. Vayda AP (1983) Progressive contextualization: methods for research in human ecology. Hum Ecol 11(3):265–281
18. Triviños ANS (1987) Introdução à pesquisa em ciências sociais: a pesquisa qualitativa em educação [Introduction to research in the social sciences: qualitative research on education]. Atlas, São Paulo
19. Maggs-Rapport F (2000) Combining methodological approaches in research: ethnography and interpretive phenomenology. J Adv Nurs 31(1):219–225
20. Gomes LJ (1997) Extrativismo e comercialização da fava d'anta (*Dimorfhandra* sp.): estudo de caso na região de cerrado de Minas Gerais. [Extractivism and the commercialization of *Dimorphandra* sp]. (Dissertação) [MA dissertation], Universidade Federal de Lavras, Lavras, Minas Gerais
21. Jesus NB (2010) Relações socioambientais no extrativismo da aroeira (*Schinusterebenthifolius*Raddi) no baixo São Francisco SE/AL [Socio-environmental relationships in the extractivism of Brazilian pepper tree (*Schinus terebinthifolius* Raddi) in the low São Francisco area SE-AL]. (Dissertação) [MA dissertation], Universidade Federal de Sergipe, Aracaju, Sergipe
22. Fiani R (2006) Teoria dos jogos: com aplicações em economia, administração e ciências sociais [Game theory: with applications to economics, administration, and the social sciences]. Elsevier, Rio de Janeiro
23. Binmore K (2009) La teoria de juegos: una breve introducción [Game theory: a short introduction]. Lopez PV (trad). Alianza Editorial SA., Madrid
24. Barash DP (2004) The survival game: how game theory explains the biology of cooperation and competition. Holt Paperbacks, New York
25. Hardin G (1968) The tragedy of the commons. Sci J 162:1243–1248
26. Albuquerque UP, Hanazaki N (2009) Five problems in current ethnobotanical research—and some suggestions for strengthening them. Hum Ecol 37:653–661
27. Begossi A (2000) Ampliando o CPR toolkit [Enlarging the CPR toolkit]. Common Property Resour Dig 55:9
28. Jesus NB, Gomes LJ (2010) Cadeia produtiva da pimenta-rosa a partir do Baixo São Francisco, Sergipe, Alagoas [The supply chain of Brazilian pepper from the Low San Francisco area, Sergipe, Alagoas]. In: Silva VA, Almeida AS, Albuquerque UP (eds) Etnobiologia e Etnoecologia: pessoas e natureza na América Latina [Ethnobiology and ethnoecology: people and nature in Latin America]. NUPEEA, Recife

Chapter 22

Methods and Techniques Applied to Ethnobotanical Studies of Timber Resources

Marcelo Alves Ramos, Patrícia Muniz de Medeiros, and Ulysses Paulino Albuquerque

Abstract

In this chapter, we discuss a number of steps that are relevant to the planning and execution of an ethnobotanical study of the domestic use of timber. We introduce the main methods used in this type of study, including their applications and limitations, allowing the reader to identify the techniques that should be implemented in your studies.

Key words Wood resources, Firewood, Ethnobotany, Traditional knowledge, Categories of use, Domestic use of timber

1 Introduction

Forest ecosystems provide a large amount of goods and services for humanity; trees are the most exploited of these products because they form the major structural component of forests. Timber is the main product obtained from woody plants, and the extraction of timber from forests is harmful to plant populations. Beyond of the large-scale commercial uses, timber can be used to supply the different needs of local populations, including the construction of houses, boats, and fences for land delimitation, the manufacture of working tools, and the use of firewood as fuel and for charcoal production.

However, despite the importance of timber resources for the subsistence of local populations and for biodiversity conservation, there is a notorious lack of ethnobotanical studies specifically addressing this subject. Most ethnobotanical studies, even those focusing on conservation, continue to address questions regarding the use of medicinal plants or plant resources in general. In the studies of general plant resources, the categories of use that include timber products are diluted and lack a more comprehensive approach.

Ulysses Paulino Albuquerque et al. (eds.), *Methods and Techniques in Ethnobiology and Ethnoecology*,
Springer Protocols Handbooks, DOI 10.1007/978-1-4614-8636-7_22, © Springer Science+Business Media New York 2014

As the subject of timber use has been largely neglected in the ethnobotanical studies, there is an lack standardization and methodological consistency. Therefore, in this chapter, we introduce the main topics that are important for the ethnobotanical study of the domestic use of timber. This information will enable the reader to (a) distinguish the main methods/techniques that are being reported in the literature, (b) know the types of data that can be acquired from each of the methods, and (c) identify the limitations of each of these tools.

1.1 Identifying the Timber Uses in a Region

An initial challenge for the investigation of the domestic use of timber is register the diverse uses that encompass the collection of this resource. As researchers, we commonly use the term "timber" because our training typically includes the conceptual distinction between timber products and non-timber products. However, these terms are not always part of the cultural universe of the studied population.

Given this scenario, imagine the following situation: a particular researcher decides to use the interview technique to determine the diversity of timber uses in a given region and asks his/her informers "what timber uses are known?" or "which uses require timber extraction?" Will the choice of words affect the research in some way? In our experience, the choice of words is mostly likely to contribute to the collection of underestimated data, which would not faithfully represent the uses that actually utilize timber in the study area.

Therefore, we emphasize the necessity of avoiding communication problems by eliminating any questions that are not fully understood by the studied population. Because a limited number of uses are commonly reported in studies that record the diversity of timber uses, investigators should be aware of this issue. An interesting alternative is to perform pretests, incorporating questions such as "what are the uses of tree branches and trunks in this community?" Another option is to conduct guided tours of houses, noting the uses of timber and questioning the informers about the names of these uses. From these preliminary analyses, the researcher can get an idea of the common uses of timber in the community and thus may conduct interviews about each use.

Another relevant point for the study of timber products is the lack of standardization of what actually constitutes timber use. For example, some studies may consider products, such as chairs, tables, and house roofs, as timber use, whereas other studies may consider the chair legs, chair seat, table legs, house beams, house lines, slats, etc., and not the chair, the table, or the house itself. These differences in the perception of what constitutes timber use may produce different results when, for example, calculating the use value of timber species. Therefore, we suggest that the classification of timber uses be performed according to the second example in which each part of a wooden structure is considered a separate use. Different tree species can be used for each different wooden part of a structure, even within a single structure (e.g., a house); therefore, these parts should be considered individually. Using the construction of a house as an example, it is possible that different species are suitable for the construction of each part of a house, such as the roof, the walls, and the structural support elements. Therefore, the researcher should identify the variety of parts that exist in a wooden structure.

2 Use Categories

Another point that should be emphasized is that the researcher should pay attention when classifying timber products in "use categories." Although it is common to find this category in ethnobotanical studies of the general uses of plants, in the case of timber products, the analysis of the types of uses is rare. Therefore, Table 1 demonstrates our suggestions for the standardization of the categories (types) of uses, and the respective definitions and examples are included to assist the researcher in this classification.

Following an extensive analysis of the literature, we determined that there are four distinct categories for the types of timber uses: fuel, construction, religious uses, and technology. Clearly, depending on the region and the studied population, an increased variety of timber use categories may exist; however, the researcher is responsible for identifying these uses during your study.

Table 1
Main categories for domestic timber use

Category	Definition	Examples
Construction	Elements that compose structures to territorial delimitation, houses and animal/object shelters	Fences, doors, windows, wall filling, roof filling, wall supports, ceiling and roof supports
Fuelwood	Wood destined to burning and heat generation. It can be used for cooking and heating	Charcoal and firewood
Magic-religious	Wood use that aims to create closer contacts between people and the spiritual world. It can be destined to liberate negative energies from the user's bodies, to establish links with gods and ancestors, or even to acknowledge them	Bonfires (used in festivities dedicated to catholic saints in Brazil), *Igoqo* e *ubuhlantii* (Used in religious rituals in some South African communities)
Technology	Products that suffer manipulations from the original raw material, but are not designated to area delimitation	Tool handles, chairs, tables, canoes

Distinguishing between the types of timber uses and the standardization of the terms is important because each of these types of uses may exhibit different dynamics of use, consumption patterns, and specific collection methods. In general, these differences are not considered in these types of studies. Walters [1] performed a study in the central Philippines and determined that for the construction category, although people tend to select timber with specific classes of diameters, diameter requirements are not important when using timber as fuel; therefore, the collection of timber for fuel is more flexible regarding the size of the timber. Medeiros et al. [2, 3] also noted huge differences in the selection patterns and timber uses between the distinct categories, noting that the fuel category encompassed large amounts of timber with a short replenishment time, and the collection was based on the branches and trunks of live or dead trees. In contrast, the construction and technology categories exhibited long replenishment times (years or decades), and the collection was primarily based on the trunks of live trees; however, the difference between these categories was reflected by the amount of timber used, which was greater for construction than for technology. Ramos [4], in a study performed in Northeastern Brazil, also considered the dynamics of the collection and consumption of timber resources between and within the distinct categories of use. The fuel category, for example, remains a common and locally widespread use because it is less restrictive in the community, allowing the use of a greater diversity of species. In contrast, a local trend of abandoning the use of timber for house construction has been observed. This abandonment has occurred in the region because timber uses are restricted; therefore, the reduced availability of raw material suppliers has been followed by the cultural adjustment of the community, which has been forced to resort to industrial material sources.

Among the timber use categories presented in this chapter, there are few records of the religious uses of timber in the literature. The religious uses have been investigated in the northeastern region of Brazil and include the exploitation of timber for building bonfires, which are used for cultural/religious purposes every June to honor three catholic saints, St. Anthony, St. John, the Baptist, and St. Peter [5, 6]. In South Africa, timber uses related to cultural/religious traditions have also been recorded, e.g., the *igoqo* and *ubuhlantii* [7–9]. For the *igoqo*, a pile of timber is placed in front of a house and is a female religious practice, whereas the *ubuhlantii* is related to male rituals. The *ubuhlantii* is an arrangement of wooden stakes set into the ground, which structurally resembles a fence; however, this use cannot be placed in the construction category because of its religious and cultural associations.

3 Methods/Techniques Employed in Studies of Domestic Timber Uses

Different techniques have been used for the study of domestic timber uses (see Table 2). A number of these techniques are common to ethnobiological and ethnoecological studies and have been

Table 2
Methods and techniques employed in study of the domestic use of timber resources

Method/technique	Main aim	References that used the method
Analysis of aerial photographs	• To identify harvesting areas • To analyze changes in forest cover over the years	[27, 28]
Footpath survey	• To identify forest exits, watch collectors, and weight their material	[22, 23]
Forest biomass	• To estimate wood biomass available in the forest	[9, 23, 27]
Interviews	• To record socioeconomic information • To identify the types of use that require wood collection • To record known, used, and preferred species • To register the collection patterns • To evaluate resource's replacement time • To evaluate collection distances • To estimate the amounts of consumed wood	[2, 3, 10, 13, 29–34]
In situ inventory	• To register plants and use categories available in the household (real use) • To measure the amount of collected/consumed wood • Record the collected diametric classes	[2, 3, 13, 17, 35, 36]
Vegetation inventory	• To evaluate the local status of useful species • To register the offer of useful species for each diametric class • To register individuals with harvesting evidences	[1, 9, 21, 37–39]
Weight and volume of collected wood	• To record what species or use categories demand more wood • To estimate the amount of consumed wood	[12, 13, 36, 40–43]
Participatory methods	• To identify collection areas (community mapping) • To build a seasonal collection calendar • To access local representations of collection and consumption issues	[2, 3, 27, 44]
Economic valuation	• To estimate the economic value of wood resources (cost/benefit relation)	[7, 27, 45]

referred to in previous chapters. The main idea we want to discuss in this topic is about the importance of selecting the most appropriate technique to be used in their research, i.e., "which tools would provide the most consistency for the questions raised in the study?"

In ethnobotanical studies, especially the study of timber resources, it is common to find biased results because of the inappropriate selection of a methodological tool. This is worrisome because ethnobotanical studies have a strong conservationist appeal. Therefore, in the following sections, we discuss these methodological tools and their applications and limitations in detail.

3.1 Interviews

In timber resource studies, the use of interviews is the most common method for data collection. However, caution should be applied when using this tool because researchers often draft questions that require the in-depth knowledge of a subject that the informers have not mastered; therefore, the questions are difficult for the informers to answer reliably. An example of this issue, which is discussed at a later stage, are questions regarding the units of measurement, such as the distances that the informers travel to collect the resource and the amount of timber they collect for a particular purpose.

We recommend that interviews be used in combination with other methodologies for the later triangulation of the results. The data obtained solely through interviews may lead to premature conclusions, especially when the authors use the list of species cited by the informants to highlight conservation issues and to indicate changes in the structure of plant populations.

3.1.1 Species Used

A proportion of the primary data obtained through interviews relates to the richness of the species used for the different categories of timber uses. These data should be collected with caution because when a given species is mentioned in an interview, it does not mean that the species was actually used; instead, it could be just part of the informant's body of knowledge.

Another point of concern is that a species could be used by several people for a particular purpose, but in small quantities and infrequently, whereas other species could be used by few people, but in large quantities. Therefore, one must be careful to avoid assuming that the most cited species are also the species under the greatest pressure.

3.1.2 Preferred Species

Another common issue in ethnobotanical studies is the recording of the preferred species for each use. This information is important for demonstrating the taxonomic groups that could be targets for greater extractivist pressure. In a given region, if the timber use is governed by matters of preference, the species will likely suffer the greater pressure.

Nevertheless, a number of studies do not differentiate between "preference" and "use," indicating that the most used species are the preferred ones, and vice versa. A species that is the target of greater use cannot always be considered the preferred species because there are several agents acting in the selection of a specific plant that can be related to cultural and ecological issues and the intrinsic characteristics of a species.

Therefore, we recommend that during the interviews, questions aimed at recording known species should differentiate between the species that are used and the species that are preferred. Miah et al. [10] and Jashimuddin et al. [11] performed studies in Bangladesh and described a method to record preferences using the practice of ranked pairs. In the ranked pairs method, the informant indicates the preferred plant between a choice of two, and the plant that is not selected is replaced with a new plant to complete the pair. Using this method, the plants are compared until each informant succeeds in choosing the preferred species.

3.1.3 Distance Travelled and the Frequency and Quantity of Consumed/Collected Timber

From a conservation perspective, interviews alone have been used to collect important data, such as (a) the distance travelled by the population to collect the timber, (b) the time dedicated to the activity, (c) the quantity of collected/consumed timber, (d) the frequency of collection, and (e) the replenishment time. The use of interviews alone likely contributes to the methodological inconsistencies observed in this field of knowledge, making it difficult to obtain information directly and accurately. Regarding distance, Bhatt and Sachan [12] travelled with the residents to the timber extraction area to measure the distance travelled and time spent on the activity; however, this is not always possible because it requires time and depends on the level of trust that the researcher has built within the community.

Estimates of timber resource consumption are often obtained solely from interviews, although it would be more rational to perform a survey of timber stocks inside the homes (*see* [13]). The use of interviews is advantageous from the perspective of time, resource limitations, and logistics [14]. In interviews, the following approaches are used: (a) the researcher directly asks the informant how much of the resource he/she consumes in a given time interval, using standard measurement units (kilograms and cubic meters) or local measurement units (head loads, wagon loads, and the quantity of timber a donkey can carry), and (b) the informant is asked to separate the amount he/she collects/consumes, which is weighed by the researcher. Although the latter option is less subjective, both have the potential to generate inaccurate data.

Although interviews are a practical alternative to quickly estimate the domestic consumption of timber, a number of limitations are evident for this type of analysis. Shankar et al. [14] highlighted the following issues: (a) the informant may exhibit a lack of interest in answering the questions, leading the researcher to record inaccurate data; (b) the lack of appropriate knowledge about the amount used, i.e., when the informant is asked to give the weight of timber he/she consumes daily, an incorrect answer may be given because the person is estimating a weight that he/she seldom measures; and (c) the intentional distortion of information, especially in studies conducted in areas where timber extraction is illegal.

In addition, the following issues have come to our attention during our investigation of this subject:

1. When the researcher is studying timber resources that have a long replenishment time, such as the construction of a house, the probability of recording inaccurate data during interviews increases. It occurs because in these situations, the replacement of the timber is performed for only one or a small number of damaged elements at a time; therefore, it could be difficult for the interviewee to indicate the amount consumed per time interval because this consumption is occasional and there is no frequent collection of the given material.
2. Estimating the quantities consumed using interviews may not allow the detailing of the quantities per species, which is even more worrisome when the researcher specifically asks the informants about the quantities consumed (or collected) per species. This analysis is even more inaccurate when the gathering events are random and the informant may have some difficulty remembering the amount collected for each species. The consumption data are also impaired because of temporal variations in species consumption.
3. Seasonal differences may affect timber consumption and consequently influence the informant responses regarding the consumed quantities. Therefore, if the studied area features strong seasonality, we recommend interviewing the same informants two or more times. McCrary et al. [15] and Pote et al. [9] considered seasonal differences in their consumption estimates. Ramos and Albuquerque [13] performed a 1-year monitoring of timber stocks in two rural communities in Northeastern Brazil and observed a strong influence of seasons on the volume and richness of the stocked species. The studies demonstrated that the use of one-time interviews, or even single inventories, did not encompass the full extent of the resource usage.

Therefore, as mentioned previously, to increase accuracy, the data collected through interviews should always be accompanied by other research methods.

3.2 In Situ Inventory

In situ inventories are used to record the plant material, types of use, species richness, and the diametric classes of timber in an informant's house. During the inventory, the researcher has the opportunity to estimate the amount of timber (volume or mass) in a house (*see* [13]). The inventory technique has the advantage of not depending on interviews to obtain information regarding the actual and current use of the resources; however, when combined with the interview data, it can be more functional (*see* [16]).

Gaugris and van Rooyen [17], in a study performed in South Africa, compared the data from "interviews" and from "in situ inventory" on the use of timber species for house construction. The study investigated the confidence level of these tools incorporated into models of the sustainable use of the resource. Through questionnaires, the study demonstrated that five species were highlighted, receiving the highest number of citations regarding use. The authors expected that the same species would be observed at a high frequency in the in situ inventory; however, only one of the highlighted species was part of the elements used in the structure of the houses. The study clearly showed that if the data from the questionnaires alone were used to aid the conservation efforts in any restoration project, the conclusions would have been misleading, generating a distorted representation of the actual situation regarding the use of the resources.

Despite the importance of the in situ inventory, as with any other tool, it has limitations that should be considered, such as the following:

1. Buildings and other timber structures may be covered with clay or rocks, which makes it difficult to identify the species used (material). If the studied area exhibits a predominance of covered materials, it is preferable to use another technique or to use inventories only to compare the categories of timber use (e.g., construction vs. fuel) and not to consider classifying the material according species. However, if the covered material is in the minority, continue using the inventory method.
2. Unfortunately, it is not always possible to know the reliability of the local information on timber identification. A reference guide would help minimize the problem of timber identification, enabling the researcher to distinguish the species. To help identify the materials found in the residences, the reference guide should contain samples of each timber species and a morphological description of each sample. Another option would be to use a checklist interview, such as that reported by Medeiros et al. [18] in which informants were shown and asked to identify timber samples. Therefore, the in situ inventory would be constructed based only on local information in the case of high correct identification rates. These complementary techniques, although very useful, are relevant only in long-term studies.

3.3 Vegetation Inventory

The data regarding the vegetation structure are crucial when the researcher is investigating the status of useful species exploited in a given forest area and verify whether the plant community could support the demands of the population (demand vs. supply). An interesting topic of investigation is whether the species used most frequently for a particular population corresponds to the species with the greatest abundance in the area. If there is no positive relationship between these variables, a group of plants will be more vulnerable to extraction, and the population structure could be highly affected over time because of the influence of selective logging.

Other contributions from ecological studies include the assessment of the distribution of useful species based on diameter (age structure) [19, 20] and the recording of individuals exhibiting selective logging (Table 2) [9]. An example of this analysis was reported in a study by Walters [1, 21] in which the individual trees that displayed evidence of logging were marked during a study of the vegetation in the natural and cultivated forests in the Philippines. The author noted that of all the trees sampled, 1/3 of the individuals in the natural areas exhibited signs of cuts on the trunks compared with only 8.9 % of species in the cultivated areas. It was suggested that the increased number of logged individuals in the natural forests is related to the increased basal area and diameter of the species. Although an interesting technique, the assessment of logging is limited in that the researcher cannot know precisely the intended use of the logged species.

The following precautions should be taken when drawing conclusions from the integration of ethnobotanical and ecological data:

1. If several human communities use the same area of vegetation inventoried and the ethnobotanical study only encompasses one of these communities, the remnant plant structure might be the result of use by all of the people from the community and not necessarily by the community targeted in the study.
2. The local status of plant species might not be a solely result of timber use; however, timber use is known to exert the greatest pressure on a species because the extraction method is extremely destructive. Therefore, we recommend performing a basic general study before beginning the specific study of timber uses, which will allow the researcher to understand the uses related to the timber uses for a given species and infer the relevance of any additional uses. In addition, when the researcher is performing the plant inventory, he/she should note, in addition to the numbers of logged and whole individuals, the plants showing signs of damage from other uses (e.g., evidence of bark removal for medicinal use). Even for studies of specific timber uses, it is important to consider the diversity of uses in the study area because a plant is often under pressure from a set of factors and not from just one type of use.

3. A species might be rare because of its natural distribution; alternatively, this actual distribution may reflect the previous use of the plant. Therefore, the researcher should use caution when attributing the scarcity of plants to the impact of current exploitation.

3.4 Biomass Aboveground

The assessment of biomass available aboveground and its relationship with the local collection needs (supply vs. demand) is seldom used (Table 2). Certain biomass calculations are labored and require the logging of the entire tree, which can limit the use of this technique.

An example of the biomass aboveground application can be found in Shankar et al. [14]. This study demonstrates that the demand for firewood in the study area appears lower than for the net production of the forest (biomass). The data indicate that the vegetation supports the local demands for the collection of timber. However, the study recommends considering the following two caveats for studies that rely on biomass production to indicate the sustainability of use:

1. There is a tendency for people to collect timber repeatedly from the same site in the forest, especially in the areas closest to settlements; therefore, the impacts of exploitation are not homogenous throughout the area.
2. People are selective regarding the species they will use, and biomass assessment considers the species populating the area as a whole.

Therefore, these limitations should be considered in studies focusing on the assessment of biomass approach and studies investigating policy-making and planning in the management of forest resources.

3.5 Mass and Volume of Collected/Consumed Timber

As mentioned previously, many studies use interviews to estimate the mass or volume of collected timber. However, a number of researchers have used less subjective methods that involve the direct measurement of the timber during the in situ inventory.

To measure stocks or bundles of firewood, most studies use the unit of mass (kg); however, using the volume of stacked timber (m^3) to measure stock is also common [13]. The volume measurements can be converted to mass, which is the most common unit for comparing daily timber consumption. For a number of timber uses, such as the construction of houses and fences, measuring the volume is recommended because the structure cannot be removed for weighing. The measurement of volume can be performed through the geometric measurement of the timber elements (e.g., the radius and height in the case of cylindrical elements).

Measuring the mass/volume provides data on the quantity of collected timber with more accuracy; however, it is noteworthy that these methods provide static measurements (punctual) and should be linked to the data collected through interviews, such as the frequency of collection and the size of the population responsible for this activity. This integration of the data allows for the estimation of the consumed quantities of the resource per unit time.

Measuring the mass/volume is a time-consuming technique; therefore, it is recommended for small communities in which the low number of participating households permits concluding the study within a reasonable amount of time. Because this technique can result in the researcher spending a lot of time with a single interviewee, it is advisable to conduct the inventory in segments and on different days to avoid exhausting the interviewee.

The use of the in situ inventory method can be extremely invasive because depending on the focus of the study, the researcher must enter the households and spend time measuring the materials. Therefore, the level of confidence that the target community has for the research group and how the people feel about this type of research being performed inside their homes should be considered.

3.6 Footpath Survey

The footpath survey technique is seldom used and is usually applied to the collection of firewood; however, the technique can be applied to many timber uses (*see* [13, 22]). The footpath survey entails the mapping of all points of exit from the forest and placing people at these sites to observe the exit of collectors and to estimate, visually or by weighing, the amount of collected timber (kg). McCrary et al. [23] advocates the use of this tool because his studies were performed in a forest reserve area where the people were engaged in illegal logging and did not cooperate with the interviews, alleging that they did not extract resources from the forest. In the study, the method enabled the documentation of the extraction of all forest products from the reserve, creating a complete pattern of use that generated geographical and temporal data and information on the type of resource driving the activity.

According to Shankar et al. [14], the data generated from footpath surveys have limited scientific value because the method does not consider seasonal variation and can be applied only in small areas of forest in which the boundaries are well defined and all entrance paths are known accurately. Therefore, this method requires a large team to monitor the forest entrances and to measure the quantity of material. It is also clear that using this method, the visual measurement of the amount of the resource collected and the species identification are subjective.

Another issue associated with footpath surveys is that the fuelwood collection is often performed in anthropogenic areas and not just in areas of native forest. In addition, wood can be bought or even collected from nearby native forest areas that are not targeted for assessment. Therefore, these studies should not assume that the amount of timber collected is the same as the amount of timber consumed by the local population.

3.7 Weight Survey

The weight survey method is used to determine the firewood consumption in kilograms. In each household visited, a stack of firewood greater than the amount consumed in one day is weighed and is placed near the kitchen. The determination of this amount must be performed with the informant's help to avoid the risk of leaving a quantity smaller than that consumed daily in the household. After weighing, the researchers ask the residents to use firewood only from the pile. After 24 h, the pile is weighed again, and the difference between the initial and final weights is the daily amount of firewood consumed. Shankar et al. [14] used a similar methodology, visiting selected households in the afternoon and weighing the entire firewood stock available in the house. In the study, care was taken to note whether the families had enough firewood to avoid collecting timber in the next 24 h. The team returned the next day and weighed the timber remaining in the stock.

This methodology is more efficient if it is applied at different times in the same residence to avoid estimating the amount of consumed firewood on an atypical day, which would influence the interpretation of the results. For example, Fox [24] performed a weight survey at four different times in each household. If the study area exhibits strong seasonality, it is important that the weighings be distributed evenly across the seasons. Shankar et al. [14] studied the firewood consumption in an area of the Mysore district, India, and distributed the weighings throughout the three local seasons (summer, monsoon, and winter).

Because the target of the weight survey approach is consumption, sources of bias should be avoided. For example, if the study involves investigating the amount consumed by targeting the local forest areas, the researcher should first investigate if all the consumed material is sourced locally or if some people have bought timber from different regions. Another issue concerns the identification of the weight per species. The weight survey method might not allow the measurement of the weight per species because the amount of firewood designated for use by the residents throughout the day is chosen randomly or in the manner most convenient for the researcher or the resident (e.g., the timber on the top of the pile). Therefore, the material is not a true sample of the diversity of the species used and does not reflect the proper proportions used.

3.8 Other Quantitative Techniques for Fuelwood Use

An alternative to performing geometric measurements for each element one-by-one is to calculate the stacked volume of fuelwood stocks existing in the households. The calculation is performed by multiplying the length by the width by the average of five heights of the stack. The unit of stacked volume is the stereo, which is equal to 1 m^3 of stacked timber and can be converted and estimated in terms of mass. However, these analyses should be performed with caution, especially when the measurements are performed in a static time. The researcher cannot assume, for example, that the larger stocks represent greater consumption because the collection events might be less frequent but intense, forming large stocks; the collections can also be frequent and less intense, with a high consumption but a less conspicuous stock (*see* [13]).

For studies of the general timber use, the use of stereo to measure certain uses and the conventional volume to measure other uses is not recommended because comparisons between the uses measured by natural volume and stacked volume are not accurate. In these cases, cubication factors can be used, which estimate the total solid volume based on the stacked volume.

Brouwer and Falcão [25] used a different approach; a sample of residences and residents were selected and were encouraged to record and weigh all of the firewood they consumed during 1 month. This method can provide the exact amount consumed, without estimates; however, the reliability of local information again must be considered.

Samant et al. [16] used a community approach in which ten bundles of fuelwood were selected randomly and the total weight and the weight per species were measured. Using data, such as the days of collection and the number of collectors per residence, the authors estimated the average amount of fuelwood consumed per day, per residence, and per species.

Abbot and Homewood [26] surveyed the flow of fuelwood for seven consecutive days per month for 11 months. On the morning of the first day, the stock was weighed and new elements that were bought or collected were weighed daily. The output material (donated or sold) was also recorded. Therefore, the stocks were weighed again on the last day of the week to determine the weekly consumption of fuelwood.

In addition to the methods/techniques outlined in this chapter, other methods, such as the economic values, participative rural diagnostics, and the use of aerial photographs, are included in Table 2. These techniques are used less often for the study of timber resources, and currently, they are used as complementary tools; however, the methods should be investigated in more detail.

As mentioned previously, the data collected through interviews should be combined with other techniques to increase credibility and to provide actual data to aid the conservation issues, guiding the development of management plans that are appropriate to each site.

4 Final Considerations

Each method used for the investigation of timber uses exhibits advantages and limitations and can be useful or not, depending on the study goals. Therefore, when the research methods are selected, the purpose of the study should be considered and should include the context of the community. Above all, the following should be considered: (a) The openness level with the community; (b) The uses that will be investigated (general timber uses, firewood, construction of houses, etc.); (c) The aspect of timber uses that will be studied (knowledge, pressure of use, patterns of collection, etc.); (d) The influence of seasonal issues in the study area; (e) The size of the community; (f) The time and resources available to perform the research.

Finally, we determined that although timber resource studies are rare, they encompass a variety of approaches and demonstrate enormous potential for developing more studies on the subject. Therefore, we recommend the careful review of the domestic use of timber, including the adoption of different strategies that enable reliable data to be obtained and that aid conservation and forest resource management policies. In addition to ensuring biodiversity conservation, the search for sustainable alternatives for local communities is needed.

References

1. Walters BB (2005) Patterns of local wood use and cutting of Philippine mangrove forests. Econ Bot 59:66–76
2. Medeiros PM, Silva TC, Almeida ALS, Albuquerque UP (2011) Pressure indicators of wood resource use in an Atlantic forest area, Northeastern Brazil. Environ Manage (New York) 47:410–424
3. Medeiros PM, Silva TC, Almeida ALS, Albuquerque UP (2012) Socio-economic predictors of domestic wood use in an Atlantic forest area (north-east Brazil): a tool for directing conservation efforts. Int J Sustain Dev World Ecol 19:189–195
4. Ramos MA (2011) Uso de produtos madeireiros na caatinga: uma avaliação das estratégias locais de seleção e coleta de recursos. Thesis, Universidade Federal Rural de Pernambuco
5. Martins PL, Baracuhy JG, Trovão DM, Costa GM, Cavalcanti ML, Almeida MV (2004) As essências florestais utilizadas nas fogueiras de São João, na cidade de Campina Grande-PB. Revista de Biologia e Ciências da Terra 4(1):1–13
6. Almeida AL, Medeiros PM, Silva TC, Ramos MA, Sieber SS, Albuquerque UP (2008) Does the June tradition impact the use of woody resources from an raea of Atlantic forest in Northeastern Brazil? Funct Ecosyst Communities 2:32–44
7. Cocks ML, Wiersum KF (2003) The significance of plant diversity to rural households in Eastern Cape province of South Africa. Forests Trees Livelihoods 13:39–58
8. Cocks ML, Bangay L, Wiersum KF, Dold AP (2006) Seeing the wood for the trees: the role of woody resources for the construction of gender specific household cultural artifacts in non-traditional communities in the Eastern Cape, South Africa. Environ Dev Sustain 8:519–533
9. Pote J, Shackleton CM, Cocks ML, Lubke R (2006) Fuelwood harvesting and selection in Valley Thicket, South Africa. J Arid Environ 67:270–287
10. Miah D, Ahmed R, Uddin MB (2003) Biomass fuel use by the rural households in Chittagong region, Bangladesh. Biomass Bioenergy 24:277–283

11. Jashimuddin M, Masum K, Salam M (2006) Preference and consumption pattern of biomass fuel in some disregarded villages of Bangladesh. Biomass Bioenergy 30:446–451
12. Bhatt BP, Sachan MS (2004) Firewood consumption pattern of different tribal communities in Northeast India. Energy Policy 32(1):1–6
13. Ramos MA, Albuquerque UP (2012) The domestic use of firewood in rural communities of the Caatinga: how seasonality interferes with patterns of firewood collection. Biomass Bioenergy 39:147–158
14. Shankar U, Hedge R, Bawa S (1998) Extraction of non-timber forest products in the forests of Biligiri Rangan hills, India. 6. Fuelwood pressure and management options. Econ Bot 5:320–336
15. McCrary JK, Walsh B, Hammet al (2005) Species, sources, seasonality and sustainability of fuelwood commercialization in Mayasa, Nicaragua. For Ecol Manage 205:299–309
16. Samant SS, Dhar U, Rawal RS (2000) Assessment of fuel resource diversity and utilization patterns in Askot Wildlife Sanctuary in Kumaun Himalaya, India, for conservation and management. Environ Conserv 27(1):5–13
17. Gaugris JY, van Rooyen MW (2006) Questionnaires do not work. A comparison of methods used to evaluate the structure of buildings and wood used in Rural Households, South Africa. Ethnobot Res Appl 4:119–131
18. Medeiros PM, Almeida ALS, Ramos MA, Albuquerque UP (2008) A variation of checklist interview method in the study of firewood plants. Funct Ecosyst Communities 1:45–50
19. Lykke AM (1998) Assessment of species composition change in savanna vegetation by means of woody plants' size class distributions and local information. Biodivers Conserv 7:1261–1275
20. Obiri J, Lawes M, Mukolwe M (2002) The dynamics and sustainable use of high-value tree species of the coastal Pondoland forest of the Eastern Cape Province, South Africa. For Ecol Manage 166:131–148
21. Walters BB (2005) Ecological effects of small-scale cutting of Philippine mangrove forests. For Ecol Manage 206:331–348
22. Ganesan B (1993) Extraction of non-timber forest products, including fodder and fuelwood, in Mudumalai, India. Econ Bot 47: 268–274
23. McCrary JC, Hammet al, Barany ME, Machado HE, García DJ, Barrios JI (2004) Illegal Extraction of Forest Products in Laguna de Apoyo Nature Reserve, Nicaragua, Carribean. J Sci 40:169–181
24. Fox J (1984) Firewood consumption in a Nepali village. Environ Manage 8(3):243–250
25. Brouwer R, Falcão MP (2004) Wood fuel consumption in Maputo, Mozambique. Biomass Bioenergy 27(3):233–245
26. Abbot JIO, Homewood K (1999) A history of change: causes of miombo woodland decline in a protected area in Malawi. J Appl Ecol 36: 422–433
27. Campbell BM, Luckert M, Scoones I (1997) Local-level valuation of Savanna resources: a case study from Zimbabwe. Econ Bot 51:59–77
28. Nkambwe M, Sekhwela MBM (2006) Utilization characteristics and importance of woody biomass resources on the rural-rural fringe in Botswana. Environ Manage 37: 281–296
29. Aguilar S, Condit R (2001) Use of native tree species by an Hispanic community in Panama. Econ Bot 55:223–235
30. Holmes CM (2003) Assessing the perceived utility of wood resources in a protected area of Western Tanzania. Biol Conserv 11:179–189
31. Kakudidi EK (2007) A study of plant materials used for house construction around Kibale National Park, western Uganda. Afr J Ecol 45:22–27
32. Gavin MC, Anderson GJ (2007) Socioeconomic predictors of forest use values in the Peruvian Amazon: a potential tool for biodiversity conservation. Ecol Econ 60:752–762
33. Ramos MA, Medeiros PM, Almeida ALS, Feliciano ALP, Albuquerque UP (2008) Use and knowledge of fuelwood in an area of Caatinga vegetation in NE Brazil. Biomass Bioenergy 32:510–517
34. Sá e Silva IMM, Marangon LC, Hanazaki N, Albuquerque UP (2009) Use and knowledge of fuelwood in three rural caatinga (dryland) communities in NE Brazil. Environ Dev Sustain 11:833–851
35. Kvist LP, Andersen MK, Stagegaard MH, Llapapasca C (2001) Extraction from woody forest plants in flood plain communities in Amazonian Peru: use, choice, evaluation and conservation status of resources. For Ecol Manage 150:147–174
36. Gaugris JY, van Rooyen MW, Bothma JP, van Der Linde MJ (2006) Hard Wood utilization in buildings of rural households of the Manqakulane community, Maputaland, South Africa. Ethnobot Res Appl 5:97–114

37. Banks DI, Griffin CM, Shackleton CM, Shackleton SE, Mavrandonis JM (1996) Wood supply and demand around two rural settlements in a semi-arid savanna, South Africa. Biomass Bioenergy 11:319–331
38. Chazdon RL, Coe FG (1999) Ethnobotany of woody species in second-growth, old-growth, and selectively logged forests of northeastern Costa Rica. Conserv Biol 13:1312–1322
39. Luoga EJ, Witkowski ETF, Balkwill K (2002) Harvested and standing wood stocks in protected and communal miombo woodlands of eastern Tanzania. For Ecol Manage 164:15–30
40. Bhatt BP, Sachan MS (2004) Firewood consumption along an altitudinal gradient in mountain villages of India. Biomass Bioenergy 27:69–75
41. Top N, Mizoue N, Kai S, Nakao T (2004) Variation in woodfuel consumption patterns in response to forest availability in Kampong Thom Province, Cambodia. Biomass Bioenergy 27:57–68
42. Top N, Mizoue N, Ito S, Kai S (2004) Spatial analysis of woodfuel suppy and demand in Kampong Thom province, Cambodia. For Ecol Manage 194:369–378
43. Nascimento VT, Sousa LG, Alves AGC, Araújo EL, Abuquerque UP (2009) Rural fences in agricultural landscapes and their conservation role in an area of caatinga (dryland vegetation) in Northeast Brazil. Environ Dev Sustain 11:1005–1029
44. Luoga EJ, Witkowski ETF, Balkwill K (2000) Economics of charcoal production in miombo woodlands of eastern Tanzania: some hidden costs; associated with commercialization of the resources. Ecol Econ 35:243–257
45. Pattanayak SK, Sills EO, Kramer RA (2004) Seeing the forest for the fuel. Environ Dev Econ 9:155–179

Chapter 23

Ethnobiological Research in Public Markets

Ulysses Paulino Albuquerque, Júlio Marcelino Monteiro, Marcelo Alves Ramos, Elba Lucia Cavalcanti de Amorim, and Rômulo Romeu Nóbrega Alves

Abstract

Fairs or traditional markets are centers of buying and selling that commercialize plant and animal products and are important for gathering, concentrating, maintaining, and diffusing empirical knowledge about these resources. In this chapter, you will first encounter a brief discussion of questions and hypotheses taken from ethnobiological research on open-air markets. The principal social factors involved in this type of investigation will be presented along with several methodological problems that researchers face while conducting ethnobiological research in centers at which medicinal resources are bought and sold. Finally, practical suggestions and recommendations for conducting rigorous research will be discussed from a scientific viewpoint.

Key words Ethnobotany, Ethnozoology, Conservation, Traditional knowledge, Medicinal plants

1 Introduction

Public and open-air markets are public places designated for the sale of various products that simultaneously function as spaces for the trade and acquisition of cultural information. Although the terms are frequently used as synonyms, they are distinguishable, as public markets have a permanent physical structure and generally daily operation, while open-air markets may have permanent or temporary structures and may operate daily or only 1 day a week [1]. These markets function as hallmarks of a particular culture or society by reproducing, on a small scale, the cultural and biological diversity of a region. In these markets, one can find specific sectors that commercialize plant and animal products for medical ends,

This chapter was compiled from transcriptions of the text originally published in the Journal of Ethnopharmacology and titled "Medicinal and magic plants from a public market in Northeastern Brazil," vol 110, pp 76–91, by Albuquerque et al. (2007), reproduced with permission from the publisher Elsevier (License No. 3071441454661).

Ulysses Paulino Albuquerque et al. (eds.), *Methods and Techniques in Ethnobiology and Ethnoecology*,
Springer Protocols Handbooks, DOI 10.1007/978-1-4614-8636-7_23, © Springer Science+Business Media New York 2014

gathering from different sources, and maintaining and diffusing empirical knowledge about these resources. The growing search for natural products, as part of a socially shared strategy, increases the importance of these traditional centers. Despite their importance, however, few ethnobiological investigations have studied the vendors in public and open-air markets [2]. For Almeida and Albuquerque [3], it was possible to obtain information in these markets about the native and exotic flora and fauna of a region, thereby aiding plans to conserve commercialized resources. These markets also feed a group of religious practices and beliefs that employ animal and plant resources in a magical-religious manner [4, 5].

Ethnobiological studies in traditional markets are scarce because the methods and approaches for collecting information in these markets are still being developed (Box 1). The studies that

Box 1 Methodology Used for Ethnobotanical Surveys in Public Markets by Bye and Linares [25] and Reproduced by Nguyen [26]

1. Visit the market early in the morning or during regular hours. This approach will facilitate your access to the vendors before they become too occupied with their clients. Regular visits allow for the observation of changes in product offerings (seasonal fruits, for example).
2. Interviews with the vendors should include: (a) the name of the plant; (b) indication(s) and use(s); (c) preparation(s); (d) collection locale; (e) whether the plant is collected or cultivated; (f) the habitat or origin; (g) price; and (h) type of vendor (e.g., collector-vendor, reseller).
3. For proper documentation and collection of the plant:
   (a) Verify if the plants were purchased and from what locale they were collected.
   (b) Preparation of voucher specimens: (1) for herbarium specimens, the material should be pressed in accordance with herbarium standards. (2) Voluminous or succulent materials should be dried or preserved in the appropriate liquids. Drying in the sun is recommended to remove excess water. After these procedures, seek specialists for proper handling of the voucher specimen.
   (c) For seeds, stems, and roots, additional specimen material should be acquired, preferentially reproductive material (flowers), to identify the plants.
   (d) Photographs of specimens.
4. Visits to the field where the products are collected for the acquisition of ethnobotanical and ecological information.
5. Identification of the specimen using specific literature, comparisons with material deposited in herbaria or by consulting specialists.
6. Proper labelling of the voucher specimen.
7. Deposit the voucher specimen, photographs, and notes concerning the plant in the appropriate herbarium.

have been completed are basically descriptive and focus on the diversity of commercialized plants and/or animals [6–13], although studies have been conducted to understand the importance and seasonality of commercialized plant species [14] or to take advantage of the market context to discuss quantitative analyses in ethnobotanical research [15, 16]. The same descriptive tendency is found among studies published in Brazil [17–22] with several forays into the origin and conservation of commercialized resources [1, 3, 23]. Martin [24] noted that such studies are essential for assessing the status and conservation of commercialized species.

2 Research Questions and Hypotheses

Generally, studies of local or traditional markets take descriptive approaches that produce a list of useful plants and/or animals (e.g., [6, 18–20]). It is rare to find articles that test hypotheses or propose practical questions and objectives for investigation (cf. Hanlidou et al. [10]). We present a few examples of questions, based on the literature (see [10, 11, 13, 16, 27, 28]) that can be used in ethnobiological studies of markets: What is the origin of the commercialized plants and/or animals? In what proportion are wild and cultivated plants or wild and domesticated animals found in the markets? What parts of the species or what products are being commercialized? Are there commercial implications to the conservation and sustainable use of the commercialized species? What quantity of a commercialized species is used per year? What natural environments produce the native commercialized species that are offered in the market? Are there differences in the proportions of commercialized species from different forms of vegetation? Does the number of commercialized species vary among respondents? If there is variation, what are the reasons? What is the range of commercialized medicinal flora or fauna in relation to the total known medicinal flora or fauna of the region? How does the composition of the native flora or fauna in each region influence the choice of commercialized medicinal species? Are there differences in the taxonomic richness of medicinal species depending on the geographic location of the markets? Does the price of a commercialized species vary, depending on demand or conservation status? Is there an overlap of the therapeutic indications for domestic and wild animals or medicinal plants? Can medicinal products derived from threatened animal species be substituted with products derived from medicinal plant species or domestic animals without placing the possible substitutes at risk? Are the merchants aware that the commercialization of wild animals is illegal? Are there differences in the understanding of the respondents? Is the knowledge system of the market open or closed? Do gender, professional occupation, time dedicated to the activity and level of education affect the knowledge about commercialized medicinal resources? What is the perception of the vendors about the status of commercialized species from a conservationist viewpoint? Do the commercialized plant and/or animal products vary according to season or some other temporal consideration? With regard to this question, for example, Monteiro et al. [16] showed that there were significant changes in the richness of commercialized medicinal plants over 4 years in the Caruaru Market in Northeastern Brazil, including an increase in unique species. A similar situation was reported for medicinal animals in the city of Recife, located in the same region of Brazil [13, 29].

In general terms, Martin [24] suggests a basic information set that is important to obtain: (1) information about the vendors; (2) the origin of the product—questions about the local where the plants and/or animals are collected; (3) the condition of the product—dried or fresh, sterile or fertile; (4) the handling and commercialization of the product—cultivated or wild, collected by the vendor or purchased from others; (5) information regarding the quantity, price, and availability; and (6) changes in demand for and/or stock of the commercialized resources.

3 Sampling and Selection of Interviewees

Identification of and contact with the interviewees is a crucial task for studies on local and traditional markets. In Brazil, there are clear distinctions among the responses of different informants based on their experiences and the functions that they serve in the system. Considering the multicultural nature of Brazil, these markets also become religious centers. The commercialization of medicinal plants and/or animals, products and objects for use in the initiation rites and magical-religious practices of Afro-Brazilian cults is not uncommon. In this sense, the presence of these resources for magical-religious use can be broad, depending on the local context in which they come to the market.

The success of a study and the fulfillment of the proposed objectives may depend on the appropriate selection of interviewees. In accordance with our experience, the following general social actors can be identified. (1) Vendors, i.e., persons who commercialize products while possessing little information about them. Sometimes individuals from the same family with a tradition of selling medicinal species will alternate running the market stand during the day. (2) "Erveiros (herb healers)," who commercialize their products and possess profound knowledge about them. (3) "Doutor Raiz (root doctors)," experienced individuals who are generally rare in each market and who function as local "doctors." These persons may or may not commercialize plants and/or animals, and they generally keep concoctions or preparations ready to treat a wide range of diseases. In Brazil, the name of these specialists may vary regionally. "Doutor Raiz" (root doctors) and "raizeiro" (root healers) are understood to be specialists in the collection and sale of medicinal plants in the markets [30]; other terms also designate these actors, such as "erveiro" or "ervateiro" (herb healer). In most cases, the same actors involved in the commerce of medicinal plants also commercialize medicinal animals, although a large portion do not admit to selling zootherapeutic products because they know that commerce in wild animals is illegal. The clandestine or semi-clandestine nature of the commercialization of wild animals is certainly one factor that contributes to the scarcity of detailed information about this subject.

Related to the selling of medicinal resources, each of these "doctor" types of interviewees can also be a collector of plant and animal products, though this role appears less common when they sell in large urban centers. Normally, these collectors receive plants and animals through intermediaries; these can be individuals with experience collecting products or persons who opportunistically collect and periodically offer plants in public markets. These intermediaries are key elements in ethnobiological investigations and are particularly important for studies with conservationist approaches because they can identify, for example, the collection sites, the quantity collected and the price that is charged to the sellers. Unfortunately, this social group is rarely accessible to studies because intermediaries, unlike vendors, are not present at the markets at regular hours or days. Recent studies have demonstrated the existence of commercial routes for medicinal plants and animals among different cities in northern and Northeastern Brazil to meet the demand for medicinal biological products (e.g., [23, 31–35]).

There are also small, local producers who supplement their income by cultivating certain species to supply the markets. It is not difficult to find multiple merchants in one market who depend on the same intermediaries. The description of "raizeiros" (root healers) by scholars of popular culture in Brazil in the second half of the twentieth century remains accurate in various locations in the Northeast, although many elements have disappeared:

> *In the various northeastern markets that we visited, the stand of the root user was always the same: a mat of chili, lying on the ground on which were placed small piles of roots, bark, seeds, vines, bird feathers, armadillo and tortoise shells, deer hooves, capybara teeth and nails, horns, snake skin, cans with powders from certain woods or seeds, lard and inevitably, a huge bull's horn, full of roasted leaves, the indispensable snuff* [30].

All of these elements must be taken into consideration for the selection of informants, especially because people can be found who only recently became involved in selling products in markets. For example, Almeida and Albuquerque [3] found that among the informants in a popular market in the State of Pernambuco, Brazil, some stood out for the quality of the information they possessed as a function of the time they had dedicated to the trade. Furthermore, public markets can be the stage for a great deal of learning by novice vendors. Additionally, Bussmann et al. [36], while studying traditional markets in Trujillo and Chiclayo in Peru, indicated that the local pharmacopoeia had changed and that the number of available species appeared to increase due to experimentation by the vendors. It is not uncommon that the markets are open systems of knowledge (see [37] for a definition of open and closed systems of knowledge), as information is frequently traded between vendors, and they can form a consensus as to the use of plants and animals to treat specific health problems (see [3]). Inter-vendor trade of products is also common, as a vendor can consult another vendor when a client does not find a desired product in his or her stall. The reputation of these specialists varies as a function of their honesty and their command of the medical and magical-religious world of the markets. For example, adherents to the Afro-Brazilian cults tend to seek out vendors with ample knowledge (or a wide range of products) such that they may purchase necessary materials to practice their rites. Dishonesty arises when a different plant is substituted for the one desired by the client. In the case of animals, it is common to sell products from domestic animals as if they were derived from threatened wild animals; for example, ox lard (*Bos taurus*) is sold as if it were derived from a manatee (*Trichechus manatus* or *T. inunguis*).

The initial contact with the world of the public market is one of the most delicate tasks because different factors can influence the willingness of individuals to participate in interviews. These factors include the following: (1) Fear that the interviewers belong to an oversight agency (because products from animals threatened by extinction are commercialized in the market) or that they are future competitors that are eager to gain information quickly and easily. (2) The flow of clients in the stands. There are specific days, known as "market days," with increased activity in the market. Visiting on market days is not recommended because there can be a dangerous competition between the client and the researcher for the attention of the vendor, and the latter will be ignored. (3) Personal vanity, when local specialists with good reputations who have appeared on television or news programmes refuse to talk to researchers; this reluctance is not uncommon. Our experience has indicated that the best way to initiate an interview is to explain the study in great detail within the previously discussed context and to obtain authorization to conduct the study. The introduction of the research by an individual who knows the vendors can make the acquisition of information easier. It is not uncommon for people to initially refuse the invitation to participate in the study (because of distrust or fear) but then to quickly become willing to collaborate when they see that their colleagues are participating in the process without great problems. Thus, acquiring the trust of a respected vendor can help to ease contact with the remaining vendors. Truthfully, establishing contact with the vendors is considered the most critical aspect of the investigation [24] for reasons that will be discussed later. The situation becomes more difficult when the study is related to the trade of medicinal animals, as the majority of commercialized species are wild, therefore entailing legal implications (especially for animals on threatened-species lists). Oversight by environmental agencies is common in the markets and results in the seizure of medicinal products derived from wild animals. Thus, the researcher can be mistaken for an undercover agent.

The receptivity of interviewees in public markets is generally notably different from that in local communities. Fear and distrust are much more apparent in markets and can lead to frustration for the researcher because time is lost without gaining the cooperation of the people. A constant presence in the market, informal conversations and the purchase of products for the researcher's own use help to bring down these barriers. The researcher must attend to these considerations because the quality of the information that he or she receives depends on maximal reduction of biases; an interviewee can respond quickly and without commitment to simply extricate himself or herself from the conversation with the researcher. If respect and patience are necessary ingredients in any ethnobiological research, in these cases, the investigator must employ a large dose of both virtues. Certain researchers adopt a strategy of purchasing plants as a way to financially compensate the interviewees for their time while simultaneously gaining their trust [11].

The source of knowledge of these specialists can be varied. It is common to observe from their speech that their information regarding plants is obtained from "mateiros" (bushmen) and intermediaries. In this case, the bushmen and intermediaries can be viewed as "generators" or primary sources of empirical knowledge about herbal medicine. Moreover, it is not uncommon to encounter herb vendors who relate the acquisition of their knowledge to reading books concerning medicinal plants; while being interviewed, they may frequently need to consult such books to remember the names, indications, and useful forms of the plants. Certain vendors claim that knowledge can be acquired together with the clients/users of the medicinal resources. These vendors may seek a species used by folk medicine to treat a sickness of which they are already aware. However, there are some vendors who learned the therapeutic uses of a plant from their parents or close relatives and who follow in the family path; still, they consult specialized literature because they have had the opportunity to attend traditional schools. New means of communication have caused the younger generation to be less interested in learning the knowledge of the older generation, leading to the erosion of traditional knowledge transmission.

Another world of information begins to be perceived, and formal education begins to remove the youth from the older generation significant periods of time, thereby reducing youthful interest in the elders' knowledge [38]. However, the researcher must be careful in making such claims. Many studies that encounter discrepancies between the knowledge of the older and younger generations explain the findings using the "erosion of knowledge" hypothesis. In most cases, this hypothesis is a misinterpretation; it is natural for the older generation to know more, as they have had more time to accumulate information [39].

Thus, one of the great difficulties in gathering data in public markets is related to obtaining a large number of interviewees. This difficulty stems from two principal problems, specifically, obtaining adequate information and finding study participants. How can someone respond to questions about this type of work in a process that requires time and the good will of the interviewees; Good sense on the part of the interviewer is required first; if it is absent, those who participate in the study with time restrictions or even against their own desires may omit information that is relevant to the study. Bearing these considerations in mind, the selection of the interviewee group must be careful, criteria-based, and patient.

4 The Collection of Information

Semi-structured interviews are common in studies of public markets [1, 3, 11, 16, 22, 23, 28]. Our experience has shown that our interviewees normally become anxious when the interviewer asks consecutive questions using a questionnaire. We have observed situations in which the interviewee becomes disengaged while waiting for the investigator to note something or ask the next question. Depending on the objectives, problems with the accuracy of the information can arise, as an interview is a strongly reactive method. The manner in which the questions are formulated is particularly important in this environment. May we ask what types of plants and/or animals you know or indicate for medicinal use? The interviewee can speak about his or her individual repertoire but may not discuss the species that are sold. We believe that highly informal situations contribute positively to the quality of the information. The technique of using a free list can be a notably useful tool in conjunction with the direct observation of commercialized products. The collection of information can be completed by asking the interviewees to give information concerning each of the products that they commercialize. In this manner, additional data concerning the price, availability, and seasonality of the commercialized species can be directly collected. The selection of a specific methodology clearly depends on the desired objectives. Successive meetings and visits with the interviewees are, in our perspective, indispensable for the collection and distillation of the information obtained [10, 11], and they allow for the acquisition of specimens for taxonomic identification [6, 26].

5 Voucher Specimens

We have no doubts that one of the greatest difficulties in the ethnobiological research of public markets is the collection of material for identification and later deposit; voucher specimens represent the principal reason for the lack of scientific rigor in many studies, especially for the most commonly cultivated plants [26, 40].

In the case of plants, Martin [24] emphasizes that the procedures are practically identical for any ethnobotanical survey. However, it is necessary to emphasize that the collected samples may be inadequate for deposit in an herbarium or may fail to meet the requirements of herbarium curators for incorporation into their collections. All of the ethnobotanical studies in specialized periodicals indicate the need to reference the number of the collector or the number of the herbarium for each plant cited. This referencing is not always possible, however, especially when traditional Brazilian markets are considered. It is also common to neglect the collection of highly common materials, which can lead to the loss of important information. For example, Albuquerque [41] described a new variety of basil (*Ocimum minimum* L. var. *religiosum*) sold in public markets in Recife in northeastern Brazil. In the Caruaru Market, also in northeastern Brazil, Monteiro et al. [16] found discrepancies in the identification of some plant materials collected before and after a 4-year interval. Plants with the same common name were identified as distinct species, genera, and even families, as in the case of the "espinheira santa" (holy thorn bush). This plant was first identified as *Maytenus* sp. (Celestraceae) but later as *Zollernia ilicifolia* (Brongn.) Vogel (Fabaceae). Many of these plants are dried, pulverized, or fragmented when they are sold, making their identification difficult. Other times, the plant product consists of only seeds, roots, leaves, or bark from the stem. In situations where the vendor is also the collector, it is possible to ask for their assistance in collecting the raw botanical material [14, 19]. This approach, however, is not always possible because (a) there are vendors who are not collectors, and (b) the vendors may not be available to go into the field. The greatest difficulty is not in the identification of the plant but in the acquisition of material to be deposited in an herbarium. Let us consider both difficulties: identification problems and problems with collection for herbarium deposit.

First, medicinal plants can be identified in the field using samples that are obtained directly from vendors (either purchased or collected with their help). For laboratory identification, specialized literature and materials from the herbarium are used for comparison, and help is sought from specialists (e.g., [6]). In the case of a plant that is sold in fragments or parts, if it is not possible by any means to collect complete, fertile samples for identification, the only remaining approaches are to provide a "taxonomic track" based on the scientific information that is locally available or to attempt to identify the sample by means of pharmacobotanical procedures [40]. For example, in our research group, we possess detailed morphological and anatomical descriptions of certain regionally common medicinal plants that we use as standards to identify other samples. As for the collection of material for herbaria, it is difficult to resolve this problem if the investigators are unable to collect adequate material. Our experience has demonstrated that purchasing samples from vendors for incorporation into herbaria as collected material [11] is not always a reliable procedure. There is no guarantee of the accuracy of information, such as the original plant collection locale, and curators may not accept the specimens. We believe that researchers should make every effort to collect samples; if this is impossible, we propose an alternative and complementary approach: to identify the samples and compare them to herbarium material that will become "reference material." Nguyen [26] cites the possibility of photographing the plant for identification and incorporation into herbaria. However, this approach should only be used if collection is impossible. Visual materials represent a good resource for quick (Internet) shipment to specialists as a substitute for traditional practices (sending material to herbaria) [26]. Moreover, photographs can be used as supplementary materials to plant voucher specimens because the exclusive use of photographs does not provide all of the necessary information. Other options for researchers include concentrating additional efforts on identifying the intermediaries who supplied the plants and creating their own mobile herbaria with reference samples; informants can then be asked whether they match the plants being sold [26]. However, it is common that interviewees do not recognize the entire fertile branch of a plant when they only sell the bark, seeds, or roots. It is important to recognize that plants called by the same name may not belong to the same species. We have found in the markets of Pernambuco (Northeastern Brazil) that the bark of the stem of two species known as "aroeira" can be offered without any distinction: *Myracrodruon urundeuva* All and *Schinus terebinthifolius* Raddi. The first species is found in the semiarid areas of the state, whereas the second occurs more commonly, whether cultivated or naturally, in areas closer to the coast.

The difficulties encountered in obtaining voucher specimens for ethnobotanical research are similar to those observed in ethnozoological research. Medicinal animals can be sold whole (in the case of smaller species), but in most cases they are sold in parts such as feathers, paws, fur, leather, teeth, lard (fat), milk, meat, horns, spine, scales, nails, blood, penis, bones, liver, meat, heart, head, testicles, bone marrow, eyes, ears, or other parts. In this form, either the entire animal or isolated parts can serve as voucher specimens, depending on the species or the part that is sold.

In the case of small animals that are sold whole, voucher specimens can be acquired from the merchants through donations or purchases. In the case of larger animals, in most cases, the commercialized parts make it possible to identify the species. Skeletal elements such as skulls allow for the identification of many species of mammals, reptiles and birds, and they are commonly accepted as voucher specimens in zoological collections. Considering that different medicinal products can be obtained from the same species, it is possible that one zootherapeutic product (fat, for instance) will not allow the rapid identification of a species, but other derivatives, such as the bones or skin, do permit identification.

Photographs of commercialized parts can also be used for identification, depending on the species. Reference images of the species can also be shown to the vendors such that they may confirm the identification of the animal products. As a last resort, a taxonomic track can be used, and for this approach, we recommend consulting both the specific literature (searching for published studies performed in the region) and specialists who work in local universities and know the composition of the regional fauna.

References

1. Lima PGC, Coelho-Ferreira M, Oliveira R (2011) Plantas medicinais em feiras e mercados públicos do Distrito Florestal Sustentável da BR-163, estado do Pará, Brasil. Acta Bot Bras 25:422–434
2. Jain SK (2000) Human aspects of plant diversity. Econ Bot 54(4):459–470
3. Almeida CFCBR, Albuquerque UP (2002) Uso e conservação de plantas e animais medicinais no estado de Pernambuco (Nordeste do Brasil): um estudo de caso. Interciencia 27(6):276–285
4. Vogel A, Mello MAS, Barros JFP (1993) A Galinha-d'angola. EDUFF, Rio de Janeiro
5. Albuquerque UP (2006) Folhas Sagradas: As Plantas Litúrgicas e Medicinais nos Cultos Afro-Brasileiros, 2nd edn. Nuppea, Recife
6. Nicholson MS, Arzeni CB (1993) The market of medicinal plants of Monterrey, Nuevo León, México. Econ Bot 47:184–192
7. Narváez A, Stauffer F (1999) Products derived from palms at the Puerto Ayacucho markets in Amazonas state, Venezuela. Palms 43:122–129
8. Williams VL, Balkwill K, Witkowski ETF (2000) Unraveling the commercial market for medicinal plant parts on the Witwatersrand, South Africa. Econ Bot 54:310–327
9. Lev E, Amar Z (2002) Ethnopharmacological survey of traditional drugs sold in the Kingdom of Jordan. J Ethnopharmacol 82:131–145
10. Hanlidou E, Karousou R, Kleftoyanni V, Kokkini S (2004) The herbal market of Thessaloniki (N Greece) and its relation to the ethnobotanical tradition. J Ethnopharmacol 91:281–299
11. Macía MJ, García E, Vidaurre PJ (2005) An ethnobotanical survey of medicinal plants commercialized in the markets of La Paz and El Alto, Bolívia. J Ethnopharmacol 97:337–350
12. Andel T, Myren B, Onselen S (2012) Ghana's herbal markets. J Ethnopharmacol 140:368–378
13. Ferreira FS, Albuquerque UP, Coutinho HDM, Almeida WO, Alves RRN (2012) The trade in medicinal animals in Northeastern Brazil. Evid Based Complement Altern Med 2012:1–20
14. Mertz O, Likke AM, Reenberg A (2001) Importance and seasonality of vegetable consumption and marketing in Burkina Faso. Econ Bot 55:276–289
15. Williams VL, Witkowski ETF, Balkwill K (2005) Application of diversity indices to appraise plant availability in the traditional

medicinal markets of Johannesburg, South Africa. Biodivers Conserv 14:2971–3001
16. Monteiro JM, Ramos MA, Araujo EL, Amorim ELC, Albuquerque UP (2011) Dynamics of medicinal plants knowledge and commerce in an urban ecosystem (Pernambuco, Northeast Brazil). Environ Monit Assess 178:179–202
17. Berg ME (1984) Ver-o-Peso: the ethnobotany of an Amazonian market. Adv Econ Bot 1:140–149
18. Costa-Neto EM (1999) Healing with animals in Feira de Santana City, Bahia, Brazil. J Ethnopharmacol 65:225–230
19. Parente CET, Rosa MMT (2001) Plantas comercializadas como medicinais no Município de Barra do Piraí, RJ. Rodriguésia 52(80):47–59
20. Nunes GP, Silva MF, Resende UM, Siqueira JM (2003) Plantas medicinais comercializadas por raizeiros no centro de Campo grande, Mato grosso do Sul. Revista Brasileira de Farmacognosia 13(2):83–92
21. Pinto AAC, Maduro CB (2003) Produtos e sub-produtos da medicina popular comercializados na cidade de Boa Vista, Roraima. Acta Amazônica 33(2):281–290
22. Monteiro JM, Araújo EL, Amorim ELC, Albuquerque UP (2010) Local markets and medicinal plant commerce: a review with emphasis on Brazil. Econ Bot 64:352–366
23. Alves RRN, Rosa IL (2007) Zootherapy goes to town: the use of animal-based remedies in urban areas of NE and N Brazil. J Ethnopharmacol 113:541–555
24. Martin GJ (1995) Ethnobotany: a people and plants conservation manual. Chapman & Hall, London
25. Bye RA, Linares E (1983) The role of plants found in Mexican markets and their importance in ethnobotanical studies. J Ethnobiol 3(1):1–13
26. Nguyen MLT (2005) Cultivated plant collections from markets places. Ethnobot Res Appl 3:5–15
27. Cunningham AB (2001) Etnobotánica aplicada: Pueblos, uso de plantas silvestres y conservación. Nordan Comunidad, Montevideo, p 311
28. Botha J, Witkowski ETF, Shackleton CM (2004) Market profiles and trade in medicinal plants in the Lowveld, South Africa. Environ Conserv 31(1):38–46
29. Silva MLV, Alves AGC, Almeida AV (2004) A zooterapia no Recife (Pernambuco): uma articulação entre as práticas e a história. Biotemas 17(1):95–116
30. Araújo AM (2004) Medicina rústica, 3rd edn. Martins Fontes, São Paulo
31. Alves RRN, Rosa IL (2010) Trade of animals used in brazilian traditional medicine: trends and implications for conservation. Hum Ecol 38:691–704
32. Alves RRN, Rosa IL, Santana GG (2007) The role of animal-derived remedies as complementary medicine in Brazil. BioScience 57:949–955
33. Alves RRN, Lima NH, Tavares MC, Souto WMS, Barboza RRD, Vasconcellos A (2008) Animal-based remedies as complementary medicines in Santa Cruz do Capibaribe, Brazil. BMC Complement Altern Med 8:44
34. Alves RRN, Léo-Neto NA, Brooks SE, Albuquerque UP (2009) Commercialization of animal-derived remedies as complementary medicine in the semi-arid region of Northeastern Brazil. J Ethnopharmacol 124(3): 600–608
35. Oliveira ES, Torres DF, Brooks SE, Alves RRN (2010) The medicinal animal markets in the metropolitan region of Natal City, Northeastern Brazil. J Ethnopharmacol 130:54–60
36. Bussmann RW, Sharon D, Vandebroek I, Jones A, Revene Z et al (2007) Health for sale: the medicinal plant markets in Trujillo and Chiclayo, Northern Peru. J Ethnobiol Ethnomed 3:37
37. Garro LC (1986) Intracultural variation in folk medical knowledge: a comparison between curers and noncurers. Am Anthropol 88:351–370
38. Amorozo MCM (1996) A abordagem etnobotânica na pesquisa de plantas medicinais. In: Di Stasi LC (ed) Plantas medicinais: arte e ciência. Editora Unesp, São Paulo, pp 47–68
39. Voeks RA, Leony A (2004) Forgetting the forest: assessing medicinal plant erosion in Eastern Brazil. Econ Bot 58:294–306
40. Ramos MA, Albuquerque UP, Amorim ELC (2005) O comércio de plantas medicinais em mercados públicos e feiras livres: um estudo de caso. In: Albuquerque UP, Almeida CFCBR, Marins JFA (eds) Tópicos em conservação, etnobotânica e etnofarmacologia de plantas medicinais e mágicas. NUPEEA, Recife, pp 127–164
41. Albuquerque UP (1999) Three new varieties in *Ocimum* L. (Lamiaceae). Braz Arch Biol Technol 42(1):1–8

Chapter 24

Techniques for Analysis of Quantitative Ethnobiological Data: Use of Indices

Valdeline Atanazio da Silva, Viviany Teixeira do Nascimento, Gustavo Taboada Soldati, Maria Franco Trindade Medeiros, and Ulysses Paulino Albuquerque

Abstract

Quantitative techniques for the analysis of collected data have become increasingly popular among ethnobiologists and ethnobotanists in particular. Since the 1990s, a number of techniques have been proposed, and many authors have adopted them in their research. However, this acceptance of quantitative techniques was not accompanied by an analysis of their limitations and weaknesses. This chapter presents a discussion of the role of quantitative techniques for the analysis of plant data and an overview showing some of the most commonly used techniques. These include examples cited in some reviews, along with more recently proposed additions and limitations for some of these examples. The techniques discussed here were extracted from ethnobotanical works but are nevertheless applicable in other fields of ethnobiology.

Key words Quantitative ethnobotany, Quantitative indices, Data analysis in ethnobiology

1 Introduction

The use of quantitative techniques is becoming increasingly popular among ethnobiologists and ethnobotanists in particular (see [1, 2]). This popularity, however, has not been accompanied by the necessary analysis of the limits and weaknesses of these procedures. Researchers need to fully understand the assumptions that are made when adopting a specific quantitative technique to analyze their data. They also need to be aware that "measuring" traditional knowledge is not an easy task, insofar as it involves different dimensions (theoretical and practical) and elements (plants, animals, soil, etc.). Because many techniques have been proposed in the literature and because a method may not be appropriate for a specific question, the aspect of knowledge measurement warrants further consideration. For example, techniques used to analyze theoretical

Ulysses Paulino Albuquerque et al. (eds.), *Methods and Techniques in Ethnobiology and Ethnoecology*,
Springer Protocols Handbooks, DOI 10.1007/978-1-4614-8636-7_24, © Springer Science+Business Media New York 2014

knowledge, such as the potential uses of a plant, do not necessarily capture the real use of the resource [3].

In 1996, Phillips [1] made an important revision to the techniques used in ethnobotanical studies by considering the degree of subjectivity (large or small) in these studies and by dividing them into three principal types: (1) informant consensus, (2) subjective allocation, and (3) aggregation of uses. Hoffman and Gallaher [4] conducted a new analysis of ethnobotanical research techniques and used the RCI (relative cultural importance) term for designating the different techniques cited by Phillips [1] as well as more recent techniques (Table 1). Although these techniques were published as components of ethnobotanical studies, they are nevertheless applicable in other fields of ethnobiology. In 2011, Medeiros et al. [2] analyzed 87 different techniques according to the three principles cited by Phillips [1].

As one example, informant consensus permits the analysis of the "relative importance of each use because it is calculated directly from the degree of consensus of the responses of the informants" [1]. A high consensus among informants indicates that a plant is well known in the community and further suggests possible efficacy or validation of the knowledge for a specific end. For example, consensus may lead to some strong candidates as a source of a particular active ingredient for future scientific investigation [5]. In this sense, the techniques based on informant consensus assume a correlation between the importance of a plant and the extent of shared knowledge of that plant within a social group. In other words, the more important a plant, the more people know and recognize it as a resource. Some of the most commonly used techniques in ethnobotany belong to this category of informant consensus. The use value (UV) technique, for example, was originally published by Phillips and Gentry [6, 7] and estimates the knowledge that people have about plant use.

The subjective allocation of the relative importance of each species reflects the mode of "seeing or understanding" of a specific culture by the investigator [1]. The studies carried out by Turner [8] and Stoffle et al. [9] are examples of this type of study that use the cultural significance index (CSI) to quantify the importance of useful species for traditional groups. These techniques assess the relative importance of each species by considering "categories" that are developed by investigators from their experience in the field and are thus very subjective. For example, while studying a specific group, the investigator may perceive that the plants or animals associated with religious rituals are highly valued locally. This could result in the creation of a "religious" category to be considered in the relative importance calculations. The values of the resources would be attributed as follows: 0 for a resource that is not used in the rituals; 0.5 if the species plays a secondary role in

Table 1
Comparison of the relative importance indices (modified from [2])

Analysis criteria	Sum of uses	Use value [31]	Cultural significance [34]	Fidelity level [11]	Use value [6, 7]	Importance value [28]	Total use value [26]	Rapid informant ranking [29]	Cultural, practical, and economic value [3]
Explicit method	No	No	Yes	Yes	Yes	Yes	Yes	Yes	Yes
Objective values	Yes	No	No	Yes	Yes	Yes	Yes	Yes	Yes
Method sensitive to sample size	Yes	Yes	No	No	No	No	No	No	No
Type of data generated	Discrete ratio scale	Discrete ordinal scale	Continuous	Continuous	Continuous	Continuous	Continuous	Continuous	Continuous
Time for data collection and analysis	Low	Low	Moderate	Moderate	High	Moderate	High	Moderate	High
Statistical analysis possible for habitat/area?	Yes	Yes	Yes	No	Yes	No	Yes	No	No
Statistical analysis possible for category of use?	Yes	Yes	Yes	Yes	Yes	No	Yes	Yes	Yes

the rituals; and 1, the maximum value, if a species is the central element in the religious traditions of the investigated group.

For the sum of uses, the indications for each resource by usage category (e.g., construction, medicinal, food) and by taxon or type of vegetation are summed to determine the importance of the species or its uses. Thus, these techniques assign greater importance to more versatile resources, i.e., those with more uses and those belonging to a larger number of categories. This technique is frequently used because it is rapid and practical. This makes it especially useful in the initial or exploratory phase of analyzing data [1]. One example is increasing the percentage of use of a species in a specific area, as in a study by Carneiro [10] that evaluated the percentage of use of 0.7 ha of forest by the Kuikuru Indians in MatoGrosso, Brazil.

A brief analysis of some of the techniques used in the literature follows. It should be noted that the majority of the techniques can be used to test use hypotheses and/or knowledge about natural resources.

1.1 Informant Consensus

1. Friedman et al. [11]—*Fidelity level* (*FL*): This technique was described for the medicinal uses of plants and considers only one use category. However, it can be adapted to any category. The technique relies on agreement among informant responses for a principal therapeutic indication.

$$FL = \left(I_p / I_u\right) \times 100\%$$

where:

FL = fidelity level

I_p = number of informants that cited the principal use of the species

I_u = total number of informants that cited the species for any purpose

Consider the calculation of FL for "species *x*." Suppose that out of 100 informants in the community, 30 cited the species for a total of five therapeutic indications: "to cure flu," "cough," "headache," "cleaning the blood," and "worms." Of these uses, "cough" had the greatest consensus. That is, it was cited by the largest number of informants, 20 in total, making this the principal use of "species *x*." The index for this species would be calculated in the following manner: FL = (20/30) × 100, resulting in a value of 66.66. The calculation of this index demonstrates that relative importance considers informant consensus on two levels: (a) the distribution of the most important information in the social group and (b) the distribution of knowledge of the species in the social group. This index has been used by various authors including

[12, 13]. The same authors also proposed a second index, rank order priority (ROP), which combines FL, the previously calculated index and the RP (relative popularity).

$$ROP = FL \times RP,$$

where:

ROP = rank order priority

FL = fidelity level (see previous calculation)

RP = relative popularity

To calculate the relative popularity (RP), the original paper [11] proposed that each species be primarily classified as popular or unpopular. For this purpose, a coordinate system is used in which the number of respondents who mention each species is plotted on the abscissa and the number of different uses for each plant is plotted on the ordinate. The intersection between the two lines defines the zone that separates popular from unpopular plants. After all species have been plotted on the graph, the stabilization number can be determined, i.e., the value at which the number of uses of the whole plant does not increase even if the number of respondents increases. To calculate the RP of unpopular species, the number of citations of the species can be divided by the stabilization number. The RP is then defined as 1 for popular species and <1 for unpopular species. A simplified calculation of RP has been proposed [14] in which the ratio of the number of informants who cite a given species to the number of informants who cite the most frequently cited species is determined.

Considering the previous hypothetical situation, "species *x*" was cited by 50 informants and was the most frequently cited species. The ROP calculation for "species *x*" would be as follows: ROP = (66.66) × (30/50), for a value of 39.99. Thus, the ROP includes the new consensus level—the distribution of knowledge of the species in relation to the richness of the resources cited in the studied use category.

2. Troter and Logan [15]—*Informant consensus factor* (*ICF*): These authors sought to identify body systems (or categories of diseases) with greater relative local importance (emphasis). Body systems are a subcategorization of the medicinal use category, as they group together diseases related to diverse systems, such as "respiratory system diseases," "genitourinary system diseases," and "digestive system diseases" [16, 17]. The body systems can be obtained from the ICD (international classification of diseases). The ICF index can nevertheless be used for any use category. For example, it could be used for the use category "timber" with the following subcategories: "timber for artifacts," "timber for domestic construction," and "timber for rural construction." The investigator should order the

popular indications as subcategories to allow for a more detailed analysis of the categories of use. Examples of the use of the ICF index include [13, 14, 18, 19].

The ICF is determined using the following formula:

$$ICF = (n_{ur} - n_t) / (n_{ur} - 1),$$

where:

ICF = informant consensus factor

n_{ur} = number of citations of uses for a subcategory

n_t = number of species used for a subcategory

The maximum value that can be assigned to a subcategory is 1. The closer a subcategory is to this value, the greater the agreement among the informants about the use of the species in the different subcategories. Imagine that there are 30 use citations for the body system "respiratory system diseases," with three therapeutic indications ("cough," "flu," and "expectorant") distributed among four species. The ICF calculation for this subcategory (respiratory system diseases) would be as follows: ICF = (30 − 4)/(30 − 1), which gives 0.8965.

3. Phillips and Gentry [6, 7]—*Use value*: The use value index is currently the most widely used technique by investigators, and examples of its use include [13, 20–23]. For this index, the relative importance of a plant reflects the number of uses assigned to it. The results using this technique are commonly interpreted as the pressure on a given resource resulting from use, considering the logic that the most well-known resources are also the most used. However, no studies have established a direct relationship between use value and the real pressure on a given resource resulting from use [24].

The use value, as conceived by the authors, is calculated using the following formula:

$$UV_{is} = \sum U_{is} / n_{is},$$

where:

UV_{is} = use value of species *s* to informant *i*

U_{is} = number of uses of species *s* mentioned in each event by informant *i*

n_{is} = number of events in which informant *i* cited species *s*

Note that this index determines the use value of the species by the informant. With the same informants, various meetings (events or interviews) are necessary for the same species. Imagine that informant A was visited five times. In three meetings, informant A cited seven uses for "species *y*" in the following way: "cure cough" at meeting 1; "cure cough," "cleanser," "stakes," "firewood," and "juice" at meeting 2; and "cure cough" at meeting 3. The use value of "species *y*" for informant A would be calculated as follows: $UV_{is} = (1+5+1)/3$, giving a value of 2.33.

Next, to obtain the total use value of "species *y*," the individual use values for the species are summed using the following equation:

$$UV_s = \sum_i UV_{is} / n_s,$$

where:

UV_s = total use value for species *s*

n_s = number of informants interviewed for species *s*

To calculate the total UV_s for a given species *y*, assuming that informant B cited 4, 3, and 5 uses in 3 respective meetings (the same number of meetings with informant A), $UV_{is} = (4+3+5)/3$, yielding a value of 4. Thus, the UV_s of the species can be calculated as $(2.33+4.00)/2$, resulting in 3.17.

Phillips and Gentry [7] also proposed comparing the relative knowledge of each informant ("relative use value," RUV_i), as certain informants tend to know less about plant resources than others. The formula for calculating RUV_i is as follows:

$$RUV_i = \left[\sum (UV_{is} / UV_s) / n_i\right],$$

where:

RUV_i = relative use value for informant *i*

UV_{is} = use value for each species *s* cited by informant *i*

UV_s = total use value estimated for each species described in the study

n_i = for each informant, the number of species with data from two or more other informants

The botanical family use value can also be calculated using the following formula:

$$FUV = \sum UV_{sf} / n_f,$$

where:

FUV = family use value, which equals the average total use value for each species in the family

UV_{sf} = use value of the species belonging to the family

n_f = number of species in the family

Consider the following hypothetical scenario: the family Anacardiaceae is represented by the three species *x*, *y*, and *z*, whose respective total use values are 1.09, 0.89, and 0.034. The family use value would be calculated in the following way: $FUV = (1.09+0.89+0.034)/3$, which gives 0.67.

Modifications to the use value calculation have been proposed by different investigators ([25, 26]—see item 6). In [25] the author calculated the use value of a species using the following formula:

$$UV = \sum U/n,$$

where:

UV = use value of an species

U = number of quotations per species

n = number of informants

This formula represents a simplification of the index calculation originally by Phillips and Gentry [6], but it should be noted that for this calculation, the repetition of interviews/events with the same informant is not necessary.

Consider the following scenario in which 57 members of the community are interviewed. During the survey, three species are mentioned: *x*, *y*, and *z*, which received 10, 20, and 5 citations, respectively. Thus, the use value for each species can be calculated as the ratio of the number of citations to the total number of respondents as follows: *x*(10/57 = 0.17), *y*(20/57 = 0.35), and *z*(5/57 = 0.09).

4. Bennett and Prance [27]—*Relative importance* (*RI*): The formula for RI is relatively easy to use and has been employed in the study of medicinal plants. Using this index, the utility of plants reflects versatility of use, e.g., a greater number of therapeutic indications or body systems to which it belongs. The index is divided into two factors, and the maximum value that can be assigned to a species is 2.

$$\text{RI} = \text{NBS} + \text{NPH},$$

where:

RI = relative importance

NBS = number of body systems

NPP = number of pharmacological properties

The two factors are calculated using the following formula:

NBS = NBS_S (the number of body systems treated by a specific species)/NBS_{SV} (number of body systems treated by the most versatile species)

NPP = NPP_S (the number of pharmacological properties attributed to a specific species)/NPSV_{SV} (number of pharmacological properties attributed to the most versatile species)

Suppose that in a traditional pharmacopeia, "species *x*" is the most versatile species, with 15 indications distributed over four body systems. "species *y*" receives seven citations for use in two body systems: RI = (2/4) + (7/15), resulting in 0.97. This index confers the greatest importance to the species with the greatest number of uses, without taking into consideration the number of people that cite the uses. To convert to a scale of 100, we multiplied the combined scores (NBS + NPH) by fifty yielding a relative importance of 48.33 for "species *y*." Like the FL index of Friedman et al. [11], the RI index can be adapted for any use category.

5. Byg and Balslev [28]: These authors proposed a set of techniques to evaluate the knowledge of the use of a species of palm trees, and the proposed indices can be applied to diverse situations. In the following description, the palm trees were substituted with species so that scholars can use the formulas more generally for plant resources (Table 2).
6. Gomez-Beloz [26]—*Reported use* (*RU*): This index designates the total number of uses reported for each plant. The author considers this index to be similar to that of [6, 7], but it differs by the fact that the informants have been interviewed only once (one event). In addition to the total RU, the author proposes additional formulas for calculating the following: the use value for each plant and part of the plant (PPV), the specific use (SU), the intraspecies use value (IUV) and the overall use value (OUV). The Table 3 contains all of the formulas for each calculation proposed by Gomez-Beloz [26]. This index assigns species value according to the total set of uses that are attributed to it; it does not consider how knowledge about the plant is distributed locally.
7. Lawrence et al. [29]—*Rapid informant ranking* (*RIR*): Informants list the ten most important taxons. The rank of the species is converted into a score. For example, the most highly ranked species receives a score of 10, the second a score of 9, the third a score of 8, and so forth. The final ranking of the plant is based on the sum of the ranking of the species by all of the informants, as follows:

$$\mathrm{RIR}_{\mathrm{taxon}} = \frac{1}{2}\left(\sum T_{\mathrm{m}} / N_{\mathrm{m}} + \sum T_{\mathrm{f}} / N_{\mathrm{f}}\right),$$

where:

$\mathrm{RIR}_{\mathrm{taxon}}$ = rapid informant ranking for a given species

T_{m} = total of the scores given by the men for a specific species

N_{m} = number of men that were interviewed

T_{f} = total of the scores given by the women for a specific species

N_{f} = number of women that were interviewed

Imagine a community in which ten men and seven women are interviewed. Four men cite "species *y*" among the ten most important

Table 2
Formulas presented by Byg and Balslev [28] with adaptations for more general use

Technique	Description	Formula
Total species diversity (SD_{tot})	Measures how many species are used and how they contribute to total use. The values vary from 0 to *n*	$SD_{tot} = 1/\sum P_s^2$, where P = the contribution of species s to the total species use (the number of times species s was mentioned divided by the total number of uses for the cited species)
Total species equitability (SE_{tot})	Measures how different species contribute to the total use, independent of the number of species used. Values vary from 0 to 1	$SE_{tot} = SD_{tot}/n$, where n = the number of species used
Importance value (IV_s)	Measures the proportion of informants that cited a species as the most important. The values vary from 0 to 1	$IV_s = n_{is}/n$, where n_{is} = the number of informants that consider species s to be the most important and n = the total number of informants
Use diversity (UD_s)	Measures for how many use categories a species is used and how evenly these contribute to its total use. Values range between 0 and number of use categories for which it is used	$UD_s = 1/\Sigma P_c^2$, where P_c = the contribution of the use category c to the total utility of species s (the number of times that species s was mentioned within each use category divided by the total number of use citations for species s among all use categories)
Use equitability (UE_s)	Measures how different uses contribute to the total use of a species, independent of the number of use categories. The values vary between 0 and 1	$UE_s = UD_s/UD_{smax}$, where UD_{smax} = the maximum possible usage diversity value for species s with uses in a given number of categories

Informant diversity (ID_s)	Measures how many informants use a species and how its use is distributed among them. The values vary between 0 and the number of informants that use a species	$ID_s = 1/\sum P_i^2$, where P_i = the contribution of informant i to the total set of knowledge about species s (the number of records of use of species s by informant i divided by the total number of records of use of species s)
Informant equitability (IE_s)	Measures how the use of a species is distributed among the informants, independent of the number of informants using it. Values vary between 0 and 1	$IE_s = ID_s/ID_{smax}$, where ID_{smax} = maximum informant diversity value for a species s which is known by a given number of informants
Use consensus (UC_s)	Measures the degree of agreement among the informants with relation to whether a species is useful or not. Values vary between −1 and +1	$UC_s = 2n_s/n - 1$, where n_s = the number of people that use species s
Proposition consensus (PC_s)	Measures the degree of agreement among the informants with respect to uses. The values vary between 0 and 1	$PC_s = \sum Pu^2/S$, where P_u = the proportional contribution of use u to the total utility of species s (the number of times that use u was recorded for species s divided by the total number of recorded uses of species s). S = the total number of types of uses for species s
Species diversity (SD_i)	Measures how one informant uses many species and how the uses are distributed among the species. The values vary between 0 and the number of species used by the informant	$SD_i = 1/\sum P_s^2$, where P_s = the contribution of species s to the total species use of informant i (the number of times that informant i mentions species s divided by the total number of responses of informant i)
Species equitability (SE_i)	Measures how one informant makes use of the plants that they know, independent of the number of plants used. Values vary between 0 and 1	$SE_i = SD_i/SD_{imax}$, where SD_{imax} = the maximum value of species diversity for informant i

Table 3
Formulas designated for the calculation of different aspects of the reported use value according to Gomez-Beloz [26]

Technique	Formula
Plant part value (PPV)	Ratio of the total number of reported uses for a given part of a plant ($RU_{(plant\ part)}$) to the sum of the reported uses for all of the parts of the plant ($\sum RU$)
Specific reported use (SU)	Corresponds to the number of times that a specific use was cited by an informant
Intraspecies use value (IUV)	Ratio of the specific use ($SU_{(plant\ part)}$) to the reported uses for the part of the plant ($RU_{(plant\ part)}$)
Overall use value (OUV)	OUV = PPV × IUV. Compares the importance of uses among a group of plants

species, allocating it to the first, third, ninth, and fourth positions in the ranking, which correspond to scores of 10, 8, 2, and 7, respectively. Two of the seven women also cite "species *y*." Following the same logic, the scores assigned are 5 and 4. Thus, the calculation for "species *y*" is as follows: $RIR = 1/2[(10+8+2+7)/10+(5+4)/7]$, equaling 1.99.

8. Castaneda and Stepp [30]—*Ethnoecological importance value (EIV)*: This proposed method estimates the relative importance for a particular habitat (type of vegetation). This index is calculated for each species in each habitat and allows for a quantitative comparison of the ethnoecological value that particular habitats have to different gender, age, or cultural groups according to their abundance of species. The following formula is used:

$$EIV = \sum (S)(n_x / N_x),$$

where:

EIV = ethnoecological importance value;

S = species salience;

N = total number of species encountered in the study;

N_x = total number of individuals of the species x encountered in all of the habitats in the study;

n_x = total number of individuals of the species x found in a habitat.

Thus, the EIV of a habitat is the sum of the ethnoecological importance of each useful species contained within it. The ethnoecological importance of each species is the product of the species cultural salience (S) times its relative abundance (n_x/N_x).

The calculation of EIV requires the following steps [30]:

1. Free lists must be made in the community calculating the salience (S) for each species mentioned. The salience can be calculated, for example, from rankings using a similar methodology to the RIR index proposed by Lawrence et al. [29], with values assigned to the species on the free list.
2. Field plots in the habitats are required to determine the abundance of the species (the total number of individuals that occur in each habitat) and the relative importance of each species from an ecological point of view.
3. The ethnoecological value of each plant is calculated for each species in each habitat.
4. The ethnoecological value for all of the species in each habitat is summed to determine the EIV.

1.2 Subjective Allocation

1. Prance et al. [31]—The calculated use value of a plant reflects the biases of the investigator. Thus, plants of lesser use, according to the field observations of the investigator, receive a value of 0.5, and the more important plants receive a value of 1.0. The total value of a plant is given by the sum of the values calculated for each of the uses of a plant. This is an extremely subjective technique that depends on the "view" of the investigator. Consider a community in which the following five uses are given for "basil": "to cure flu," "cough," "dispel the evil eye," "make tea," and "garnish." The investigator considers the medicinal and religious applications as the most important and assigns points accordingly: 1 ("to cure flu"), 1 ("cough"), 1 ("dispel the evil eye"), 0.5 ("make tea"), and 0.5 ("garnish"). The calculation would be as follows: $UV = 1 + 1 + 1 + 0.5 + 0.5$, totaling 4.0.
2. Turner [8]—Subjective allocation predominates among the techniques that analyze the value or importance of a species. The CSI proposed by Turner [8] aims to record the role of plants in a culture. The term cultural significance had already been used by Hunn [32] to indicate the importance or role that a taxon plays in a culture. Berlin et al. [33] understood cultural significance as the practical value of biological knowledge to a given culture. The CSI, initially proposed by Turner [8], and later modified by Stoffle et al. [9] without substantial alterations, was severely criticized for preestablishing use values for species according to their category of use, once again reflecting the "view" of the investigator.

The CSI, according to Turner [8] and Stoffle et al. [9], is represented in the following manner:

$$\text{CSI} = \sum CSI(q \cdot i \cdot e),$$

where:

CSI = cultural significance index

q = quality of use

i = intensity of use

e = exclusivity of use

The variation for the values attributed by the investigator are $q = 5$ to 1; $i = 5$ to 1; and $e = 2$ to 0.5.

This index was altered by Silva et al. [34] in the following way:

$$\text{CSI} = \sum (i \cdot e \cdot c) \cdot CF,$$

where:

CSI = cultural significance index

i = management of the species

e = preference of use

c = frequency of use

CF = correction factor

The values of the variables "i," "e," and "c" are either 2 or 1 and are determined by each citation of use. These modifications lend a more objective character to the formula while diminishing its subjectivity.

Management of the species (i) refers to the impact of the plant on the daily life of the community [8].

Preference of use (e) represents the preference of use for one species in relation to another for a specific function [8].

Frequency of use (c) considers the currently used plants. In agreement with the values proposed by Stoffle et al. [9], it attributes a value of 2 to plants that are currently known and used and a value of 1 to plants that are rarely cited.

The *correction factor* (CF) takes into consideration the consensus among the informants. The value comes from the number of informants that cited the species divided by the number of informants that cited the most cited species.

In short CSI equals the sum of each value ($i \cdot e \cdot c$) for a given species—for each use that it may have—and this sum is multiplied by the CF value.

Below is an example of a CSI calculation:

The "species *x*" is an important plant cited as: treatment for kidney problems, firewood, and making stakes. Thus, CSI "species *x*" = $\sum$(species *x*'s value as kidney problems) + (species *x*'s value as a firewood) + (species *x*'s value as a making stakes)·CF. "Species *x*" it is important to the community, it is assigned a value of 2 for "management of the species"; because it is preferred for the kidneys problems, it receives the value of 2 for medicinal use, and receives a value of 1 for firewood use and making stakes. Because it suffers a constant use impact, the "species *x*" receives a value of 2 for "frequency of use" in each use. Finally, the value of the correction factor is 0.75 because the "species *x*" is cited by 37 informants, while the most cited species is cited by 49 informants. Thus, the index calculation is as follows: CSI "species *x*" $= [(2\times2\times2)+(2\times1\times1)+(2\times1\times2)]\times0.75$, resulting in a value of 12.08.

1.3 Sum of Uses

For this index, the uses are simply summed by category, vegetation type, or taxon, which is useful for the first analysis of the data [31, 35, 36].

2 Final Considerations

Studies are still necessary for the evaluation of the quality and adequacy of the large quantity of currently available techniques for their proposed objectives (see [2, 3]). Furthermore, the constant publication of new techniques impedes "standardization" among investigators and studies. This does not mean that new techniques should not be developed but that there should be some criteria for their creation. How do you compare data and establish relations between studies and communities if there is a new technique with each new publication? At the same time, there is a need to establish a procedure that can "translate" or "measure" the importance of plant resources to a given community.

With regard to the relative importance of species, another aspect that should be considered is that knowledge of the use of a plant by an individual or a group of people does not necessarily imply that the plant is being used effectively. Many of the discussed techniques emphasize the potential for use. The difference between these two "moments" must be observed at the time of study.

Despite the succinctness of this exposition, the reader is invited to reflect on the techniques that are used in his or her own studies, as observed by Hoffman and Gallaher [2]. It should be emphasized that techniques based on informant consensus are more robust and function better for specific types of inferences and analyses.

References

1. Phillips O (1996) Some quantitative methods for analyzing ethnobotanical knowledge. In: Alexiades M (ed) Selected guidelines for ethnobotanical research: a field manual. The New York Botanical Garden, New York, pp 171–197
2. Medeiros MFT, Silva PS, Albuquerque UP (2011) Quantification in ethnobotanical research: an overview of indices used from 1995 to 2009. SientibussérieCienciasBiológicas 11(2):211–230
3. Reyes-García V, Huanca T, Vadez V et al (2006) Cultural, practical, and economic value of wild plants: a quantitative study in the Bolivian Amazon. Econ Bot 60:62–74

4. Hoffman B, Gallaher G (2007) Importance indices in ethnobotany. Ethnobot Res Appl 5:201–218
5. Amiguet VT, Arnason JT, Maquim P et al (2005) A consensus ethnobotany of the Q' Eqchi' Maya of Southern Belize. Econ Bot 59:29–42
6. Phillips O, Gentry AH (1993) The useful plants of Tambopata, Peru: I. Statistical hypothesis tests with a new quantitative technique. Econ Bot 47:15–32
7. Phillips O, Gentry AH (1993) The useful plants of Tambopata, Peru: II. Additional hypothesis testing in quantitative ethnobotany. Econ Bot 47:33–43
8. Turner NJ (1988) "The importance of a rose": evaluating the cultural significance of plants in Thompson and Lillooet interior Salish. Am Anthropol 90(2):272–290
9. Stoffle RW, Halmo DB, Evans MJ et al (1990) Calculating the cultural significance of American Indian plants: Paiute and Shoshone ethnobotany at Yucca Mountain, Nevada. Am Anthropol 92:416–432
10. Carneiro RL (1978) The knowledge and use of rain forest trees by the Kuikuru Indians of central Brazil. In: Ford RI (ed) The nature and status of ethnobotany, Anthropological papers. University of Michigan, Ann Arbor, pp 202–216
11. Friedman J, Yaniv Z, Dafni A et al (1986) A preliminary classification of the healing potencial of medicinal plants, based on a rational analysis of an ethnopharmacological field survey among bedouins in the Negev desert, Israel. J Ethnopharmacol 16:275–287
12. Albuquerque UP, Andrade LHC (2002) Uso de recursos vegetais da caatinga: o caso do agreste do estado de Pernambuco. Interciencia 27:336–346
13. Amaral CN, Guarim-Neto G (2008) Os quintais como espaços de conservação e cultivo de alimentos: um estudo na cidade de Rosário Oeste (Mato Grosso, Brasil). Bol Mus Par Emílio Goeldi 3(3):329–341
14. Amorozo MCM, Gély A (1988) Uso de plantas medicinais por caboclos do Baixo Amazonas. Barcarena, PA, Brasil. Bol Mus Par Emílio Goeldi Sér Bot 4:47–131
15. Vendruscolo GS, Mentz LA (2006) Estudo da concordância das citações de uso e importância das espécies e famílias utilizadas como medicinais pela comunidade do bairro Ponta Grossa, Porto Alegre, RS, Brasil. Acta Bot Bras 20:367–382
16. Troter R, Logan M (1986) Informant consensus: a new approach for identifying potentially effective medicinal plants. In: Etkin NL (ed) Indigenous medicine and diet: biobehavioural approaches. Redgrave Bedford Hills, New York, pp 91–112
17. Albuquerque UP, Medeiros PM, Almeida ALS et al (2007) Medicinal plants of the caatinga (semi-arid) vegetation of NE Brazil: a quantitative approach. J Ethnopharmacol 114: 325–354
18. Teklehaymanot T, Giday M, Medhin G et al (2007) Knowledge and use of medicinal plants by people around DebreLibanos monastery in Ethiopia. J Ethnopharmacol 111:271–283
19. Akerreta S, Cavero RY, López V et al (2007) Analyzing factors that influence the folk use and phytonomy of 18 medicinal plants in Navarra. J Ethnobiol Ethnomed 3:16
20. Torre-Cuadros MLA, Islebe GA (2003) Traditional ecological knowledge and use of vegetation in southeastern Mexico: a case study from Solferino, Quintana Roo. Biodivers Conserv 12:2455–2476
21. Lucena RFP, Araújo EL, Albuquerque UP (2007) Does the local availability of woody *caatinga*plants (Northeastern Brazil) explain their use value? Econ Bot 61:347–361
22. Molares S, Ladio A (2009) Ethnobotanical review of the Mapuche medicinal flora: use patterns on a regional scale. J Ethnopharmacol 122:251–260
23. Thomas E, Vandebroek I, Dammea PV et al (2009) The relation between accessibility, diversity and indigenous valuation of vegetation in the Bolivian Andes. J Arid Environ 73:854–861
24. Albuquerque UP, Lucena RFP, Monteiro JMM et al (2006) Evaluating two quantitative ethnobotanical techniques. Ethnobot Res Appl 4:51–60
25. Rossato SC, LeitãoFilho H, Begossi A (1999) Ethnobotany of Caiçaras of the Atlantic Forest Coast (Brazil). Econ Bot 53:387–395
26. Gomez-Beloz A (2002) Plant knowledge of the WinikinaWarao: the case for questionnaires in ethnobotany. Econ Bot 56:231–241
27. Bennett BC, Prance GT (2000) Introduced plants in the indigenous pharmacopoeia of Northern South America. Econ Bot 54: 90–102
28. Byg A, Balslev H (2001) Diversity and use of palms in Zahamena, Eastern Madagascar. Biodivers Conserv 10:951–970
29. Lawrence A, Phillips OL, Ismodes AR et al (2005) Local values for harvested forest plants in Madre de Dios, Peru: towards a more contextualised interpretation of quantitative ethnobotanical data. Biodivers Conserv 14: 45–79
30. Castaneda H, Stepp JR (2007) Ethnoecological importance value (EIV) methodology: assessing the cultural importance of ecosystems as sources of useful plants for the Guaymi people of Costa Rica. Ethnobot Res Appl 5:249–257

31. Prance GT, Balée W, Boom BM et al (1987) Quantitative ethnobotany and the case for conservation in Amazônia. Conserv Biol 1:296–310
32. Hunn ES (1982) The utilitarian factor in folk biological classification. Am Anthropol 84:830–847
33. Berlin B, Breedlove DE, Laughlin RM et al (1973) Culturalsignificance and lexical retention in Tzeltal-Tzotzilethnobotany. In: Edmonson MS (ed) Meaning in Mayan languages. Mounton, The Hague
34. Silva VA, Andrade LHC, Albuquerque UP (2006) Revising the cultural significance index: the case of the Fulni-ô in Northeastern Brazil. Field Methods 18:98–108
35. Balée W (1986) A etnobotânica quantitativa dos índios Tembé (Rio Gurupi, Pará). Bol Mus Par Emílio Goeldi Sér Bot 3: 29–50
36. Balée W, Gély A (1989) Managed forest succession in Amazonia: the Ka'apor case. Adv Econ Bot 7:129–158

Chapter 25

Ecological Methods in Ethnobotanical and Ethnobiological Research: Using Diversity Measurements and Richness Estimators

Nivaldo Peroni, Helder Farias Pereira Araujo, and Natalia Hanazaki

Abstract

This chapter discusses the application of diversity measures for ethnobiological and ethnobotanical data sets. Diversity measures are based on the assessment of the heterogeneity of information, with two main components: richness and abundance. In ecological studies, this information is usually equivalent to heterogeneity in the distribution of individuals of different species in a given area. In studies in which we are dealing with local knowledge about species—i.e., vernacular names, ethnogenera or ethnospecies—information often corresponds to the quotations from each respondent about the perceived items. In ethnobiological research these analyses allow us to broaden the discussions on the evaluation of sampling effort; on the comparability between data sets obtained from different regions; on the objective analysis of the distribution of knowledge within a given human group; and on the possibility of integration ethnobiological data with ecological and biological information.

Key words Diversity, Richness, Evenness, Rarefaction, Sampling effort

1 Introduction

Anthropology has played a central role on influencing the origin and development of ethnobiology and ethnobotany. The growth of these fields is reflected in the development of their methodology, which has incorporated increasingly quantitative approaches, to complement the more qualitative approaches. An important contribution came between the 1980s and 1990s, with the publication of manuals discussing methodological issues in ethnobotany [1–3], and articles that dealt with quantitative approaches [4–7].

In the Brazilian context, there are few publications with emphasis on methods in ethnobotany and ethnobiology (see, e.g., [8–10]). In particular, the application of ecological methods in ethnobotanical and ethnobiological research can be illustrated by article [7, 11–13]. It is also important to remember authors who have used ecological methods, like the use of diversity indexes as

Ulysses Paulino Albuquerque et al. (eds.), *Methods and Techniques in Ethnobiology and Ethnoecology*, Springer Protocols Handbooks, DOI 10.1007/978-1-4614-8636-7_25, © Springer Science+Business Media New York 2014

estimators of food niche breadth (e.g., [14–16]). Commonly used taxonomic and ecological methods, such as multivariate analysis, also have a wide applicability in studies of this nature (e.g., [17, 18]).

Among the various ecological methods which may be applied in ethnobotanical research, this article discusses the application of diversity measurements and richness estimators for ethnobotanical and ethnobiological data sets, based on surveys conducted in the southeastern region of the Brazilian Atlantic coast [12, 19, 20] and on mangrove-estuarine complexes on the northern coast of the state of Paraíba [21]. The potential application of ecological methods in ethnobotanical and ethnobiological research lies in some interrelated factors, such as, for example: evaluation of sampling effort; comparability between data sets obtained in different regions; the ability to objectively analyze the distribution of ethnobotanical and ethnobiological knowledge; the applicability of ethnobiological and ethnobotanical studies in conservation programs; the possibility of integrating ethnobotanical and ethnobiological data with ecological and biological data; and the possibility of evaluating patterns of biological and cultural variation.

2 Measurements of Diversity

Diversity simply reflects the variety that exists in a certain place, and refers to the number of categories that can be differentiated, and the proportions (or relative abundance) of the objects in each category. Biological diversity, or biodiversity, includes diversity among species (diversity of organisms) within species (genetic diversity) and diversity of communities or ecosystems (ecological diversity) [22].

A more intuitive measure of diversity is the number of individuals in a given area. A count of the number of species present in a community expresses its species richness, which corresponds to a simple measurement of diversity. However, only obtaining the number of species present in a community does not provide information regarding species abundance, or rather, which species are rare and which are common.

The second important component is the evenness of diversity. To ensure that diversity measurements confer different weights to rare and common species, it is necessary to consider the equitability among species in the sample, in other words, its abundance. Figure 1 illustrates two hypothetical communities, both with eight species and 32 individuals. Despite having the same species richness, the abundance of the eight species is more evenly distributed in community "b," where all species have the same chance of being sampled. In community "a" there are some more abundant species (e.g., the fish and white flowers), while other species are less

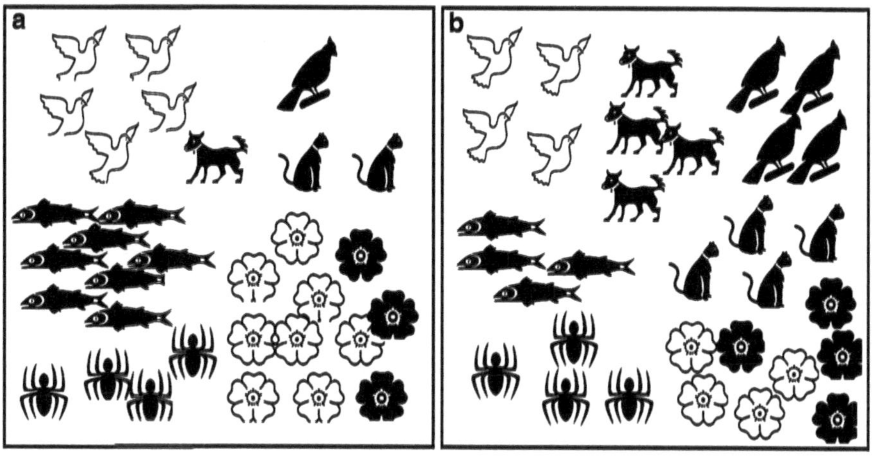

Fig. 1 Two hypothetical communities "a" and "b," with eight species and 32 individuals each

common (e.g., the dog or black bird). When comparing these two communities, considering their richness and abundance, community "b" is more diverse than community "a."

Since diversity measures mostly evaluate the heterogeneity of the information, for ethnobotany and ethnobiology these concepts must be translated into questions like, for example: what is the richness of species or item citations in a sample of users within a community? How do groups between and within communities differ based on species diversity and/or items cited? How many species/items could still have been cited if the sample size (number of respondents) were increased? Is there any relationship between patterns of known species and species availability in the environment?

2.1 Analyzing Citation Diversity When Considering the Components of Richness and Abundance

To measure the diversity of a community considering both richness and abundance, relatively simple indices may be used, such as the Simpson diversity index. This index is calculated by determining, for each species, the proportion of individuals (or biomass) that contributes to the total of the sample. The Simpson index ($1/D$) corresponds to the inverse of dominance in a community (D), which is calculated by Eq. 1:

$$D = \sum pi^2 \quad (1)$$

where p is the proportion of the total number of species to the ith species. In ethnobotany and ethnobiology, the number of individuals of each species can be replaced by the information that is being measured: for example, the number of citations for a given animal or plant.

Since the Simpson diversity index is a measure opposed to dominance, the complementary value to dominance, or $1 - D$, can also express the index. This index reflects the probability of any two individuals caught at random belonging to the same species. It is also a measurement of dominance, known as a measure of the concentration of individuals into a few species.

Another index that is often used is the Shannon index, also known as Shannon–Wiener index (H). This index expresses the probability of correctly identifying an individual picked at random from a sample, and is given by Eq. 2:

$$H = -\sum pi \log pi \quad (2)$$

where p is the ratio of the total number of species to the ith species.

Although it has been used extensively, the Shannon–Wiener index is highly sensitive to sample size. Some of these measures are discussed in the following section and can be adjusted to ethnobotanical and ethnobiological studies, see for example [23, 24]. In order to analyze the diversity of a data set, whether it is species, items, or citations, it is recommended to use different measures of diversity. For a more in-depth approach to the topic, consult [25, 26].

Each index has peculiar characteristics and may be more appropriate for specific situations. For example, in the case of registration of all species in a given location, rather than a sample, it is recommended to use the Brillouin index (HB), given by Eq. 3:

$$HB = \frac{\log_2 N! - \sum \log_2 n_i!}{N} \quad (3)$$

where N is the total number of species and n_i is the number of occurrences for each species.

2.2 Analyzing the Diversity of Citations Considering the Richness of Species Citations

Often we have to resort to the use of sampling, since it may be very difficult and expensive interviewing all residents of a community, or determining all the species within the community's gardens, for example. However, when sampling, it is important to take into account the influence of the sample size on the number of items mentioned or present in the samples. It is also important to be careful when drawing comparisons between locations with different sample sizes. Moreover, often times the interest of this work lies only in the analysis of species citation richness, independently of their distribution abundances. Although it is easy to compute the number of species cited in a sample, there is a possibility of computing more species citations if our sample size or our sampling effort were increased.

Therefore, we can then make use of "extrapolation," or so-called "richness estimators," which will provide an estimate of the number of items (e.g., species) that could have been computed at a given location, without necessarily taking into consideration the quantity (abundance) of individuals or number of citations per species. The literature on ecological richness estimators continues to increase, for example the application of this approach in ethnobotany [13, 27, 28]. For an in-depth theoretical work, consult [29–32]. It is also important to reinforce that some of the methods presented assume that each sample unit represents a set of items in a single sample unit, where curves are "based on samples" [31]. In a guided tour, for example, where informants identify species, there is a sequential accumulation of individual species. Other approaches could also be used for curves "based on individuals," for example, [33] and classical rarefaction [34], which is discussed later in this chapter.

Species richness computed in interviews will depend on the intrinsic characteristics of the informants and the community where the study is being conducted, as well as the sampling effort for data collection. A simple way to evaluate the sample effort is to adapt the species accumulation curves, widely used in ecological studies, to assess accumulation of cited species from interviews.

Accumulation curves can be useful even in the early stages of data collection when the botanical or zoological species has not been identified. In this case the analysis of accumulation curves using citations of popular names allows for the evaluation of the sampling effort, allowing preliminary assessments and possible redirection of data collection strategies.

A recent approach developed recommends the use of measurements and comparisons of species richness through the use of sample randomization in accumulation curves [29, 31]. When accumulation curves are constructed by plotting only the addition of new species to sampling units (e.g., gardens or interviews), steps within the curves appear, indicating the heterogeneity of the samples, or even any particular change. For example, this may occur, when a species survey is conducted where a large garden is surveyed followed by smaller ones, or even an interview where the informant knows a greater number of species compared to others already sampled (Fig. 2). In this hypothetical example, beginning from informant number 14 there is a significant addition to the set of species previously examined, showing a pattern change from the inclusion of this informant in the sample.

In cases like the hypothetical example of Fig. 2, the shape of the curve can vary depending upon the order of sample unit addition. The method proposed is to build a curve that expresses an average of the calculated richness values, beginning with the randomization of a sample's position [29, 31]. Thus, average values of richness are calculated for random combinations of one sample, two samples, three samples, and so on until completing the total value of samples. Then the average curve may be plotted, which shows a "smooth" form (Fig. 3). The curve may then be compared with other accumulation curves, as well as with curves based on

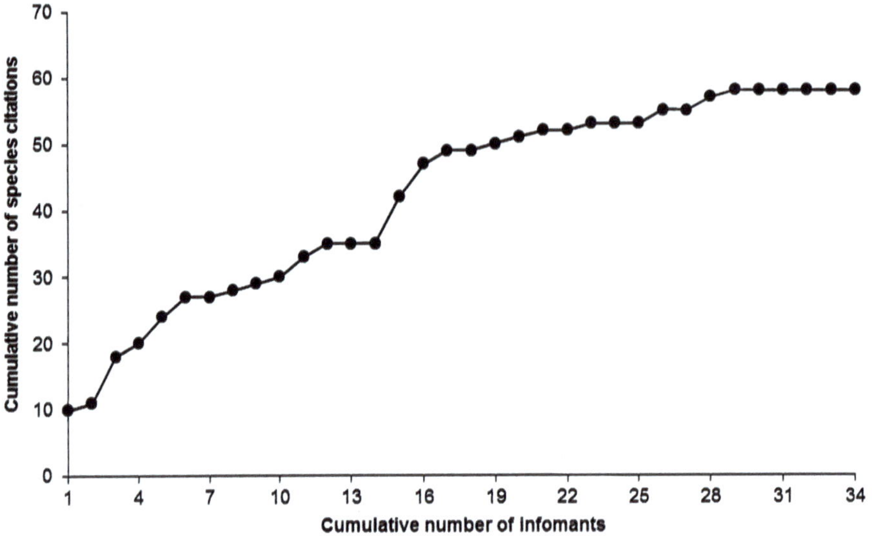

Fig. 2 Hypothetical example of a species citation accumulation curve from a set of 34 informants and 58 species citations

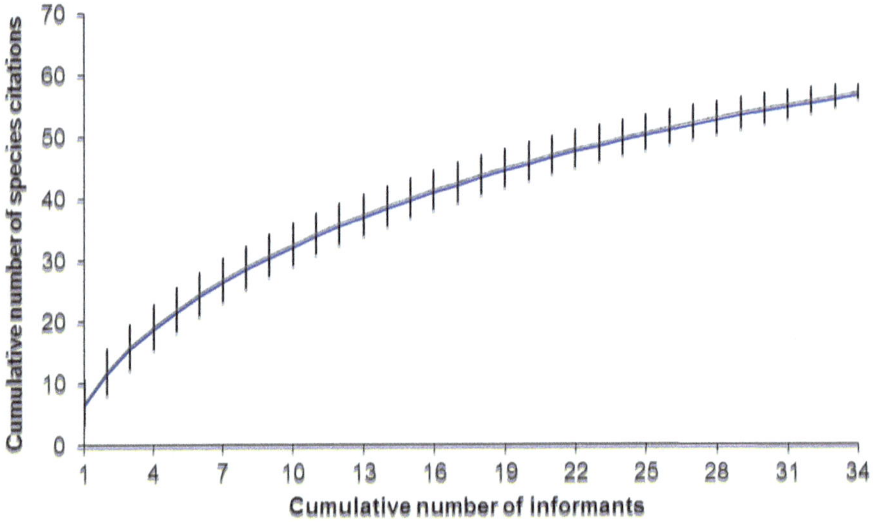

Fig. 3 Example of a hypothetical accumulation curve of species citations between 34 informants. The *line* represents an average curve calculated from 500 curves with randomized addition of samples. The *vertical lines* indicate the possible deviations between the calculated curves

calculations from specific indices, known as "richness estimators." Particularly for this type of analysis, calculations are facilitated by the use of the software EstimateS [35].

The simplest calculation in this approach rests on the total number of species observed in each sample (S_{obs}), which is represented by a cumulative species curve, plotted on a graph showing the sampling effort. One can then compare the plotted curve with other similar sample-sized curves or choose a diversity estimator to produce a curve which will be compared to the observed accumulation curve (S_{obs}).

The estimation methods discussed here belong to the category known as nonparametric richness estimates [26, 36]. Among the various estimators discussed [32], just a few that have been tested with ethnobotanical and ethnobiological data and showed satisfactory results will be discussed (e.g., [13, 21, 27, 28, 37]). These indices can be divided into three groups according to the type of estimate: (a) those where the number of species accumulated in a curve is estimated; (b) those that indicate the estimation of species richness based on rare species (citations) shared between sample sets; for this some estimators include Jackknife 1 and 2, Chao 1 and 2, and Bootstrap; and (c) those where the estimate is based on the richness of rare species shared among groups of samples (ICE, incidence-based coverage estimator and ACE, abundance-based coverage estimator) (see [32, 36]). The formulas for the estimators may be found in Appendix 1.

2.2.1 Jack 1 and Jack 2

The Jackknife technique is a procedure used to reduce the underestimation of the true number of species in a sample, based on the number of items in a sample. A first-order Jackknife estimator is based on the presence of "uniques" (species, or citations of species present in a single sample unit). A second-order estimator (Jack 2) relies on the "uniques" plus "duplicates" (species present in exactly two sample units) [29].

2.2.2 Chao 1 and Chao 2

The Chao 1 estimator is based on abundance, however it utilizes the relationship between the number of "singletons" (species represented by only one individual) and "doubletons" (species represented by exactly two individuals) to estimate the number of species not observed [38, 39]. In this estimator, the greater the number of species, the greater the difference between the observed and estimated richness. Chao 2 is a diversity estimator based on the incidence of species (presence or absence) from data expressing rarity, "uniques" and "duplicates," that is the number of species found in only one or two samples, respectively [39]. Similarly to Chao 1, the greater the number of "uniques" and "duplicates" species, the greater the difference in estimations.

2.2.3 Incidence-Based Coverage Estimator and Abundance-Based Coverage Estimator

These estimators of species richness are modifications of [40] and are discussed in [29, 41, 42]. ACE is based on species with less than ten individuals in the sample (thus considers its abundance). The corresponding ICE is also based on the species found in less than ten sample units (considers its incidence) [41]. These are richness estimators based on the statistical concepts of sample cover, which refers to the sum of the probabilities of finding the species observed within the total amount of species present.

In other words, it assumes that the most abundant and widely distributed are more likely to be found in any sample, but were not observed [32].

2.2.4 Bootstrap

This method is based on the species incidence (presence or absence) [41, 43]. Thus only requires incidence data (presence/absence).

2.3 Analyzing the Diversity of Citations Considering a Distinct Number of Sampling Units Based on Individuals (Abundance)

Rarefaction is applied in comparative studies, where there is an interest in analyzing the diversity present in different communities with different numbers of interviews, that is, different sample sizes. It is also very common to compare the diversity of cited items by men and women from the same community, or even compare classes of different ages (e.g., adults and youth). In these cases one must resort to strategies of analysis where diversity is compared taking into account a standardized sample size, and use techniques such as rarefaction analysis.

Rarefaction is a technique that estimates the expected number of species in each sample if all samples have the same standard size [33, 34]. The expected richness for each point on the curve is given by Eq. 4:

$$E(s) = \sum f(x) = \left\{1 - \left[\binom{N - Ni}{n}\right] \Big/ \left[\binom{N}{n}\right]\right\} \quad (4)$$

where $E(S)$ is the expected number of species in the rarefied sample (expected richness given by n), n is the standardized sample size, N is the number of species (richness), and N_i is the number of samples where the ith species occurs. The rarefaction curve is obtained by calculating the $E(S)$ for a sequence of points, wherein the highest point corresponds to the total richness measured for that community.

3 Applications in Ethnobotany and Ethnobiology Measures of Diversity and Rarefacation

A parallel can be drawn between the number of species present in an area with the number of species known or used by a particular human group. Thus, instead of using the number of species in sample units, the number of citations of each ethnospecies is used from interviews that make up a sample of a human community [12].

This parallel was utilized before to compare the diversity and sampling effort of several published ethnobotanical studies, also utilizing rarefaction measurements [7]. Rarefaction allows the comparison between subsamples of the same size, from species richness and abundance [25]. The same procedure was used by others authors [44] when comparing ethnobotanical data sets in three study areas: Rio Negro, Rio Araguaia, and the Atlantic Forest.

Table 1
Diversity indices of plant use among the Caiçara communities of southeastern Brazil

Location	*R*	*1/D*	*H′*	Citations	References
Picinguaba	216	77.12	2.06	1,552	Rossato et al. [11]
Praia Camburí	162	60.23	1.98	541	Hanazaki et al. [12]
Praia Almada	152	70.35	2.00	433	Hanazaki et al. [12]
Sertão Puruba	140	55.87	1.92	525	Rossato et al. [11]
Praia Puruba	124	58.14	1.92	414	Rossato et al. [11]
Casa de Farinha	108	55.98	1.85	393	Rossato et al. [11]
Gamboa	100	25.14	1.65	558	Figueiredo et al. [47]
Praia Calhaus	75	23.81	1.53	482	Figueiredo et al. [48]
Ilha da Vitória	57	32.14	1.61	195	Rossato et al. [11]

R richness, 1/*D* Simpson, *H′* Shannon–Wiener

When analyzing multiple ethnobotanical data sets from *caiçara* communities in the Atlantic Forest, some general patterns regarding diversity are established (Table 1). The largest differences were found between communities located on the continent when compared to the diversity found on islands (Gamboa, Calhaus and Ilha da Vitória). This trend is discussed by others authors [11] based on the theory of island biogeography, since islands are expected to have a lower diversity of biological species. In some situations, a higher number of ethnospecies' citations does not correspond to a greater diversity of species when abundance is considered by the Simpson index and Shannon–Wiener index (see, e.g., inversions between Camburi and Almada and between the Praia Puruba and Sertão Puruba in Table 1). In this sense, Praia Almada and Praia Puruba show a similar knowledge distribution, when these data sets are compared to Praia Camburi and Sertão Puruba, respectively.

Analyzing data sets of Praia Almada and Praia Camburi, both on the northern coast of São Paulo, the distribution of knowledge among men and women were compared, and between older and younger people in three broad categories of plant use: edible plants, medicinal herbs, and manufacturing plants (Fig. 4) [12]. There is no significant difference when comparing the diversity of plants used in each category in both locations [12]. However, when comparing ethnobotanical diversity according to the gender of the respondents (Fig. 4), men and women have differentiated knowledge, especially with regards to manufacturing plants. Older and younger informants also have different knowledge regarding medicinal and manufacturing plants, but not edible plants (for further discussion, see [12]).

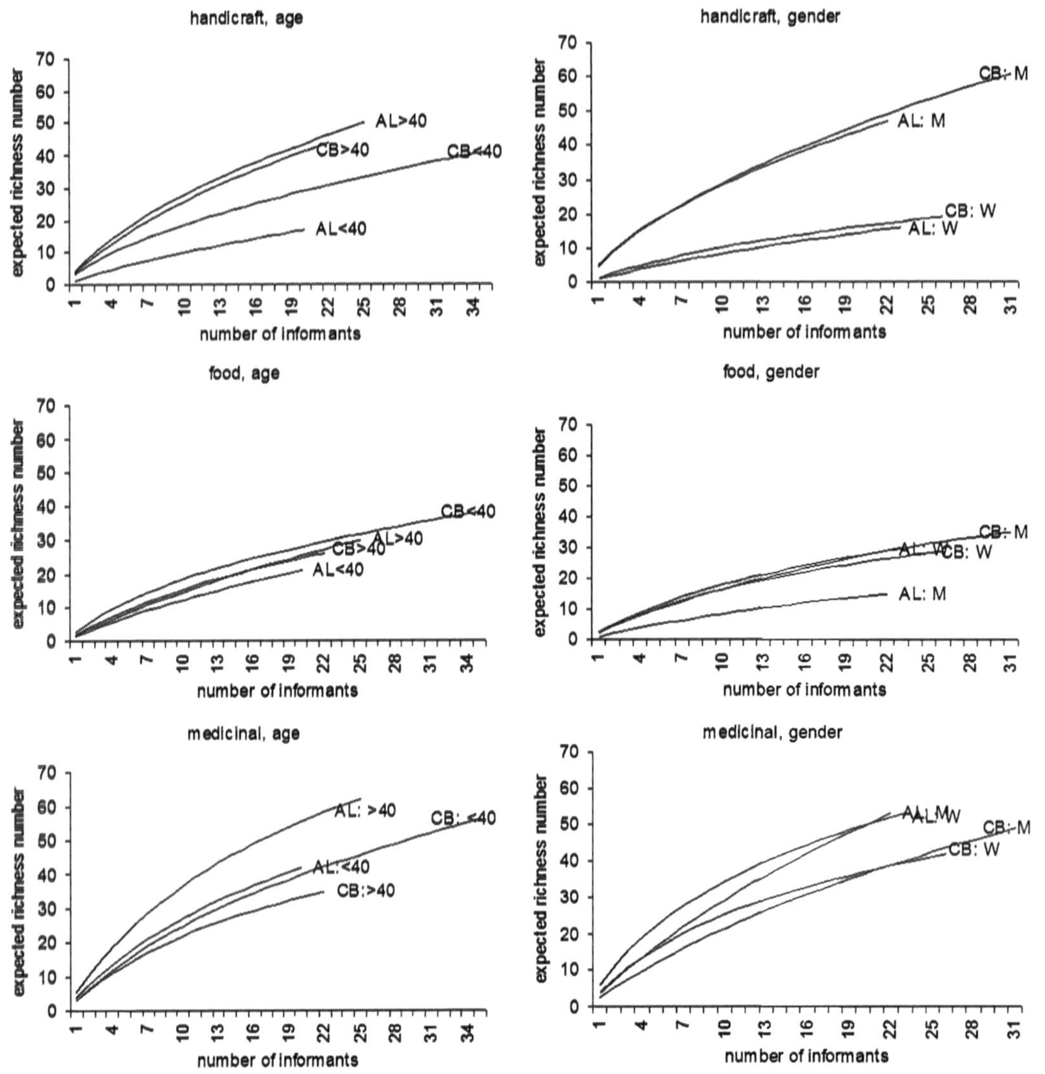

Fig. 4 Rarefaction curves based on number of informants per plant in each community. *AL* Praia da Almada, *CB* Praia do Camburí, *W* women, *M* men, <40 = informants between the ages of 18 and 40 years, >40 = informants 40 years or more (adapted from [12])

Another measurement of diversity, the Brillouin index, was used by others authors [20] to compare the knowledge and management of varieties among farmers in the southern coastal region of São Paulo. In this case, rather than using sampling, every species grown by the farmer was recorded (Fig. 5). Farmers who manage a greater diversity of varieties are important subjects in the regional system of shifting cultivation [20]. These farmers manage more varieties and a greater number of species, and tend to have better knowledge regarding the agricultural systems. Moreover, these farmers have ownership of their farming land and can grow more

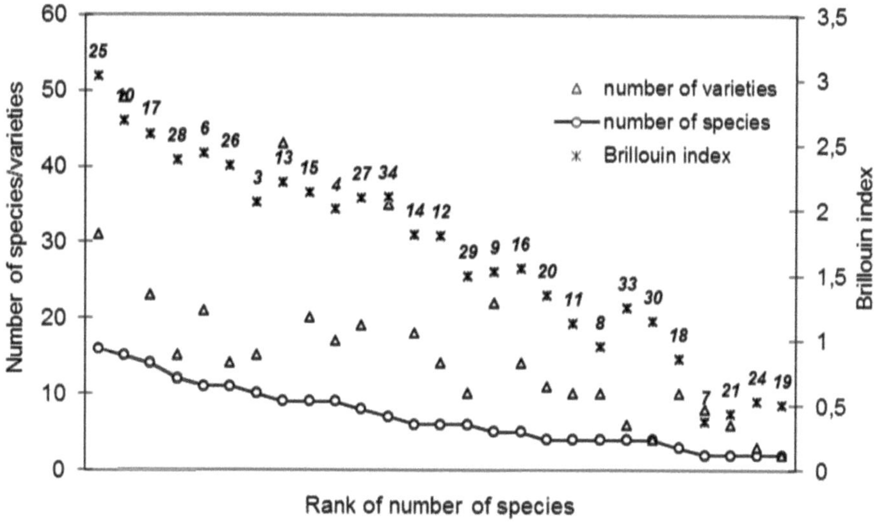

Fig. 5 Distribution of farmers according to the number of cultivated species, from highest to lowest. The numbers in *italics* identifies each farmer (modified from [20])

than one field every year, rotating the fields on their properties from one year to another. Farmers who manage low diversity are those that tend to simplify their farming systems, handling fewer species and fewer varieties.

3.1 Richness Estimators

Species accumulation curves have been successfully used to evaluate sampling effort and to compare diversity indices [13, 27]. The use of richness estimators [29, 31] has not been thoroughly tested to prove their efficiency with ethnobotanical and ethnobiological data. However, these methods have a large potential, especially since these methods enable the development of comparative analysis based on sample data sets.

In order to construct cumulative diversity curves, some authors [27] have used diversity measurements in combination with accumulation curves. The goal of these authors was to determine a viable sample size for the use of diversity indices in specific data from markets, fairs and individual merchants of medicinal plants, or "herbalists." The authors compared 22 measurements of diversity in six data sets with a sample of 50 merchants and 100 informal street vendors from traditional medicinal plant markets in Johannesburg and a seventh data set from merchants on the western boundary of Kruger National Park, South Africa. After comparing the indices, the authors concluded that a sampling in the informal market could not be less than 35 sampling units, whereas a sampling within the market would have to include at least 20 samples ("merchants"). In another study [13], the authors studied the previously mentioned 50 merchants, by using rarefaction, they showed that the sampling effort ($N=50$) was sufficient, where 511 species and 6,285 citations were registered, and each "merchant" added 1.3 new species, at a rate of one new species for every 100 species citations.

In a study on the composition of the avifauna known by artisanal fishermen in two mangrove-estuarine complexes in the state of Paraíba [21], the authors used the Chao 2 estimator to verify the number of species that would increase the richness observed from 60 informants. The fishermen cited a total of 94 species and the estimated number of species known by the local population was 96 (+3). The authors evaluated smaller sizes than the number of samples (informants), at intervals of 15, which would allow for similar estimations. With these numbers of interviews, the Chao 2 curves did not stabilize and showed significant differences with the estimation of 60 interviews, while, with the evaluation of 45 informant interviews, the estimator obtained a stable performance and no significant difference with the total estimated. The stability of the estimator is one form of measuring its accuracy [45]. Therefore, 45 informants was a sufficient number to assess known birds among the artisanal fisherman communities.

A survey about live bird trade in Brazil revealed that at least 295 bird species belonging to 177 genera and 56 families are illegally traded as pets in Brazil [37]. However, with estimates derived from these data, using two estimators, the authors point to a total of more than 400 species (Fig. 6)—about 23 % of the extant bird species in the country [37].

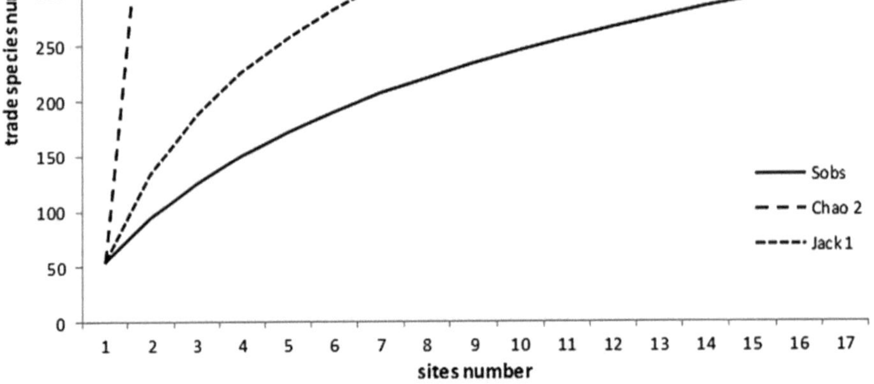

Fig. 6 Trade bird species number (S_{obs}) derived from review of 17 sites in Brazil and estimates number (Chao 2 and Jack 1) [37]

Bias and accuracy along with precision are essential variables to evaluate performances of estimators (see [45]). However, studies assessing bias and accuracy of estimators using ethnobiological data are not yet available in the literature. Some authors founded that Michaelis–Menten Means, an estimator not mentioned here, based on extrapolations of accumulation curves, was the most accurate or most stable among the tested ethnobotanical data from Johannesburg and Mpumalanga [28]. However, it was not as accurate as the Jack 1 estimator, albeit the bias and accuracy were not statistically tested. Such studies would require the stabilization of rarefaction curves, which in many cases does not occur easily. Even with the lack of studies that evaluate and determine ideal estimators for ethnobiological data analysis, the use of estimation methods enable the development of comparative analysis that assist in several studies, as discussed and exemplified.

There are two predispositions in using estimation methods from interview data: the broad spatial distribution of information in a community and the interviews with a maximum number of key informants in each community [21]. In the first case, the amount of information of species may be restricted to certain places in certain communities, either for cultural reasons or proximity to certain biologically distinct environments. In the second case, usually, some informants have a comprehensive and/or specific knowledge and may be responsible for changes in the pattern of species accumulation curves, as observed of informants 14–16 in hypothetical example of Fig. 2. Consequently, these changes may influence the predisposition of estimators, which use the behavior of rarefaction curves derived from the species accumulation curves.

4 Final Considerations

Other diversity estimates can be used to compare the patterns of diversity in ethnobotanical and ethnobiological data. Parallel to the richness estimators presented here, there are other types of estimators [36], that are adaptable to ethnobotanical and ethnobiological data, such as, similarity measures that can reflect beta diversity [46] and may be used as measures of difference between data sets from different areas.

Ecological methods permit the issues raised by ethnobiologists and ethnobotanists to be answered objectively, fostering an even closer link between ethnobotany and conservation. Using these methods, frequently seen as overly "quantitative," allows a shift in the focus of ethnobotanical and ethnobiological studies, going beyond merely listing species to identifying patterns of local knowledge. Importantly, values and numbers alone provide little information, because the measurements of diversity have a comparative value. Their applications in ethnobiology and ethnobotany are also essential to understanding the context of diversity. Thus, measurements of diversity inform nothing if they were disconnected to well-described qualitative data regarding the human populations who are the users and holders of this knowledge about diversity.

Appendix 1: Formulas for Species Richness Estimators (Adapted from [32, 41])

Estimators
Jack 1: $S_{\text{jack 1}} = S_{\text{obs}} + Q_1(m-1)/m$
Jack 2: $S_{\text{jack 2}} = S_{\text{obs}} + \{[Q_1(2m-3)/m] - [Q_2(m-2)^2/m(m-1)]\}$
Chao 1: $S_{\text{chao 1}} = S_{\text{obs}} + F_1^2/2F_2$
Chao 2[a]: $S_{\text{chao 2}} = S_{\text{obs}} + Q_1^2/2Q_2$
ACE: $S_{\text{ace}} = S_{\text{abund}} + (S_{\text{raro}}/C_{\text{ace}}) + (F_1/C_{\text{ace}})\Upsilon_{\text{ace}}^2$
ICE: $S_{\text{ice}} = S_{\text{freq}} + (S_{\text{infreq}}/C_{\text{ice}}) + (Q_1/C_{\text{ice}})\Upsilon_{\text{ice}}^2$
Bootstrap: $S_{boot} = S_{obs} + \sum_{K=1}^{S_{obs}} (1 - p_k)^m$
Where
S_{obs}: total number of species observed in a set of samples
S_{abund}: number of abundant species (with more than ten individuals)
S_{raro}: number of rare species (with ten or less individuals)
S_{freq}: number of most frequent species (present in more than ten samples)
S_{infreq}: number of less frequent species (present in ten or less samples)
m: total number of samples
Q_1: number of "singletons" (species represented by exactly one individual)
Q_2: number of "doubletons" (species represented by exactly two individuals)
F_1: number of "uniques" (species present only in one sample)
F_2: number of "duplicates" (species present in exactly two samples)
C_{ace}: abundance-based coverage estimator
C_{ice}: incidence-based coverage estimator
p_k: proportion of samples that count species k
Υ_{ace}^2: estimation of variation coefficient F_1 for rare species
Υ_{ice}^2: estimation of variation coefficient Q_1 for less frequent species

[a]There are special cases which may be consulted: http://viceroy.eeb.uconn.edu/EstimateS/EstimateSPages/EstSUsersGuide/EstimateSUsersGuide.htm#AppendixA

References

1. Given DR, Harris W (1994) Techniques and methods of ethnobotany. The Commonwealth Secretariat Publications, London
2. Martin GJ (1995) Ethnobotany. Chapman and Hall, London
3. Alexiades MN (1996) Selected guidelines for ethnobotanical research: a field manual. New York Botanical Garden, New York
4. Prance GT, Baleé W, Boom BM, Carneiro RL (1987) Quantitative ethnobotany and the case for conservation in Amazonia. Conserv Biol 1:296–310
5. Phillips O, Gentry AH (1993) The useful plants of Tambopata, Peru I: statistical hypothesis testing with a new quantitative technique. Econ Bot 47:15–32
6. Phillips O, Gentry AH (1993) The useful plants of Tambopata, Peru II: additional hypothesis testing in quantitative ethnobotany. Econ Bot 47:33–43
7. Begossi A (1996) Use of ecological methods in ethnobotany: diversity indices. Econ Bot 50(3): 280–289
8. Amorozo MCM (1996) A abordagem etnobotânica na pesquisa de plantas medicinais. In: Di Stasi LC (ed) Plantas medicinais: arte e ciência. Editora da UNESP, São Paulo
9. Amorozo MCM, Ming LC, Silva SMP (eds) (2002) Métodos de coleta e análise de dados em etnobiologia, etnoecologia e disciplinas correlatas. UNESP/CNPq, Rio Claro
10. Albuquerque UP, Lucena RFP (eds) (2004) Métodos e técnicas na pesquisa etnobotânica. Editora Livro Rápido/NUPEEA, Recife
11. Rossato SC, Leitão-Filho HF, Begossi A (1999) Ethnobotany of caiçaras of the Atlantic forest coast (Brazil). Econ Bot 53(3):377–385
12. Hanazaki N, Tamashiro JY, Leitão-Filho HF, Begossi A (2000) Diversity of plant uses in two Caiçara communities from the Atlantic forest coast, Brazil. Biodivers Conserv 9(5):597–615
13. Williams VL, Balkwill K, Witkowski ETF (2000) Unraveling the commercial market for medicinal plants and plant parts on the Witwatersrand, South Africa. Econ Bot 54(3):310–327
14. Hardesty DL (1975) The niche concept: suggestions for its use in human ecology. Hum Ecol 3(2):71–85
15. Begossi A, Richerson PJ (1993) Biodiversity, family income and ecological niche: a study on the consumption of animal foods on Búzios island (Brazil). Ecol Food Nutr 30:51–61
16. Hanazaki N, Begossi A (2000) Fishing and niche dimension for food consumption of Caiçaras from Ponta do Almada (Brazil). Hum Ecol Rev 7(2):52–62
17. Höft M, Barik SK, Lykke AM (1999) Quantitative ethnobotany: applications of multivariate and statistical analyses in ethnobotany. Plants and people working paper 6. UNESCO, Paris
18. Peroni N (2002) Coleta e análise de dados quantitativos em etnobiologia: introdução ao uso de métodos multivariados. In: Amorozo MCM, Ming LC, Silva SMP (eds) Métodos de coleta e análise de dados em etnobiologia, etnoecologia e disciplinas correlatas. UNESP/CNPq, Rio Claro
19. Begossi A, Hanazaki N, Tamashiro JY (2002) Medicinal plants in the Atlantic forest (Brazil) knowledge, use, and conservation. Hum Ecol 30(3):281–299
20. Peroni N, Hanazaki N (2002) Current and lost diversity of cultivated varieties, especially cassava, under swidden cultivation systems in the Brazilian Atlantic Forest. Agric Ecosyst Environ 92:171–183
21. Araujo HFP, Nishida AK (2007) Conhecimento de pescadores artesanais sobre a composição da avifauna em estuários paraibanos: uma contribuição para conservação. Ser Cienc Biol 7(1): 67–77
22. Harper JL, Hawksworth DL (1996) Preface. In: Hawksworth DL (ed) Biodiversity measurement and estimation. The Royal Society/ Chapman and Hall, London
23. Merétika AHC, Peroni N, Hanazaki N (2010) Local knowledge of medicinal plants in three artisanal fishing communities (Itapoá, Southern Brazil), according to gender, age, and urbanization. Acta Bot Bras 24(2):386–394
24. Sousa RS, Hanazaki N, Lopes JB, Barros RFM (2012) Are gender and age important in understanding the distribution of local botanical knowledge in fishing communities of the Parnaíba delta environmental protection area? Ethnobot Res Appl 10:551–559
25. Magurran A (1988) Ecological diversity and its measurement. Croom-Helm, London
26. Magurran A (2004) Measuring biological diversity. Blackwell, Oxford
27. Williams VL, Witkowski ETF, Balkwill K (2005) Application of diversity indices to appraise plant availability in the traditional medicinal markets of Johannesburg, South Africa. Biodivers Conserv 14:2971–3001
28. Williams VL, Witkowski ETF, Balkwill K (2006) The use of incidence-based species richness estimators, species accumulation curves and similarity measures to appraise ethnobotanical inventories from South Africa. Biodivers Conserv 16:2495–2513
29. Colwell RK, Coddington JA (1994) Estimating terrestrial biodiversity through extrapolation. Philos Trans R Soc Lond B Biol Sci 345: 101–118

30. Bunge J, Fitzpatrick M (1993) Estimating the number of species: a review. J Am Stat Assoc 88:364–373
31. Gotelli N, Colwell RK (2001) Quantifying biodiversity: procedures and pitfalls in the measurement and comparison of species richness. Ecol Lett 4:379–391
32. Colwell RK (2007) User's guide to EstimateS7.5: statistical estimation of species richness and shared species from samples, Version 8.0. http://www.purl.oclc.org/estimates. Accessed 10 Aug 2007
33. Sanders HL (1968) Marine benthic diversity: a comparative study. Am Nat 102:243–282
34. Hurlbert SA (1971) The nonconcept of species diversity: a critique and alternative parameters. Ecology 52(4):577–586
35. Colwell RK (2007) EstimateS: statistical estimation of species richness and shared species from samples, Version 8.0. http://www.purl.oclc.org/estimates. Accessed 10 Aug 2007
36. Santos AJ (2003) Estimativas de riqueza em espécies. In: Cullen L Jr, Rudran R, Valladares-Pádua C (eds) Métodos de estudo em biologia da conservação e manejo da vida silvestre. Editora da UFPR, Curitiba
37. Alves R, Lima JR, Araujo H (2012) The live bird trade in Brazil and its conservation implications: an overview. Bird Conserv Int 23:53–65. doi:10.1017/S095927091200010X
38. Chao A (1984) Non-parametric estimation of the number of classes in a population. Scand J Stat 11:265–270
39. Chao A (1987) Estimating the population size for capture-recapture data with unequal catchability. Biometrics 43:783–791
40. Chao A, Lee SM (1992) Estimating the number of classes via sample coverage. J Am Stat Assoc 87:210–217
41. Chazdon RL, Colwell RK, Denslow JS, Guariguata MR (1998) Statistical methods for estimating species richness of woody regeneration in primary and secondary rain forests of NE Costa Rica. In: Dallmeier F, Comiskey JA (eds) Forest biodiversity research, monitoring and modeling: conceptual background and Old World case studies. Parthenon, Paris
42. Chao A, Hwang WH, Chen YC, Kuo CY (2000) Estimating the number of shared species in two communities. Stat Sin 10:227–246
43. Smith EP, Van Belle G (1984) Nonparametric estimation of species richness. Biometrics 40:119–129
44. Begossi A, Hanazaki N, Peroni N (2001) Knowledge and use of biodiversity in Brazilian hot spots. Environ Dev Sustain 2(3–4):177–193
45. Walther BA, Moore JL (2005) The concepts of bias, precision and accuracy, and their use in testing the performance of species richness estimators, with a literature review of estimator performance. Ecography 28:815–829
46. Whittaker RH (1972) Evolution and measurement of species diversity. Taxon 21:213–251
47. Figueiredo GM, Leitão-Filho HF, Begossi A (1993) Ethnobotany of Atlantic forest coastal communities: diversity of plants uses in Gamboa (Itacuruçá island, Brazil). Hum Ecol 21(4):419–430
48. Figueiredo GM, Leitão-Filho HF, Begossi A (1997) Ethnobotany of Atlantic forest coastal communities II: diversity of plant uses at Sepetiba bay (SE Brazil). Hum Ecol 25(2):353–360

Chapter 26

Principles of Inferential Statistics Applied to Ethnobiology and Ethnoecology

Ulysses Paulino Albuquerque, Patrícia Muniz de Medeiros, and Alyson Luiz Santos de Almeida

Abstract

A brief discussion about the role of inferential statistics in ethnobiological and ethnoecological research is presented in this chapter, as is an exposition of some simple and coherent tools for data analysis. We discuss some issues related to study design, as well as the *p* value, data transformation, and some simple statistical tools, such as mean comparisons, correlation and regression models, and tests for contingency tables among others.

Key words Quantitative tools, Study design, Regression analysis, Quantitative ethnobotany

1 Introduction

Statistical methods can be employed to interpret a wide variety of ethnobiological and ethnoecological data, such as the relative importance of plants or vegetation units to different social or ethnic groups [1]. In this chapter, we will briefly introduce some important concepts that will help us understand the logic behind statistical tests. In a completely unpretentious way, we will also offer some examples of how different statistical tests can be applied to answer our questions. Therefore, this chapter is not intended as a statistical guide for ethnobiological investigations because the authors consider themselves to be students of this subject. This chapter provides a quick view of how these tools can be useful in analyzing qualitative and quantitative data. We will essentially adopt the hypothetic-deductive perspective, but it is important to note that ethnobiology and ethnoecology are interdisciplinary in nature and comprise different epistemological orientations. It means that statistics are not always considered important to researches in these areas, especially in areas that are rooted in the social sciences and take an epistemological orientation, as opposed

Ulysses Paulino Albuquerque et al. (eds.), *Methods and Techniques in Ethnobiology and Ethnoecology*,
Springer Protocols Handbooks, DOI 10.1007/978-1-4614-8636-7_26, © Springer Science+Business Media New York 2014

to a positivist one. Our perspective is also based on common inferential statistics, but there is an increasing tendency to adopt Bayesian approaches (for more information, see [2]).

2 Why Use Inferential Statistics?

Not all research is geared toward the formulation of hypotheses because studies may be developed under different epistemological orientations. However, let us imagine, that you are interested in identifying which factors influence knowledge about plants and animals in an indigenous community. Let us also suppose that you have collected medicinal plant use data from interviews in communities of the same ethnic origin and you want to examine whether distance from the closest urban center affects the knowledge of species richness in each community. You postulate that communities that are farther from the urban center will present higher richness of known species than communities that are closer to urban centers because the latter group would have greater access to modern medical facilities or allopathic remedies. Supposing that you find differences favoring your hypothesis, what would assure us that the results could not be explained by random effects or by variability due to sampling? Because we are concerned with a sample of interviewees within a sample of communities, we cannot be sure that the differences found are real without use of appropriate statistical tests. When posing a question or facing a problem, we should ask ourselves if our hypothesis can be tested. If the answer is yes, then Doria Filho [3] suggests following six sequential steps:

1. Establish an experimental hypothesis (H_1)
2. Establish a null hypothesis (H_0)
3. Determine sample size
4. Collect data
5. Perform statistical analysis to determine the probability that the null hypothesis is true
6. Reject (or do not reject) the null hypothesis

In the example in the previous paragraph, the alternative or experimental hypotheses (H_1) could be that (a) indigenous communities that are farther from the urban center have a different level of knowledge of species richness compared with communities that are closer to the urban center, or that (b) indigenous communities that are farther from the urban center have a higher level of knowledge of species richness than communities that are closer to the urban center. There is a difference between the two alternative hypotheses postulated above: the first one is bilateral, while the second is unilateral. It means that for the first alternative hypothesis to be considered valid, there must be a difference between communities that live close or far from the urban center, regardless of which group presents a higher level of knowledge of species richness. Instead, for the second alternative hypothesis to be considered valid, the group of communities that are far from the urban center must have greater knowledge of species richness. Any other result (either no difference between groups or greater knowledge of species richness for the communities that are close to the urban center) would support the null hypothesis in the latter case.

Thus, in the first case, the null hypothesis would be that there are no differences in knowledge of species richness between communities living closer or farther from the urban center. In the second case, the null hypothesis would be that communities living farther from the urban center have less knowledge of species richness than those living closer to the urban center.

The type of the hypothesis to be tested depends on the information available about the subject of the researcher's evaluation. In this specific case, there are a number of publications indicating that communities far from urban centers usually maintain their botanical knowledge in a more successful way. Thus, it would be reasonable to choose the unilateral hypothesis.

Regardless of the type of hypothesis chosen, all hypotheses need to be falsifiable. When we decide whether to reject the null hypothesis, we can make two types of errors: Type I errors and Type II errors. The Type I error occurs when we decide to reject the null hypothesis when it should not be rejected. The probability of committing this type of error is known as α (significance level). When the null hypothesis is false and we do not reject it, we commit a Type II error. The probability of committing this type of error is denoted by β. The researcher must establish an acceptable degree of error by adopting a significance level (the most used significance levels in ethnobiological studies are 5 and 1 %). The p-value corresponds to the statistical significance of the results.

The next step in the process recommended by Doria Filho [3] is to determine the sample size. This is a very important step; chapter one of this book introduced some basic concepts regarding sample size. Problems and errors may occur when a hypothesis is being tested, and these errors can be due to sample variability. There are variations in all types of phenomena, especially in biological ones. Because of this intrinsic variability, the larger the sample is, the higher the probability of representativeness and the greater the precision of results. However, some practical limitations in working with large samples must be carefully evaluated by the researcher. In the previous example, we have two sampling decisions to make: (1) the number of communities to sample and (2) the number and profile of people to be interviewed. These decisions cannot be arbitrary and should be based on the researcher's experience, as well as on information from the literature.

If researchers do not use objective criteria to select communities and/or people, the differences that are found may be due to sample variability.

The data collection step also challenges researchers because there is no statistical test that can save an ineffectively collected dataset (or, as they say, "Garbage in, garbage out"). Inadequately formulated questions and unorganized methods of data collection, for example, can compromise data quality, rendering the information collected inadequate.

3 The *p*-Value

Interpreting the meaning of the *p*-value is one of the greatest difficulties that a beginning student of statistics can face. The *p*-value is important in deciding whether to accept or reject a hypothesis. When we find a statement in a scientific article such as, "All statistical tests were conducted considering a *p*-value of <0.05 as statistically significant," it means that the authors accept that the probability of obtaining a test statistic at least as extreme as that observed is lower than 5 %.

The *p*-value is often misinterpreted. It is not the same as the alpha value (significance level: probability that the results were not found by chance), although one traditionally rejects the null hypothesis when the *p*-value is less than or equal to the significance level.

4 Study Design

Establishing a good design can help the researcher more accurately decide whether to reject the null hypothesis. There are no formulas to be followed because study designs vary to the same degree that the potential investigation questions do. Therefore, we emphasize that there are a number of design models to choose from. Below, we offer some suggestions of steps to follow in the design process, some of which have been modified from Magnusson and Mourão [4]:

1. Determine your object of study. Try to define your investigation problem in a clear and objective way. Identify the dependent variable.
2. Make sure that your dependent variable can be objectively measured.
3. Build a scheme to elicit which variables influence your dependent variable, as well as the relations between independent variables.
4. Pay attention to design problems, such as lack of replicates, sample size, sample representativeness and variable manipulation.

5. Select an adequate scenario to develop your study and select your sample.
6. Ensure that you have an adequate statistical background to design the study; otherwise, ask for the support of someone experienced in the area.
7. Finally, decide how the results will be applied: will they be used to determine the existence of an effect, to determine the magnitude of an effect, or to make predictions?

5 Defining Study Variables

Without a doubt, one of the most important steps in planning a study is defining the variables that describe the "object" or phenomenon to be evaluated. In ethnobiology, we certainly address complex phenomena that are not easily measurable, and the choice of variables to explain or measure these phenomena or objects depends on theoretical and methodological knowledge of the subject and the possible variations in these phenomena or objects. The different possible variations should be previewed by the ethnobiologist when the investigation is being designed to avoid potential errors. While formulating the design, the researcher can consider variables of both a quantitative and a qualitative nature. Quantitative variables usually have greater statistical and comparative power than qualitative variables. Therefore, the nature of the variable influences the type of statistical tests that can be used to answer the investigative questions. Table 1 presents a classification of variables, with examples relevant to ethnobiology. When we deal with natural phenomena that are "amplified" by human presence, it is important to consider that other variables will usually be related to the phenomena that we are measuring.

Obviously, the choice of a given type of variable depends on the study to be developed, as well as on the theoretical and practical possibilities for measuring this variable. For example, we may be interested in comparing the medicinal plant knowledge of men and women in a given region. We can define this variable based on the number of medicinal plants cited in the interviews. We have to consider that "measuring" someone's knowledge is a very difficult task, and any measure we adopt will only cover part of this knowledge. Thus, we must make a methodological choice and attempt to overcome the interpretative limitations by choosing easily manageable variables.

We can further classify variables into two other types: dependent and independent. The researcher may be interested in investigating a cause-and-effect relation. Continuing with the example above, we may be interested in testing whether age influences the number of plants cited as medicinal by people in the community.

Table 1
Classification of variables according to their nature

Type of variable	Classification	Description
Quantitative	Discrete	Integers. Examples: the number of medicinal species cited by an interviewee; the number of use citations; the number of people living in a household
	Continuous	Numbers that can assume different values, including fractional numbers. Examples: a plant's height; an animal's size; a resource's use value; the weight of a firewood pile
Qualitative	Categorical-nominal	Independent categories. Examples: ethnicity; gender (male, female); knowledge status (specialist, nonspecialist)
	Categorical-ordinal	Categories that are related by a ranking relation. Example: age ranges (0–20, 21–40, 41–60...)

The independent variable, also known as the predictor variable, is the one that causes the phenomenon or effect (in this case, age). The dependent variable, on the other hand, is the one that measures the phenomenon that is being explained (in this case, the number of plants cited as medicinal).

Magnusson and Mourão [4] emphasized the importance of the quality of information used in adequately answering a question. As they state, "*Generally, it is expected that the amount of available information increases with the number of observations, but it is not always like that. When a new observation provides only the same information that we already had in previous observations, it does not increase the total amount of available knowledge and it can confuse us and make us believe that we have more information than we really do. Since this information in not a real repetition to what concerns providing more information, Hubert called it 'pseudo-replication', which means false repetition.*" In studying the relation between richness of medicinal plant knowledge and distance from an urban center, for example, we would have to be sure that each of the studied communities could be considered an independent observation. We must assure that the communities are "replications." For example, if we could assume that each human population developed its own expertise with its surrounding plants, we would have a virtually insurmountable barrier against the problem of independence in the development of an adequate methodological design. Therefore, the premises that we assume considerably influence the design of our investigation.

"Pseudo-replications" are often very difficult to avoid in ethnobiological investigations. The causes behind variations in the knowledge of human communities are not easy to elicit, and in the above-cited case, we assume that the effects of modernity are so obvious and that access to modern medicine is so widespread that we tend not to question this cause-and-effect relation. Although there are significant results supporting these assumptions, we believe that factors influencing traditional knowledge must be studied not only from a cultural perspective but also from economic, ecologic, and evolutionary perspectives. Accordingly, we believe that studies must be replicated in different contexts to determine which factors really affect people's experiences with natural resources.

6 Data Transformation

The most robust statistical analyses (those with higher predictive power) are generally parametric. Despite their strength, parametric tests may not always be used because they have some specific requirements, such as data normality. Normal data obey a normal distribution (values that are closer to the mean are more frequent, and frequency decreases with the distance from the mean, so that extreme numbers occur fewer times and intermediate numbers occur more times in a sample; see Chart 1). When our data do not have a normal distribution, we do not need to immediately give up on parametric tests because we can transform the data to approximate a normal distribution.

Chart 1
Example of a normal distribution for the number of medicinal plants cited by interviewees from a hypothetical community

Informer	N° species	
Inf.1	2	Low values (low frequency)
Inf.2	4	
Inf.3	6	Intermediate values (high frequency)
Inf.4	6	
Inf.5	6	
Inf.6	7	
Inf.7	7	
Inf.8	7	
Inf.9	10	High values (low frequency)
Inf.10	11	

There are various types of transformations, and each one is suited to a type of dataset. However, in many cases, data cannot be adjusted to the shape of a normal distribution, despite attempts to transform them. In these cases, we can use non-parametric tests. Almost all parametric tests have a non-parametric equivalent. We comment below the main types of data transformation.

6.1 Logarithmic Transformation

A logarithmic transformation "compacts" the values, decreasing the discrepancies between the lower and higher values. However, this type of transformation should not be used for datasets that contain zero or negative values. In the cases of a dataset that includes zeros, a log $X+1$ transformation can be used instead of log X, so that the log of the number's successor will be calculated. This method makes data transformation possible because a "log 0" does not exist. Logarithmic transformations can be done in various bases, but it is most common to calculate a logarithm in base 10 or base e ($\approx$2.718, known as the natural logarithm).

6.2 Nth Root Transformation

The most common nth root transformation is the square root, but the cubic and fourth roots are also commonly used. This transformation is recommended for datasets with values mostly higher than 10, and it is preferably applied to integers (discrete numbers).

6.3 Angular Transformation

The angular transformation is recommended for proportions and discrete numbers. It is often applied to complementary data in which values are part of a defined total. This transformation is performed by putting the data into percentages and then calculating the arcsine of these percentages. This transformation is especially useful for vegetation inventory datasets, such as the number of individuals from a given species out of the total number of individuals in a vegetation sample.

Some statistical packages can simultaneously transform a dataset using a variety of transformations and indicate the most adequate for that specific case. However, sometimes even the most adequate transformation indicated by those packages may not normalize the data.

Currently, power transformation is widely used in statistics. It is a family of functions that preserves the data's ranking but leaves the data as normal-like as possible. Power transformations can be performed in a number of statistical packages. However, one must be careful because those transformations can deeply change the former data.

7 Main Statistical Tests Applied to Ethnobiology and Ethnoecology

7.1 Mean Comparisons

Tests of mean comparisons are employed to verify that there are differences between two or more samples. These tests can be performed with related samples (paired tests) or independent samples.

The most used tests for comparing two independent samples are Student's *t*-test for parametric data and the Mann–Whitney test for non-parametric data. The *t*-test is only applicable to normal data and is not robust in small samples, so its application to samples with $n < 10$ is not recommended. Another requirement of the *t*-test is homogeneity of variance. Some statistical packages indicate whether the variances are homogeneous or not and consequently if the *t*-test can be used. When the variances are heterogeneous, the *F*-test can alternatively be used.

A non-parametric equivalent to the *t*-test for comparing two independent samples is the Mann–Whitney test. This test does not require data normality or homogeneity of variance. For comparisons of more than two samples, the most common parametric test is the ANOVA, and its non-parametric equivalent is the Kruskal–Wallis. When we run a test for comparing more than two samples, we can find significant differences, but we need to localize where these differences are. For example, if we are comparing four samples, maybe sample A is different from sample B, but all other combinations present no significant differences; this is why some a posteriori tests are employed. Such tests include, for example, the Tukey test (used after ANOVA) and the Dunn test (used after Kruskal–Wallis).

To compare two dependent samples, the paired *t*-test is often chosen for parametric data and the Wilcoxon test is a common non-parametric option. The requirements of the paired *t*-test are the same as those of the *t*-test for independent samples. Examples of possible applications of these tests can be seen in Chart 2.

Let us suppose that we are going to test whether there are differences in the total firewood consumption of communities A and B. We will measure the wood volume used by each household every month for 1 year. If we did not apply a paired test, the difference between the communities could be "camouflaged" because there could be greater variation in consumption in a single community, depending on the month under consideration. In the case that the sample units are heterogeneous, a simple test for independent variables might not detect differences between samples, especially if those differences are small. To avoid this problem, this type of information should be analyzed in a way that

Chart 2
Examples of ethnobiological analyses based on comparisons of related samples

- Comparison between the number of known and the number of effectively used species in a community
- Comparison of monthly firewood consumption between two communities or species

Table 2
Total firewood consumption in two hypothetical communities

Month	Volume (m^3) for community A	Volume (m^3) for community B
Jan	16	14
Feb	14	13
Mar	11	6
Apr	11	5
May	12	6
Jun	13	9
Jul	16	7
Aug	22	17
Sep	23	16
Oct	23	15
Nov	25	13
Dec	20	16

Paired *t* test:
$t = 6.6$; $p < 0.05$
Interpretation: community A consumed a volume of firewood that was significantly greater than community B

relates the samples, diminishing the influence of internal variation. Therefore, firewood consumption values should be analyzed by relating the two communities for each month in which wood volume was measured, as shown in Table 2.

7.2 Correlation

Correlation tests are used to determine whether two variables are related, i.e., their values vary concurrently (Chart 3). When the joint variation goes in the same direction—high values for variable X and high values for variable Y—we call this a positive or direct correlation. Conversely, a negative or inverse correlation is found when low values of variable X are related to high values of variable Y.

The most employed parametric test for correlating samples is the Pearson correlation coefficient. Its result is called an r-value, which ranges from −1 (perfect inverse relation) to 1 (perfect direct relation). Values close to zero tend to indicate that no correlation between variables can be found. However, the r-value is not enough to assure that the correlation is significant: for this purpose, we need to consider the p-value. Thus, we can find a small but significant correlation when r is not close to −1 or 1 and p is

Chart 3
Examples of ethnobiological analyses based on correlation analyses

- Relation between measures of relative importance (e.g., use value and relative importance)
- Relation between knowledge of a household's male and female chiefs
- Relation between native and exotic plant knowledge
- Relation between a plant's medicinal use value and its general use value (considering all use categories)

Modified from Hoft et al. [1]

Table 3
Correlation between the use value (UV) and the relative importance (RI) of 36 species native to the *caatinga* (seasonally dry tropical forest)

Species	UV	RI
Myracroduon urundeuva Allemão	1.85	1.70
Schinopsis brasiliensis Engl.	1.38	1.65
Chorisia glaziovii (O. Kuntze) E. Santos	0.10	0.38
Cordia trichotoma (Vel.) Arráb. ex Steud.	0.30	0.87
Cordia globosa (Jacq.) Humb., Bompl. & Kunth	0.08	0.57
Commiphora leptophloeos (Mart.) J. B. Gillet	0.5	1.55
Bauhinia cheilantha (Bong.) Steud.	0.78	0.92
Caesalpinia pyramidalis Tul.	1.21	1.07
Capparis jacobinae Moric.	0.12	1.01
Capparis hastata L.	0.68	1.17
Clusia sp.	0.02	0.33
(...)		

Data source: Albuquerque et al. [6]
Spearman correlation coefficient:
$rs = 0.75$; $p < 0.05$
Interpretation: the importance values of the species as obtained by the two measurement techniques are strongly related

lower than 0.05. A low but significant correlation can be found due, for example, to a large sample size.

The non-parametric equivalent to the Pearson correlation is Spearman's correlation coefficient. It is represented by the *rs* value (Spearman coefficient), which also ranges from −1 to 1. Table 3 presents one example of this test.

Chart 4
Examples of ethnobiological analyses based on regressions

• Relation between people's age and plant/animal knowledge
• Relation between socioeconomic factors and plant use (e.g., wood volume) or hunting intensity
• Relation between the household-forest distance and the amount of forest products harvested in a given period
• Relation between household distance from an urban center and plant/animal use
• Relation between time spent on hunting events and the amount of captured animals
• Relation between species' fuel value indices and their preferable use as firewood
• Relation between plant/animal species' local availability and their local use

Modified from Hoft et al. [1]

In some cases, even if there is a relation between two variables, it may not be detected by a given statistical test. For example, a correlation may not be detected when the relationship is not linear. In such cases, a data transformation can be used in order to remove a scale effect that prevents the identification of the correlation.

Before introducing the regression analysis, it is important to clarify the difference between correlation and regression. A correlation analysis does not specify a cause-and-effect relation but rather indicates whether there is or is not a joint variation in the data.

7.3 Regression

In regression analysis, we are dealing with one dependent variable (*Y*) and one or more independent variables (*X*) (Chart 4). This type of analysis is used when there are reasons to suppose a cause-and-effect relation between variables. With regression, a model can be created to predict a dependent variable's value based on the independent variables' values. We can also test the extent to which this model can be trusted, by means of the test's significance.

It must be clear that the terms "cause" and "effect" as used in statistics do not always correspond to real causal relations. That the relation between two variables is considered to be significant does not mean that one value really explains the other in nature. Therefore, the dependent and independent variables to be tested are carefully chosen by the researcher based on experience. A simple exercise demonstrates the inability of statistical tests to detect explanatory variable(s): in the case that there is a relation between two variables, turn your dependent variable into the independent one and vice versa. You will most likely find a significant result again. In this case, which one is the independent variable? The answer depends on the context of the research and on the researcher's experiments and observations.

There are several types of regression analyses, such as linear regression, polynomial regression, log-linear regression, logistic regression, and generalized linear model. However, the most commonly used is doubtless the linear regression, although it is not always easy to attain data that are adequate for its requirements. An indication of the most adequate type of regression analysis for the specific type of data to be analyzed is available in some statistical packages.

A simple linear regression can be described by the expression "$Y = a + bX$," where "Y" is the dependent variable, "X" is the independent variable, and "a" and "b" are constants. More specifically, "a" represents the intercept (value of "Y" when "X" equals zero), and "b" represents the slope of the regression, which measures both the direction and the magnitude of the relationship between the variables. Thus, when the slope is negative, the expression becomes "$Y = a - bX$." When the slope is negative, the variables are inversely related, whereas a positive coefficient shows a direct, positive relation between variables. Simple regressions search for the relationship between the dependent variable and only one independent variable, as shown in Table 4.

Despite its wide use, some criticisms have been made of simple regression. The main criticism is that it only explains a phenomenon according to one point of view, i.e., a single variable. Biological and ecological phenomena, as already mentioned, are often explained by distinct factors. In addition, what about ethnobiological phenomena? For example, what could better explain the higher or lower wood consumption of people living in a given area? The association between several variables can explain such consumption more accurately than one variable considered alone.

Multiple regressions are used to find a set of independent variables that can explain variation in the dependent variable. These regressions also output equations that allow for predictions of new elements based on the values of the explanatory (independent) variables. In nature, a phenomenon is often not explained by a single variable, but by a combination of variables, highlighting. This elicits the importance of using multiple regressions. In general, the higher the predictive power of the independent variables, the better the model, indicating that the causes of a given phenomenon have most likely been discovered.

Since multiple regressions often deal with several explanatory variables, it is common to find correlations between these independent variables. In these cases, it is recommended to ascertain the degree of multicollinearity (autocorrelation between independent

Table 4
Linear regression analysis between the use value (UV—a measure of relative importance in ethnobotany) and the importance value (IV—phytosociological index) of useful species sampled in an area of *caatinga* vegetation in Caruaru, Pernambuco

Species	UV	IV
Myracroduon urundeuva Allemão	1.85	0.56
Schinopsis brasiliensis Engl.	1.38	18.67
Chorisia glaziovii (O. Kuntze) E. Santos	0.10	0
Cordia trichotoma (Vel.) Arráb. ex Steud.	0.30	1.57
Cordia globosa (Jacq.) Humb., Bompl. & Kunth	0.08	4.66
Commiphora leptophloeos (Mart.) J. B. Gillet	0.5	11.57
Bauhinia cheilantha (Bong.) Steud.	0.78	10.11
Caesalpinia pyramidalis Tul.	1.21	36.49
Capparis jacobinae Moric.	0.12	1.94
Capparis hastata L.	0.68	20.68
Clusia sp.	0.02	0
(…)		

Data source: Lucena et al. [7]
Dependent variable: use value
Independent variable: importance value
Regression:
$R^2 = 0.186$, $p < 0.01$
Interpretation: In the case above, the importance value of a species explains its use value in the community. The value of R^2 (coefficient of determination) indicates the explanatory power of the independent variable on the dependent variable. In other words, it is the fraction of the variance that is shared by the two variables. The closer to one, the more strongly the independent variable explains the dependent variable, but this relationship can be positive or negative (as indicated by the *b*-value)

variables), and if two independent variables are more than 80 % autocorrelated, one of them is usually discarded. In such cases, high autocorrelation may indicate that one of the variables formerly considered to be independent is actually influenced by another independent variable and does not properly explain the behavior of the dependent variable. Some statisticians do not agree with this procedure, as it can eliminate variables that are actually important [5].

It is also common to perform a "stepwise" procedure in which all possible combinations of independent variables are analyzed to facilitate selection of the combination that best explains the variance in the dependent variable (i.e., the combination with the higher r^2 value).

In multiple linear regression analysis, linear relationships between variables are detected. This method's limitation lies in its imprecise detection of nonlinear relationships. Linear regressions require a normal distribution of residuals (errors). In a linear regression, it is expected that the behavior of a dependent variable is linear with respect to the other independent variables. The residuals are thus values that indicate the distance between the real value of an element and the ideal value that is predicted by the regression model.

In logistic regression, the dependent variable is categorical, being expressed in terms of 0 and 1 to indicate two possible attributes (e.g., 0 for community members who do not use animals for therapeutic ends and 1 for those who do use animals for this purpose). The independent variables are usually quantitative (e.g., income). Thus, by applying logistic regression, you can find variables that influence the categorical-dependent variable. In the above example, if the relationship between variables was significant, the monthly income of the informant would be influenced by the presence or absence of animals used for therapeutic purposes. Because it incorporates qualitative variables, logistic regression does not require residual normality.

Generalized linear models (GZLM) can be used in cases in which the independent variables are of different types (i.e., numeric and categorical) or when there are different relations between independent variables and dependent variables (e.g., linear for some variables and log-linear for others). GZLM accepts non-normal residuals and can be used for a wide range of situations. GZLM differs from general linear models (GLM), which are used especially when we are dealing with more than one dependent variable.

In the case of firewood use, the household income, the number of householders, and the average age of family chiefs, among others, may together be explanatory factors, as observed in Table 5.

8 Contingency Tables

We often need to analyze nominal data, i.e., data that are qualitative or ordinal (see Chart 5). These data have no normal distribution. Therefore, contingency tables help the investigator identify possible associations between two variables that classify the same data. There are cases in which the two variables are qualitative (e.g., age and sex), and there are cases in which one of them is qualitative (e.g., plant origin) and the other is ordinal (e.g., ranges of a given resource's extracted volume).

The chi-square and the *G*-test are certainly the most popular tests for contingency tables. The chi-square test is a nonparametric choice that is extremely popular and widely applied in various fields

Table 5
Firewood consumption and socioeconomic information from 18 households in a rural community

Number of householders	Monthly family income (US$)	Average age of family chiefs (men and women)	Annual volume of firewood consumed (m^3)
1	0	59.5	0
2	20	36.5	1
3	95	70	1
3	50	72.5	1.1
3	0	68	1.2
7	20	44.5	1.2
4	95	52.5	1.4
4	400	61.5	1.4
3	300	53	1.5
5	250	39	1.5
5	380	73	1.5
4	380	38	1.7
5	190	58	1.7
5	380	33	1.8
5	400	53.5	2
4	570	55	2.5
8	600	31	3
10	850	27.5	3.5

Dependent variable: annual firewood consumption
Multiple linear regression
Multiple coefficient of determination (R^2) = 0.89
p (general) < 0.05
p Values for each factor:
Number of householders: $p < 0.05$—direct relationship
Monthly family income: $p < 0.05$—inverse relationship
Average age of household chiefs: $p > 0.05$
Interpretation: in this example, notice that annual firewood consumption is explained by the number of inhabitants and the monthly family income but is not explained by the average age of the household chiefs

of study. For the chi-square test to work adequately, contingency tables may not contain observed values lower than 5. When this happens, an interesting option is to apply Monte-Carlo simulations to simulate the p-value and avoid the bias of the low values.

Chart 5
Example of ethnobiological data suitable for contingency tables

• Relation between plant/animal species' local availability
• Different groups of resource users using different parts of the plant or animal
• Different groups of resource users earning different amounts of money by selling their products
• Effectiveness of an animal- or plant-derived drug to combat a certain disease
• Number of occasions on which a plant or animal is mentioned for curing a certain disease

Modified from Hoft et al. [1]

Table 6
Contingency table with two qualitative variables: citation of *M. urundeuva* Allemão (aroeira) for its anti-inflammatory purposes (hypothetical data) in two communities (A and B)

	A	B
Number of people who cited the aroeira for anti-inflammatory purposes.	40	35
Number of people who did not cite the aroeira for anti-inflammatory purposes.	90	30

Chi-square in a 2×2 contingency table
$\chi^2 = 9.75$ $p < 0.05$
Interpretation: citations of aroeira for anti-inflammatory purposes were proportionally higher in community B than in community A. It can be seen that, although community A had a higher total number of people citing the aroeira as anti-inflammatory, a higher proportion of people in community B cited the species for the abovementioned purpose. It becomes clear that the important value in this type of test is the proportion rather than the absolute value

An alternative to the chi-square, the *G*-test, is also quite popular for contingency tables (see example in Table 6). It does not require the observed values to be higher than 5.

9 Final Remarks

Finally, we would like to offer a short list of studies that have employed a wide array of ethnobiological approaches. They are example applications of some of the statistical tests presented in this chapter (see Table 7).

Table 7
Key statistical tests applied in some ethnobiological studies

These tests...	Can be found in...
t test	Fox [8], Hanazaki et al. [9]
Kruskal–Wallis	Phillips and Gentry [10, 11], Phillips et al. [12], Salick et al. [13], Ghimire et al. [14], Albuquerque et al. [15, 16], Naughton-Trevis et al. [17]
Man Whitney	Salick et al. [13], Holmes [18], Ghimire et al. [14], Naughton-Trevis et al. [17], Thomas et al. [19]
Paired *t* test	Gaugris and van Rooyen [20], Ndangalasi et al. [21]
Wilcoxon	Phillips et al. [12], Ramos et al. [22]
Spearman correlation	Hanazaki and Begossi [23], Lykke [24], Albuquerque et al. [6], Estomba et al. [25], Ramos et al. [26]
Pearson correlation	Ghimire et al. [14]
Regression	Galeano [27], Nagothu [28], Holmes [18], Fusari and Carpaneto [29], Gavin and Anderson [30], Voeks [31], Lucena et al. [7]
Chi-square (χ^2)	Holmes [18], Ferraz et al. [32], Gaugris and van Rooyen [20], Estomba et al. [25], Ladio et al. [33], Hanazaki et al. [34]
G-test	Almeida [35], Monteiro [36], Lucena et al. [7], Albuquerque et al. [16]

References

1. Hoft M, Barik SK, Lykke AM (1999) Quantitative ethnobotany—applications of multivariate and statistical analyses in ethnobotany. People and plants working paper 6. UNESCO, Paris
2. Lee PM (2012) Bayesian statistics: an introduction, 4th edn. Wiley, West Sussex
3. Doria Filho U (1999) Introdução à bioestatística para simples mortais. Negócio Editora, São Paulo
4. Magnusson WE, Mourão G (2005) Estatística [sem] matemática—a ligação entre as questões e a análise. Editora Planta, Londrina
5. James F, McCulloch C (1990) Multivariate analysis in ecology and systematics: panacea or pandora box? Annu Rev Ecol Syst 21: 129–166
6. Albuquerque UP, Lucena RFP, Monteiro JM et al (2006) Evaluating two quantitative ethnobotanical techniques. Ethnobotany Res Appl 4:51–60
7. Lucena RFP, Araújo EL, Albuquerque UP (2007) Does the local availability of woody caatinga plants (Northeastern Brazil) explain their use value? Econ Bot 61:347–361
8. Fox J (1984) Firewood consumption in a Nepali village. J Environ Manage 8:243–250
9. Hanazaki N, Tamashiro JY, Leitão-Filho HF et al (2000) Diversity of plant uses in two Caiçara communities from the Atlantic forest coast, Brazil. Biodivers Conserv 9:597–615
10. Phillips O, Gentry AH (1993) The useful plants of Tambopata, Peru II: additional hypothesis testing in quantitative ethnobotany. Econ Bot 47:33–43
11. Phillips O, Gentry AH (1993) The useful plants of Tambopata, Peru I: statistical hypotheses tests with a new quantitative techinique. Econ Bot 47:15–32
12. Phillips O, Gentry AH, Reynel C et al (1994) Quantitative ethnobotany and amazonian conservation. Conserv Biol 8:225–248
13. Salick J, Mejia A, Tood A (1995) Non-timber forest products integrated with natural forest management, Rio San Juan, Nicaragua. Ecol Appl 5:878–895
14. Ghimire SK, McKey D, Aumeeruddy-Tomas Y (2005) Conservation of Himalayan medicinal plants: harvesting patterns and ecology of two threatened species, Nardostachys grandiflora

DC. and Neopicrorhiza scrophulariiflora (Pennell) Hong. Biol Conserv 124:463–475
15. Albuquerque UP, Medeiros PM, Almeida ALS et al (2007) Medicinal plants of the caatinga (semi-arid) vegetation of NE Brazil: a quantitative approach. J Ethnopharmacol 114:325–354
16. Albuquerque UP, Monteiro JM, Ramos MA et al (2007) Medicinal and magic plants from a public market in northeastern Brazil. J Ethnopharmacol 110:76–91
17. Naughton-Trevis L, Kammen DM, Chapman C (2007) Burning biodiversity: woody biomass use by commercial and subsistence groups in western Uganda's forests. Biol Conserv 134: 232–241
18. Holmes CM (2003) Assessing the perceived utility of wood resources in a protected area of Western Tanzania. Biol Conserv 111:179–189
19. Thomas E, Vandebroek I, Van Damme P (2007) What works in the field? A comparison of different interviewing methods in ethnobotany with special reference to the use of photographs. Econ Bot 61:376–384
20. Gaugris JY, van Rooyen MW (2006) Questionnaires do not work! A comparison of methods used to evaluate the structure of buildings and wood used in rural households, South Africa. Ethnobotany Res Appl 4:119–131
21. Ndangalasi HJ, Bitariho R, Dovie DBK (2007) Harvesting of non-timber forest products and implications for conservation in two montane forests of East Africa. Biol Conserv 134: 242–250
22. Ramos MA, Medeiros PM, Almeida ALS et al (2008) Use and knowledge of fuelwood in an area of caatinga vegetation in NE Brazil. Biomass Bioenergy 32:510–517
23. Hanazaki N, Begossi A (2000) Fishing and niche dimension for food consumption of Caiçaras from Ponta do Almada (Brazil). Hum Ecol Rev 7:52–62
24. Lykke AM (2000) Local perceptions of vegetation change and priorities for conservation of woody-savanna vegetation in Senegal. J Environ Manage 59:107–120
25. Estomba D, Ladio A, Lozada M (2006) Medicinal wild plant knowledge and gathering patterns in a Mapuche community from northwestern Patagonia. J Ethnopharmacol 103: 109–119
26. Ramos MA, Medeiros PM, Almeida ALS et al (2008) Can wood quality justify local preferences for firewood in an area of caatinga (dryland) vegetation. Biomass Bioenergy 32: 503–509
27. Galeano G (2000) Forest use at the Pacific coast of Choco Colombia: a quantitative approach. Econ Bot 54:358–376
28. Nagothu US (2001) Fuelwood and fodder extraction and deforestation: mainstream views in India discussed on the basis of data from the semi-arid region of Rajasthan. Geoforum 32: 319–332
29. Fusari A, Carpaneto GM (2006) Subsistence hunting and conservation issues in the game reserve of Gile, Mozambique. Biodivers Conserv 15:2477–2495
30. Gavin MC, Anderson GJ (2007) Socioeconomic predictors of forest use values in the Peruvian Amazon: a potential tool for biodiversity conservation. Ecol Econ 60:752–762
31. Voeks RA (2007) Are women reservoirs of traditional plant knowledge? Gender, ethnobotany and globalization in northeast Brazil. Singap J Trop Geogr 28:7–20
32. Ferraz JSF, Meunier IMJ, Albuquerque UP (2005) Conhecimento sobre espécies lenhosas úteis da mata ciliar do Riacho do Navio, Floresta, Pernambuco. Zonas Áridas 9: 27–39
33. Ladio A, Lozada M, Weigandt M (2007) Comparison of traditional wild plant knowledge between aboriginal communities inhabiting arid and forest environments in Patagonia, Argentina. J Arid Environ 69: 695–715
34. Hanazaki N, Alves RRN, Begossi A (2009) Hunting and use of terrestrial fauna used by Caiçaras from the Atlantic forest coast (Brazil). J Ethnobiol Ethnomed 5:36
35. Almeida CFCBR, Silva TCL, Amorim ELC et al (2005) Life strategy and chemical composition as predictors of the selection of medicinal plants from the caatinga (Northeast Brazil). J Arid Environ 62:127–142
36. Monteiro JM, Lins Neto EMF, Albuquerque UP et al (2006) Use patterns and knowledge of medicinal species among two rural communities from Northeastern Brazil semi-arid region. J Ethnopharmacol 105:173–186

Chapter 27

Problems and Perspectives in the Publication of Ethnobiological Studies

Ulysses Paulino Albuquerque, Natalia Hanazaki, and Joabe Gomes de Melo

Abstract

Ethnobiology attempts to understand the relationship between man, the environment, and the biota and has significantly contributed to the knowledge and conservation of biodiversity and the search for new molecules with biological action, especially in such disciplines as ethnopharmacology and ethnobotany. However, many publications in the field have ethical, methodological, and theoretical problems and poor and limited results. Thus, the purpose of this chapter is to review several key aspects to ensure reliable data collection and the quality of publications, especially of ethnobotany and ethnopharmacology studies.

Key words Ethics in ethnobiology, Scientific whiting, Science policy, Data collection

1 Introduction

Ethnobotany and ethnopharmacology have proven to be key tools for knowledge on biodiversity and powerful approaches in the search for natural substances with therapeutic action. Nevertheless, several limiting factors of such approaches can be mentioned, including the following: the difficulty of collecting reliable data on people; the consistent correlation of plant use in different cultures, to a greater or lesser degree, with magical-religious components; the existence of ethical issues regarding access to traditional knowledge related to biodiversity use. Undoubtedly, many studies have been developed outside of these discussions, which has led to such issues as publications with ethical, methodological, and theoretical problems and with poor and limited results as to their application.

Thus, the purpose of this chapter is to review several key aspects to ensure reliable data collection in ethnobiological research and

This chapter is a revised version from transcriptions of the text originally published by the authors in Brazilian Journal of Pharmacognosy 16(Supl.): 678–689, 2006.

Ulysses Paulino Albuquerque et al. (eds.), *Methods and Techniques in Ethnobiology and Ethnoecology*, Springer Protocols Handbooks, DOI 10.1007/978-1-4614-8636-7_27, © Springer Science+Business Media New York 2014

provide an overview of elements that have been considered essential for ensuring the quality of publications. Moreover, certain aspects of ethical and legal issues that guide ethnopharmacological and ethnobotanical studies are briefly introduced. Therefore, a number of the terms and concepts used in this chapter, which emphasizes studies that focus on the medicinal use of resources, must be clarified. The term traditional has different understandings [1, 2] but will be applied to two contexts: traditional communities and traditional know-how or knowledge (using traditional communities as a reference). The term "local knowledge" has been proposed as an alternative to the term "traditional knowledge," as the set of "knowledge, practices, and beliefs, developed by adaptive processes and passed on through the several generations by cultural transmission, on the relationships between living beings with each other and with their environment" [3, 4]. Of note, not all ethnobotanical studies on traditionally used plants are directed a priori toward contributing to the discovery of new drugs of medical or pharmaceutical interest.

The main contribution of many studies is to introduce a list of species (animals or plants) with data on their uses by local communities [5-7]. Several of these approaches, which are essentially descriptive, continue to follow the initial approach of ethnoscientific research, strongly marked by an ethnographic influence [8]. However, several studies that focus on medicinal plants, despite lacking the initial goal of contributing to the discovery of new drugs, show potential to strengthen that field of research. Most contributions, however, fail to indicate or are not directed toward answering certain basic questions. For example: how do people select plants and animals for medicinal use? What criteria are involved in these choices? These questions have not been answered, although several studies have begun to indicate the existence of patterns [9-16], which, if confirmed, will enable us to predict which groups of plants may show interesting and significant biological activity. One of these possible patterns regards the people's "preference" for selecting plants from "disturbed" environments for medicinal use [11]. Accordingly, we propose that researchers seek to link the traditional ethnobotanical survey to hypothesis testing.

2 Common Problems in Ethnobiological Studies

The identification of research problems that are crucial for the advancement of the field and improvement in the quality of publications are only possible with the use of scientific literature that reflects all accumulated knowledge on the matter. This requirement is one of the reasons for the poor quality of studies. Referring to ethnobotany studies in general, McClatchey [17] emphasizes that the poor quality of science presentation indicates that authors lacked writing skills, fluency in the language used, positive models of good-quality writing, and access to basic research methods. On the one hand, the evaluation of certain publications in Brazil seems to support McClatchey's view. On the other hand, Brazilian researchers now have easy access to worldwide scientific literature, which contradicts the lack of accuracy in data collection and analysis.

Many problems that commonly lead to the rejection of articles in journals with a board of expert reviewers are related to data compatibility with the theory or hypotheses tested. We suggest that studies show a clear and detailed description of methods, especially studies that introduce new approaches, and cite literature references for those methods. In Table 1, we introduce a proposal for structuring articles within the scope of ethnobotany/ethnopharmacology that can be extrapolated to other areas of ethnobiology. Of note, this framework proposes recommendations aimed at guiding the structuring of scientific articles; this suggestion does not imply that they should always be followed in the manner proposed. There are always exceptions, and the particularities of each study should be considered, as ethnopharmacological and ethnobotanical studies examine various environmental situations and sociocultural contexts.

Several elements must always be included in ethnopharmacological and ethnobotanical studies. In Table 2, we introduce a simple guide as a basic proposal for the evaluation of publications based on what is commonly accepted in scientific approaches involving ethnobotany and ethnopharmacology. Appropriate methods enable the best measurement of a phenomenon and conclusions based on the findings, if the questions are the main guidelines of a research study. There is an increasing trend of including quantitative techniques and statistical analyses in studies, especially since the 1990s. However, quantitative methods and statistical analyses do not replace a careful and thorough data collection because they are only tools. Pertinently, Bernard [20] emphasizes the significance of separating the concepts of quantification and science, remembering that any science is based primarily on qualitative data.

2.1 Problems with Data Collection

There are many means to obtain data from the community on plants and/or animals used for therapeutic purposes. However, before beginning a research study, it is necessary to meet people to know their living conditions, culture and social organization. Many research manuals that provide guidelines to methodologically support research are currently available [21, 22], including in Portuguese.

It is not always possible to obtain all data concerning a group's practices because many are tied to strong cultural values. For example, in certain indigenous tribes or African-Brazilian communities, specific pieces of information are zealously guarded secrets and passed on only to direct successors [23-25]. Often, the goal is not to simply deny the researcher's access to data that he or she seeks, but to preserve knowledge that is tied to cultural, social, and even sentimental values. During fieldwork, the researcher should seek to

Table 1
Proposal for structuring scientific articles in ethnobotany and ethnopharmacology

First stage
1. Avoid starting by writing the title, abstract, introduction, or literature review. Leave these to the end of the article writing
2. Begin by writing the hypothesis and/or research goals. Be as clear as possible. Prepare a list of references related to your hypothesis and/or goals
3. Accurately characterize the study area, studied population, and sample universe
4. Describe the methods used judiciously, linking them with your hypotheses and/or goals. Assess whether the same methods were used and/or reported by other authors and cite them properly. Sometimes it is interesting to explain the context of application of the methods of other researchers. When using new methods, describe them in detail and clearly
5. Be clear on the methods used. When possible, argue about the choice of methods, their potential, and limitations
Second stage
6. Organize the results based on the methods used. Organize tables, figures, and ethnographic descriptions of results. Statistical analysis of results must be included and clearly referenced
7. Write the results in a simple way using sentences to describe the relationships between the methods used, tables, and figures of results. In many cases, this consists of a short and simple sentence. Prepare a list of references that will help you discuss the results
Third stage
8. Interpret the results in light of the hypotheses and goals. Make a list of references related to your research
9. Write full references of the citations you use. It is very easy to later forget the material that you used
10. Write clear discussions of the results. The first discussion should be in the order of your presentation of the results. Do not repeat data from the results in the discussion
11. Rate your hypotheses clearly and specifically, indicating how your results support or reject them
Fourth stage
12. General theories on the people/plants relationship may arise when hypotheses are tested with effective methods. Arrange paragraphs in a logical order
13. Abstract. Write one or two sentences that summarize the introduction of the research. Introduce hypotheses or goals in one sentence. Write one or two sentences that summarize the methods. Write one or two sentences that summarize the discussion of results. Add one or two sentences for the conclusions
14. Create a title that is descriptive of the research conducted. Some authors have adopted titles in the form of questions that indicate the content of the publication to the reader. Other times, they use statements, for example: "The distance from urban centers affects the knowledge on medicinal plants of the Indians Piaroa"
15. Prepare a list of keywords. Do not repeat words from the title
16. Ask a colleague who is not directly involved in the research to conduct a critical reading of your material. Hand him or her a copy of this guide
17. Do not forget to check the bibliographic standardization and completeness of references used in the text and cited in the reference list

Modified from McClatchey [17]

identify himself or herself with the interviewee to understand why many secrets cannot be revealed [26]. The researcher in ethnobiology is essentially a data extractor and compiler. Therefore, respect for local social institutions is an important position for the researcher.

The process of data cataloging may seem to be an easy task to a beginning observer in ethnobiological research, as the data

Table 2

Guide for evaluation of scientific studies in the field of ethnobotany and ethnopharmacology, based on criteria presented by Lima et al. [18] and Castro [19] for scientific papers, thesis, and dissertations and McClatchey [17] for the problems found in ethnobotany publications

Defining the problem and objectives (A) Is the study area or topic clearly defined? (B) Is the research problem well defined and delimited? (C) Are the goals clear and well formulated?
Introduction and theoretical framework (A) Are the theoretical elements necessary to support the problem presented? (B) Are the updated and consistent data on the variables, the key factors, and methods to study the problem presented? (C) Are the hypotheses outlined? Are they discussed in light of findings from the literature? (D) Are relationships between different variables established? (E) Is the wording of the text clear, concise, interrelated, structured, and organized?
Methods (A) Is the workplace characterized? If so: size and condition of the area, infrastructure, environmental issues, physical and biological environment, agriculture, extractivism, tourism and demography, among others (B) Is the sampling design consistent with the goals? (C) Is the population defined? (D) Are the studied communities deeply characterized? Ethnicity, customs, history of space occupation, marked traits of culture, population size, religion, health and education, among others (E) Are the informants/interviewees described in detail? (F) Is the unit of observation and sampling described? (G) Are the sample selection criteria used described? (H) Does it state whether interviewees provided formal authorization to conduct the research? (I) Are sample size and type of sampling defined? (J) Are the data to be collected described? (K) Are the procedures used for data collection described? (L) Are the techniques and equipment well described and suitable? (M) Are the statistical methods of the analyses described and suitable? (N) Were the plants and/or animals collected and identified using the usual procedures? Identification using dictionaries or lists, on rare occasions, is acceptable (O) A good ethnographic description of therapeutic systems and concepts of health and disease is a key item, for studies aimed at discovering new drugs. It is also essential to understand the names of diseases/symptoms and their equivalents in terms of Western/urban medical systems, in addition to plant names and their botanical synonyms
Results (A) Are they presented in a clear and concise manner? (B) Are the tables, diagrams, graphs, and figures presented in a clear and concise manner? (C) Are all illustrations, tables, and graphs necessary? (D) Are all species correctly cited? (E) Were the cited species adequately sampled, with the record and deposit of reference material in herbaria or appropriate collections?
Discussion (A) Does the author establish comparisons with data from the literature? (B) Are the results discussed in a clear manner in relation to the proposed goals? (C) Do the results confirm or reject the hypotheses proposed? (D) If the study presents a list of plants and/or animals used as medicinal, is there a contextualized discussion on their use? (E) Do conclusions, when included, contribute to a further understanding of relationships between people and medicinal resources?

(continued)

Table 2
(continued)

Bibliography
(A) Are the bibliographic references cited written according to standards?
(B) Is the wording correct?
(C) Does it reflect the commitment of researchers to enrich the study?
Supplementary data
(A) Is the title appropriate?
(B) Does it conform to legal requirements in the case of research studies on humans?
(C) Is the chronogram, when applicable, well described with feasible steps within the time allotted?
(D) Is the language acceptable?

provided during an interview are transcribed into a notebook. Despite this initial appearance, there is a whole set of guidelines and procedures for collecting and recording qualitative data [27]. More than a simple scientific technique, interviewing is an art that is little understood by those outside of ethnobiological studies. Many problems that affect the quality and reproducibility of publications result from the lack of techniques that facilitate the initial contact with interviewees, poorly designed questions, inadequate or misguided selection of interviewees, insufficient sampling, and lack of theoretical knowledge supporting the methodological decisions. In combination with these factors, there are numerous data collection techniques that can replace or complement the traditional interviews; the selection of such methods depends on the proposed goals. Questionnaires, extremely common tools in ethnopharmacological surveys, may not be as effective given the cultural context [28]. Accordingly, Edwards et al. [28] stress that questionnaires may inhibit the informants because they are an extremely formal tool and include questions that the interviewees find difficult to understand (considering the cultural and environmental contrasts) and trigger hostility, depending on the research focus.

Many other precautions are needed in this type of research. For example, symbolism is found in the practices of different ethnic groups and is linked to belief in deities and the performance of rituals required for the preparation of drugs and other forms of healing [24]. In this sense, a people's worldview dictates the way the people relate to the environment in which they live (including plants and animals) and their perception of the world and the relationship between its components. It is the researcher's task to decode the data gathered, seeking to join emic observations and etic explanations. For example, there are several ways of perceiving and treating diseases. This diversity may result in the challenge of accepting the cultural and linguistic differences and an interpretation effort [26], which does not always have a satisfactory result. Furthermore, certain data are highly significant and, in certain

Table 3
Type of data that must be collected in ethnopharmacological and ethnobotanical studies

1.	Common (local) name of the plant
2.	Health conditions for which the plant is referred
3.	Plant parts used in preparations
4.	Other parts or substances used mixed into preparations
5.	Special collection requirements deemed necessary for the effectiveness of plants (season, time of day)
6.	Complete methods and preparation
7.	Complete forms of drug administration
8.	Amounts (based on the native systems of measurement) of all ingredients used
9.	Dosage (with special reference to the patient's age, gender, and health state)
10.	Presumed healing principles of each constituent (often organoleptic properties)
11.	Desired effects produced by each ingredient
12.	Treatment duration
13.	Special behavioral requirements to be followed by the patient during treatment (dietary restrictions, restrictions in regular activity)

Source: Berlin and Berlin [29]

cases, indispensable, especially if the study is focused on bioprospecting (Table 3). Bermúdez et al. [30] argue that ethnobotanical studies of medicinal plants should involve the following: record on medicinal plants related to the studied community through the collection of all possible data (see Table 3); quantitative analysis of the cultural significance or level of use of different species; assessment of the variation pattern of traditional knowledge and its relation to the social factors that affect it; analysis of the strategies employed by the population to make the most out of medicinal plants; assessment of the abundance, distribution, and diversity of medicinal plants used; evaluation of the extractivist impact on the structure and diversity of natural ecosystems; design of projects for sustainable use or conservation strategies of resources considering the traditional knowledge and practices; development of mechanisms toward public recognition of the studied populations' intellectual rights; and the development of strategies to compensate them for their participation in research studies. This is proposal is an ideal approach that depends on an interdisciplinary effort for its success.

2.2 Problems with Data Presentation and Analysis

One of the difficulties in ethnodirected research is the inaccuracy of published data. Waller [31] provides the example of "inflammation" because it can occur in practically the entire body. The causes may differ, such as trauma, infections, allergies, or asthma.

The ethnobotanical literature, for example, has many references of anti-inflammatory activity with little or no etiological evidence supporting future decisions on the most appropriate pharmacological models. Thus, the simple term "anti-inflammatory activity" does not enable the selection of the pharmacological evaluation model; detailed descriptions of the disorder are required [31]. For an ethnobotanical or ethnopharmacological study to be considered useful, certain data regarding disease etiology, plant preparation, mode of administration, therapeutic goals, and other details that enable researchers to understand the local culture should be properly collected and clearly presented (refer to Table 3) [32].

Several techniques and methods can reduce subjectivity in the presentation of data on medicinal plants and pharmacological testing while concomitantly offering a decision-making parameter regarding which plan is or is not worthy of a more detailed study. For example, using traditional techniques in ethnobotanical studies (free listing and semi-structured interviews), Canales et al. [33] aimed to collect data on the uses of plants by a community in Puebla (Mexico) while selecting plants for biological activity studies using the informant consensus factor. The informant consensus factor is a quantitative technique that indicates the plants and/or diseases that enjoy the broadest consensus among people. This method is widely used by researchers [6, 7, 34, 35]. In the study by Canales et al. [33], informants reached broad consensus when indicating plants for treating problems with possible bacterial origin. Therefore, the researchers selected 16 species for testing antibacterial activity, and 75 % of these showed such activity. Quantitative techniques have also been used to select plants with hypoglycemic activity [36, 37].

Using the informant consensus logic, Berlin [38] analyzed the lexical variation in names given to medicinal plants among 14 Tzotzil and Tzeltal municipalities in Mexico. Species with the broadest consensus in their use and therefore the lowest lexical variation were supposedly those with the highest pharmacological potential in contrast to species with the broader uses for many health conditions, which had a higher lexical variation.

2.3 Lack of Integration Between Data Retrieval, Theory, and Reproducibility

There is nearly a complete absence of use or development of theories that encompass the observed phenomena and more rapidly advance a theory of the relationship between people and natural resources. However, there is a great need for ethnopharmacological and ethnobotanical studies to enable the recovery and conservation of knowledge on medicinal plants native. Such research should aim to select plants for pharmacological, toxicological, and phytochemical studies [39]. The development or use of a theory is essential for any type of scientific approach intended to answer questions and predict phenomena. Several factors affect people's knowledge about the medicinal plants that they use, including ecological, economic, social, and cultural factors, which act in different ways in different cultures. Similarly, there is a nearly complete lack of reproducibility of studies with different groups and plant formations, even among studies that adopt a comparativist approach. Such reproducibility is desirable to test a hypothesis with similar methods in different contexts, which would enable us to understand further the context of relationships between people and the plants and/or animals that they use in medicine. In this sense, these studies must be made more predictive, which implies increasing the accuracy of research studies.

2.4 Legal and Ethical Issues that Permeate Ethnopharmacological and Ethnobotanical Studies

As stated by Etkin [32], ethnobotanical studies, which focus on medicinal plants (and especially ethnopharmacological studies) have the following aims: bioprospecting and the advancement of pharmaceutical science with the discovery of new drugs and their inclusion in biomedical pharmacopoeia; conservation and preservation of biodiversity; promotion of local use of plants in combination with previously known drugs and other biomedical technologies; and use of the knowledge of local communities while respecting their intellectual property. It appears that these goals are now somewhat contradictory because bioprospecting and the advancement of pharmaceutical science are driven by an underlying capitalist mechanism linked to the individual accumulation of profits or restricted to small groups or large corporations. By contrast, biodiversity conservation is not always driven by such motives, much less the concern for local intellectual property, which is often diffuse and grounded in a collective rather than individual logic. This observation may seem slightly partisan, but the systems of individual protection of knowledge (and consequently of financial profits from research studies related to bioprospecting and pharmaceutical science) do not seem to benefit local populations through the mechanisms that are currently used.

Any area of research that involves humans (including ethnobiology) must comply with the ethical and legal issues in a global and regional context. The global context is determined by organs of international reach, such as the guidelines of the UN [United Nations] and the International Society of Ethnobiology, for example. Conversely, the local sphere is restricted to the territorial borders of nations.

The Convention on Biological Diversity was a significant achievement regarding the protection of traditional knowledge. Signed at the United Nations Conference held in 1992 in Rio de Janeiro, the Convention recognizes the need to share benefits from the use of traditional knowledge and genetic resources in addition to addressing biological conservation and sustainable use. Furthermore, the way of life of traditional communities is recognized as relevant to the conservation and sustainable use of biodiversity; therefore, we should "… respect, preserve and maintain the knowledge, innovations and practices of Indigenous Peoples and other traditional, land-based communities…" [40]. The Bonn Guidelines on access to genetic resources and fair and equitable sharing of benefits from their use were subsequently published. The Bonn Guidelines have several goals, including "to provide the parties and stakeholders a transparent goal to facilitate access to genetic resources and ensure the fair and equitable sharing of benefits" [41].

The International Society of Ethnobiology (ISE) Code of Ethics is a useful reference on ethical aspects with international reach that must be obeyed. This code includes guidelines for the desirable conduct of surveys of ethnobiological nature, including diligence, acknowledgment, confidentiality, and the need for informed consent, among others [42].

There are two aspects regarding who owns the intellectual property rights [43]. The first was recognized at the Convention on Biological Diversity and states that countries are entitled to their biological resources within their legal borders and that native groups/communities have the right to protect and seek compensation for the knowledge that they have developed throughout generations based on local biodiversity [44]. The second aspect concerns the worldwide patent system, which is based on the legal concept of intellectual property rights (according to the World Trade Organization). In this system, individuals or groups claim that they have discovered or invented something, and a monopoly on the commercial development of his, her, or their "invention" or "discovery" is granted for a limited period of time (typically 20 years) [45]. The problem with this last issue is that in certain cases, the technological development is based on traditional knowledge without compensation and recognition of the group(s) or community(ies) that has the knowledge.

Below we focus on the legal issues required for certain ethnobiological studies in traditional communities of Brazil.

2.4.1 Legal Issues of Ethnobiological Research: The Brazilian Case

A current recurring issue refers to the protection of local knowledge on medicinal plants and animals. There is no consensus, although several researchers adopt certain procedures during their research, especially in the publication of results. In the case of medicinal plants, for example, several researchers do not disclose interesting plants found in their studies until the design of a tool that guarantees to the community's proper rights should the plant lead to a drug [46, 47]. According to Platiau and Varella [48], Brazil was one of the pioneering countries in international negotiations in favor of national sovereignty over genetic resources, but it created an inefficient and bureaucratic structure to manage access requests through Provisional Measure No. 2186-16 of August 23, 2001, as regulated by Decree No. 3945 of September 23, 2001. In Brazil, the protection of traditional knowledge has been widely discussed in light of this measure, which aims to regulate access to traditional knowledge related to genetic resources, implying monitoring procedures and access authorizations, leaving this role to the Genetic Heritage Management Council (Conselho de Gestão do Patrimônio Genético—CGEN), an agency linked to the Ministry of Environment.

However, there are several unforeseen practical problems that must be solved to promote the development of this research field in Brazil. There is a consensus that benefits should be shared through the signing of contracts (for which there is no agreement) of target-resources use. The issue of bioprospecting is under heated debate in Brazil because researchers are unanimous regarding benefit sharing but request justice and flexibility to ensure that laws do not impede the progress of research. Moreover, the idea of property rights requires extensive discussion. McClatchey [17] stresses that considering "the best of paternalistic intentions," there is the idea that all cultures perceive property rights in the same way, which is a significant misconception. A number of people are apprehensive regarding the idea of "owning knowledge," whereas others strongly defend their knowledge property rights [49].

The government measures designed to regulate access to biodiversity and associated traditional knowledge also aim to establish mechanisms for sharing the benefits that are generated by research studies of this nature. However, we run the risk of falling back into an ineffective bureaucratic structure. The return of benefits from ethnodirected research studies should be primarily viewed as the researcher's own ethical and moral premise, rather than a mere legal requirement. Elisabetsky [50] suggests that the sharing of benefits can occur either through monetary compensation (for example, when there is an immediate payment for the informant's work days or a payment per plant sample collected or through transfer of percentage of the royalties generated by the knowledge) or non-monetary means of benefit sharing, including contributions to the local/municipal economy, improvements in the local people's training regarding the sustainable use of natural/genetic resources and administrative training, exchange of staff and training, technology transfer, and increase in local scientific capacity with the participation of local experts.

Regardless of the abovementioned discussions, research studies that involve access to people's knowledge must also comply with the provisions of Resolution No. 196 of 10/10/1996 of the National Health Council (Conselho Nacional de Saúde—CNS), which defines research involving human beings as research that "involves human beings, either directly or indirectly, in its entirety or partly, including the management of data and materials." According to the resolution, the research project must be previously approved by the Committee on Research Ethics (Comitê de Ética em Pesquisa—CEP) of the institution. The informed (or free and informed) consent has been considered one of the cornerstones of modern research ethics [51, 52]. This consent form aims to provide information about the nature and purpose of the study, procedures, and/or techniques used in the research study (for example, whether there will be interviews and/or the researcher's constant observation of local medicine practices, among others), ensure the protection of the study subject's identity and privacy, among others.

The precepts and guidelines of the Provisional Measure No. 2186-16 of August 23, 2001 that "*adopts provisions concerning the assets, rights, and obligations related to access to genetic heritage components present in the national [Brazilian] territory, on the continental shelf and exclusive economic zone for scientific research purposes, technological development or bioprospecting; access to traditional knowledge related to genetic resources, relevant to the conservation of biological diversity, the integrity of the Country's [Brazil] genetic heritage and the use of its components; fair and equitable sharing of benefits from the exploitation of genetic heritage components and associated traditional knowledge; and access to technology and technology transfer for the conservation and use of biological diversity*" must also be followed. This provisional measure stipulates that access to genetic resources and/or access to traditional knowledge existing in the country [Brazil] will only be granted upon authorization from the Union (refer to Table 4). Actions or omissions that violate the Provisional Measure No. 2186-16 of 08/23/2001 are inclusively considered offenses against the genetic heritage or associated traditional knowledge, which should be ascertained according to an adequate administrative proceeding by the competent authority through sanctions, including warning, suspension or cancellation of registration, patent, license or permit, and fines (Refer to decree No. 5459 of June 7, 2005).

Table 5 shows a template of an informed consent form for research that involves no financial return. It should be emphasized that the data suggested in Table 5 should be tailored to different study cases and studied groups to ensure their clarity and understanding by the people who will participate in the study. Previous contacts with community leaders must almost invariably be established in addition to meetings and group discussions and other forms of communication.

3 Final Considerations

A recent editorial in the *Journal of Ethnopharmacology* [54], which is one of the most significant journals in the field, makes recommendations to authors to minimize the rejection of their studies by avoiding the lack of the following: ethnographical data of the study or regarding the methods used; the theoretical framework constituting

Table 4
Suggested steps to ensure the ethics and legality (under Brazilian law) of research studies involving access to traditional knowledge associated to genetic resources

Step 1: Research project submission to the research ethics committee of the institution
This committee is responsible for evaluating and approving, or not, the research project before it is conducted with humans. The researcher is usually requested to complete and submit an application form to the ethics committee, in addition to the research project. The process is typically fast and approval results may be released within 30 days.
Step 2: Research project submission to CGEN (Conselho De Gestão Do Patrimônio Genético/Genetic Heritage Management Council) or IPHAN (Instituto Do Patrimônio Histórico E Artístico Nacional/ National Historical and Artistic Heritage Institute)
Access to genetic resources and associated traditional knowledge, for scientific research purposes, bioprospecting and technological development requires authorization because the Brazilian genetic heritage belongs to the union. The researcher must consider the following aspects: 1. Access authorizations can only be requested by a legal entity, that is, the institution to which the researcher is associated. For example, for the researcher who is associated to a university, authorization should be requested by the university, and the legal representative is the president or the person formally appointed by him. 2. The institution should access the CGEN website (http://www.mma.gov.br/port/cgen) and identify the appropriate form for its need, complete it, and duly signed, submit it along with the documentation listed there. Regarding an authorization for scientific research purposes, this can also be requested from IPHAN. 3. The time for processing submitted applications varies greatly, so it is recommended that the request be made with a considerable advance to fieldwork.
Step 3: Benefit sharing agreement
In case there is potential for economic use of a genetic heritage resource, with the prospect of commercial use through a product or process (the development of a drug, for example), a contract must be signed between the parties involved (including the community that has the knowledge) for the sharing of possible benefits.
Step 4: Obtaining the free and informed consent
The nature of the study and each of the actors that participate in it should be conveyed, through meetings with community leaders and all those who will participate in the research study, to obtain the consent of the community and/or people involved in the study. Although the term used by Resolution No. 196, of 10/10/1996 of CNS (Free and Informed Consent) differs from that used by Provisional Measure No. 2186-16 of 08/23/2001, of CGEN (Prior Consent), both have similar functions. The latter is "the document evidencing that the provider of the genetic heritage component and/or the associated traditional knowledge effectively understood the project to be conducted and, furthermore, agreed to perform the activity" [53], whereas the former is a "consent of the research study subject and/or his legal representative, free of vices (simulation, fraud or error), dependency, subordination or intimidation, after complete and detailed explanation of the nature of the research, its objectives, methods, predicted benefits, potential risks and discomfort that it can cause, set forth in a consent form, authorizing his or her voluntary participation in the research" (Resolution No. 196, of 10/10/1996, of the CNS).

Table 5
Prior consent form template

Project name:
Principal Investigator's name:
The study you are about to participate in is part of a series of studies on the knowledge you have and the use you make of the plants from your region whether for feeding, construction, firewood, medicinal purposes, among others, and seeks no financial benefit for researchers or any other person or institution. It is a comprehensive study, which has many participants and is coordinated by the Laboratory (...) of the University (...). The study uses techniques of interviews and informal conversations, in addition to direct observations, without the risk of causing injury to participants, except a possible embarrassment because of our questions or presence. In case you agree to take part in this study, you will be invited to participate in various tasks, including interviews, listing the plants you know and use from the region, help researchers collect those plants, show, when appropriate, how you use them in your daily life. All data collected through your participation will be organized to protect your identity. Upon completing the study, there will be no way to connect your name with the data you provided. Any information on the study results will be provided when it is completed. You have complete freedom to withdraw from the study at any time. In case you agree to participate, please sign your name below, indicating that you have read and understood the nature of the study and all your doubts were clarified.
Date:__/___/__
Participant's signature or fingerprint
Name:
Address:
Researcher(s)' signature(s):
Witness(es)' signature(s):

the basis of the study; data on disease diagnosis and practices related to the specific medical use of a plant; and data on the protection and the rights of indigenous peoples or local government [55]. Although Pieroni [55] criticizes the wording of several of the above items as unclear and confusing, it is certain that the commented text reflects a trend of increasing rigor in the acceptance of ethnopharmacological and ethnobotanical studies. This can also be observed in another recent editorial in *Economic Botany* [56], which is a reference journal in ethnobotany.

The considerations regarding the ethical implications of research, in the context of current discussions about access to knowledge related to Brazilian biodiversity and return from research results, are more significant than considerations about the methodological rigor of ethnodirected research studies. Current Brazilian legislation regulating such matters remains limited. The researcher must constantly inquire about the benefits of his or her studies or, according to Posey and Dutfield [57], whose knowledge is researched and to whom it benefits.

Scientific research involving the application of local knowledge concerning the use of medicinal plants and animals is undergoing a propitious and productive period. The time when this knowledge was underestimated has passed, marking a new era of cooperation of knowledge. However, approaches toward producing a science that will truly meet the needs of local communities, society as a whole and the scientific community must advance to respond to questions and social needs. This change is not an easy task with a predefined recipe. In this chapter, we raised several questions with the hope of reflecting on our needs and the current scenario, which requires rapid changes and advances.

One of the most critical goals of bioprospecting of biodiversity is to help preserve the great diversity of languages, cultures, peoples and other organisms that inhabit the earth. One of the challenges anthropologists, ethnobotanists, entrepreneur physicians, and development professionals face is to creatively use bioprospecting of biodiversity as one of many tools to maintain and manage the fertile, but fragile, diversity of people, plants, cultures and ecosystems that are under constant threat of extinction [58].

References

1. Diegues AC, Arruda RSV (eds) (2001) Saberes tradicionais e biodiversidade no Brasil. Ministério do Meio Ambiente/USP, Brasília/ São Paulo
2. Alves AGC, Albuquerque UP (2005) Exorcizando termos em etnobiologia e etnoecologia. In: Alves AGC, Lucena RFP, Albuquerque UP (eds) Atualidades em etnobiologia e etnoecologia, vol 2. NUPEEA/ Sociedade Brasileira de Etnobiologia e Etnoecologia, Recife
3. Gadgil M, Berkes F, Folke C (1993) Indigenous knowledge for biodiversity conservation. Ambio 22:151–156
4. Berkes F (1999) Sacred ecology. Taylor & Francis, Philadelphia
5. Almeida CFCBR, Albuquerque UP (2002) Uso e conservação de plantas e animais medicinais no estado de Pernambuco (Nordeste do Brasil): um estudo de caso. Interciencia 27:276–285
6. Gazzaneo LR, Lucena RFP, Albuquerque UP (2005) Knowledge and use of medicinal plants by local specialists in a region of Atlantic Forest in the state of Pernambuco (Northeastern Brazil). J Ethnobiol Ethnomed 1:9
7. Alves RRN, Rosa IL (2006) From cnidarians to mammals: the use of animals as remedies in fishing communities in NE Brazil. J Ethnopharmacol 107:259–276
8. Albuquerque UP (2005) Introdução à etnobotânica. Editora Interciência, Rio de Janeiro
9. Stepp JR, Moerman DE (2001) The importance of weeds in ethnopharmacology. J Ethnopharmacol 75:19–23
10. Stepp JR (2004) The role of weeds as sources of pharmaceuticals. J Ethnopharmacol 92: 163–166
11. Voeks RA (2004) Disturbance pharmacopoeias: medicine and myth from the humid tropics. Ann Assoc Am Geogr 94:868–888
12. Almeida CFCBR, Silva TCL, Amorim ELC et al (2005) Life strategy and chemical composition as predictors of the selection of medicinal plants from the Caatinga (Northeast Brazil). J Arid Environ 62:127–142
13. Albuquerque UP, Lucena RFP (2005) Can apparency affect the use of plants by local people in tropical forests? Interciencia 30: 506–511
14. Monteiro JM, Albuquerque UP, Lins Neto EMF (2006) Use patterns and knowledge of medicinal species among two rural communities in Brazil's semi-arid northeastern region. J Ethnopharmacol 105:173–186

15. Estomba D, Ladio A, Lozada M (2006) Medicinal wild plant knowledge and gathering patterns in a Mapuche community from North-western Patagonia. J Ethnopharmacol 103: 109–119
16. Araújo TAS, Alencar NL, Amorim ELC et al (2008) A new approach to study medicinal plants with tannins and flavonoids contents from the local knowledge. J Ethnopharmacol 120:72–80
17. McClatchey W (2006) Improving quality of international ethnobotany research and publications. Ethnobot Res Appl 4:1–9
18. Lima CP, Jeckel-Neto EA, Baracat EC et al (1999) Analisando teses e dissertações. Femina 27:805–808
19. Castro AA (2002) Avaliação crítica da literatura: roteiro para a sessão de artigos de revista. AAC, Maceió. http://www.metodologia.org. Accessed 7 Aug 2006
20. Bernard HR (1995) Research methods in anthropology: qualitative and quantitative approaches. Altamira, London
21. Martin GJ (1995) Ethnobotany: a people and plants conservation manual. Chapman & Hall, London
22. Alexiades M (1996) Selected guidelines for ethnobotanical research: a field manual. New York Botanical Garden, New York
23. Albuquerque UP (2001) The use of medicinal plants by the cultural descendants of African people in Brazil. Acta Farm Bonaerense 20:139–144
24. Albuquerque UP (2006) Folhas Sagradas—As Plantas Litúrgicas e Medicinais nos Cultos Afro-Brasileiros. NUPEEA, Recife
25. Ikuta ARY (2001) Práticas fitotécnicas de uma comunidade indígena mbyá guarani, Varzinha, Rio Grande do Sul: da roça ao artezanato. Thesis, Universidade Federal do Rio Grande do Sul
26. Amorozo MCM (1996) A abordagem etnobotânica na pesquisa de plantas medicinais. In: Di Stasi LC (org.) Plantas medicinais: arte e ciência. Unesp, São Paulo
27. Viertler RB (2002) Métodos antropológicos como ferramenta para estudos em etnobiologia e etnoecologia. In: Amorozo MCM, Ming LC, Silva SMP (orgs.) Métodos de coleta e análise de dados em Etnobiologia, Etnoecologia e Disciplinas correlatas. I Seminário de Etnobiologia e Etnoecologia do Sudeste, Rio Claro
28. Edwards S, Nebel S, Heinrich M (2005) Questionnaire surveys: methodological and epistemological problems for field-based ethnopharmacologists. J Ethnopharmacol 100:30–36
29. Berlin EA, Berlin B (2005) Some field methods in medical ethnobiology. Field Methods 17:235–268
30. Bermúdez A, Oliveira-Miranda MA, Velázquez D (2005) La investigación etnobotánica sobre plantas medicinales: una revisión de sus objetivos y enfoques actuales. Interciencia 30:453–459
31. Waller DP (1993) Methods in ethnopharmacology. J Ethnopharmacol 38:189–195
32. Etkin NL (2001) Perspectives in ethnopharmacology: forging a closer link between bioscience and traditional empirical knowledge. J Ethnopharmacol 76:177–182
33. Canales M, Hernández T, Caballero J et al (2005) Informant consensus factor and antibacterial activity of the medicinal plants used by the people of San Rafael Coxcatlán, Puebla, México. J Ethnopharmacol 97:429–439
34. Amiguet VT, Arnason JT, Maquin P (2005) A consensus ethnobotany of the Q'eqchi' Maya of southern Belize. Econ Bot 59:29–42
35. Almeida CFCBR, Amorim ELC, Albuquerque UP et al (2006) Medicinal plants popularly used in the Xingo region—a semi-arid location in Northeastern Brazil. J Ethnobiol Ethnomed 2:15
36. Leduc C, Coonishish J, Haddad P et al (2006) Plants used by the Cree Nation of Eeyou Istchee (Quebec, Canada) for the treatment of diabetes: a novel approach in quantitative ethnobotany. J Ethnopharmacol 105:55–63
37. Andrade-Cetto A, Becerra-Jiménez J, Martinez-Zurita E et al (2006) Disease-consensus index as a tool of selecting potential hypoglycemic plants in Chikindzonot, Yucatán, México. J Ethnopharmacol 107:199–204
38. Berlin B (1999) Lexical reflections on the cultural importance of medicinal plants among Tzotzil and Tzeltal Maya. In: Gragson TL, Blount BG (eds) Ethnoecology: knowledge, resources and rights. The University of Georgia Press, Athens
39. Di Stasi LC (2005) An integrated approach to identification and conservation of medicinal plants in the tropical forest—a Brazilian experience. Plant Genet Resour 3:199–205
40. Ministério do Meio Ambiente (2000) Cópia do Decreto Legislativo número 2, de 5 de junho de 1992. http://www.mma.gov.br/estruturas/sbf_chm_rbbio/_arquivos/cdbport_72.pdf. Accessed 20 Sept 2012
41. Secretaría del Convenio sobre la Diversidad Biológica (2002). Directrices de Bonn sobre Acceso a los Recursos Genéticos y Participación Justa y Equitativa en los Beneficios Provenientes de su Utilización. Secretaría del Convenio sobre la Diversidad Biológica, Montreal

42. International Society of Ethnobiology (2006) International Society of Ethnobiology Code of Ethics (with 2008 additions). http://ethnobiology.net/code-of-ethics. Accessed 20 Sept 2012
43. Cunha MC (1999) Populações tradicionais e a conservação da biodiversidade. Estudos Avançados 13:147–164
44. Cordell GA (2000) Biodiversity and drug discovery—a symbiotic relationship. Phytochemistry 55:463–480
45. Tedlock B (2006) Indigenous heritage and biopiracy in the age of intellectual property rights. Explore 2:256–259
46. Rodrigues E, Carlini E (2003) Levantamento etnofarmacológico realizado entre um grupo de quilombolas do Brasil. Arq Bras Fitomed Cientif 1:80–87
47. Rodrigues E, Carlini E (2003) Possíveis efeitos sobre o sistema nervoso central de plantas utilizadas por duas culturas brasileiras (quilombolas e índios). Arq Bras Fitomed Científ 1:147–154
48. Platiau AFB, Varella MD (2004) Diversidade biológica e conhecimentos tradicionais. Coleção Direito Ambiental. Del Rey, Belo Horizonte
49. McClatchey W (2005) Exorcing misleading terms from ethnobotany. Ethnobot Res Appl 3:1–4
50. Elisabetsky E (2003) Direitos de propriedade intelectual e distribuição equitativa de benefícios no contexto de inovação tecnológica. In: Anais do I Simpósio de Etnobiologia e Etnoecologia da região Sul: aspectos humanos da biodiversidade. Sociedade Brasileira de Etnobiologia e Etnoecologia, Florianópolis
51. Spata AV (2005) Métodos de pesquisa: ciências do comportamento e diversidade humana. Livros Técnicos e Científicos Editora S.A, Rio de Janeiro
52. Ministério do Meio Ambiente, Brasil (2008) Diferenças entre a MP 2.186-16 e o Anteprojeto de Lei. http://www.mma.gov.br/ascom/ultimas/index.cfm?id=3774. Accessed 3 Feb 2008
53. Lavratti PC (2004) Acesso ao patrimônio genético e aos conhecimentos tradicionais associados. http://www.museu-goeldi.br/institucional/i_prop_documen.htm. Accessed 17 Feb 2008
54. Verpoorte R, Houghton PJ, Heinrich M et al (2006) Editorial. J Ethnopharmacol 103:309–310
55. Pieroni A (2006) Journal of Ethnobiology and Ethnomedicine: achievements and perspectives. J Ethnobiol Ethnomed 2:10
56. Moerman DE (2005) An open letter to SEB members, and readers of Economic Botany, from the editor in chief. Econ Bot 59: 309–310
57. Posey DA, Dutfield G (1996) Beyond intellectual property: toward traditional resource rights for indigenous peoples and local communities. International Development Research Center, Ottawa
58. Moran K, King SR, Carlson TJ (2001) Biodiversity prospecting: lessons and prospects. Annu Rev Anthropol 30:505–526

Chapter 28

"Returning" Ethnobiological Research to the Communities

Ulysses Paulino Albuquerque, Thiago Antônio de Sousa Araújo, Gustavo Taboada Soldati, and Lucia Regina Rangel Moraes Valente Fernandes

Abstract

The "return" of ethnobiological investigations to the communities involved in the data collection has been a recurring theme on the world stage. In this chapter, we present a brief discussion on this subject. Rather than trying to indicate certain paths, the idea of this chapter is to provoke a discussion on the subject. We will address proposals for the "return" of research that are divided into actions that do not depend on third parties and actions that necessarily depend on support from organizations and/or institutions, especially those involved in research and development.

Key words "Benefit", "Give back", "Return", Local knowledge, Traditional knowledge

1 Introduction

The "return" of ethnobiological investigations to the communities involved in the data collection is a recurring theme on the world stage [1]. When considering the ethical commitments and implications of ethnobiology [2], researchers are (or should be) conscious of a formal or informal "pact" with the communities. However, an ethical and moral dilemma exists because most often this pact is not fulfilled, not from lack of interest but usually from not knowing how to fulfill it. At first glance, this task appears to be easy, and the most appropriate action of "return" will depend on the researcher's personal insight and his/her comprehension of the reality of the problems of the community that he/she visited. With rare exceptions [3], our literature discusses how this "return" can be made. The works that we have at our disposal describe in detail how to collect data and the best way to obtain and analyze information, but this description is always from a unilateral perspective [3]. According to Bridges, some initiatives to fill this gap can be taken during the preparation of our scientific publications: as

Ulysses Paulino Albuquerque et al. (eds.), *Methods and Techniques in Ethnobiology and Ethnoecology*,
Springer Protocols Handbooks, DOI 10.1007/978-1-4614-8636-7_28, © Springer Science+Business Media New York 2014

authors, we might include brief notes in our articles that describe how our research "contributed" to the community studied; as reviewers of articles, we should maintain these contributions in the manuscripts; and, as editors, we should promote the recognition of this contribution.

The idea of a mandatory "return" appears to create, in our opinion, a paralyzing dilemma for the researcher, given that from the beginning of his/her academic career, he/she is constantly challenged by the notion of an "ethical obligation." The paralysis to which we refer here begins because this concern for "another" is strange given that it is not part of a researcher's professional training, either in university courses or in scientific practice, and as a result of the current contextualization of universities and research and teaching institutions before society, with the key point being the necessary discussion about the process of outreach. In addition, when we think of "return," we associate it with a grandiose project that exceeds the capacity of the researcher and the amount of time projected for the research term. In this chapter, we hope to initiate a debate concerning the possible forms of "return" for research projects that do not involve economic benefits. According to Shanley and Laird [1], "return," as an invaluable manner of participating in the benefits of a research study, means "devolver los datos en formas relevantes para los grupos locales y para la conservación aplicada [to return the data in relevant ways to the local groups and to applied conservation]." Regardless of any initiative and the proportions of an intended approach, some information that may help with this definition includes the following: (1) knowledge of the problems and difficulties of the community; (2) knowledge of the collective interests; (3) the identification of potential protection of intangible assets of collective use intellectual property; and (4) an awareness that this contribution, which may appear simple in our view as researchers, could be very important to the community or trigger a beneficial process for local transformation.

To discuss the "return" of investigations is to think about science, epistemological foundations and the values that guide our practice of being scientists and citizens. Ethnobiology and ethnoecology follow this logic; however, given their epistemological particularities, such as contact with other cognitive systems for the construction of a hybrid knowledge, they assume a very privileged ethical position of reflection about the practices of the researcher, reflections that will directly influence the "return" activities.

The following reference questions guide these reflections: "how," "why," "for whom," and "with whom," which should be made during the design of the research project. Lastly, the idea of process contributes to the idea of return defended in this text; the understanding of all of the development of research activities as an interdependent continuum in which the "return" does not only happen at the end of activities but also during all of these steps of contact, being related to all of these spheres that were constructed among the stakeholders in the research.

In light of the reflections stated above, we understand return to encompass the following: (a) a political and ethical activity that should be inherent to all researchers in this area; (b) an activity that is conceived and constructed in a dialectic form between the research stakeholders; (c) an activity that aims to contribute to local development, that is, emancipation or empowering of the social group associated with the research; (d) an activity that considers the problems, difficulties, and potential of the community, along with collective interests; and (e) a constant activity that is performed on a daily basis and not just at the end of the research.

For practical purposes, we divided the "return" initiatives into two groups: actions that can be developed individually and actions that depend on organizations or institutions, considering the limitations and difficulties of execution. To aid our discussion, Chart 1 suggests some methods of returning data to the local communities.

2 Initiatives That Do Not Depend on Third Parties

2.1 Our Products

The most simple and evident possibility of "return" is perhaps the presentation of the constructed product (dissertations, thesis, articles, etc.) to the social group as well as the effects of these products (development of theories, offering of information to public administrators, etc.). There are no apparent difficulties in executing this initiative; however, the researcher should exercise caution in the manner in which he/she conducts this process and should consider the appropriate translation of the content and methods used. These initiatives should consider the local context, such as the level of instruction, way of life, customs, and beliefs (see Chart 1). The researcher should also consider social relationships and hierarchies that structure the community.

2.2 Transfer of Knowledge and Experiences

Because the contribution of the community to our work occurs in the "transfer of knowledge and experiences" and in view of a bilateral relationship, we as ethnobiologists can provide the scientific knowledge that we possess in an attempt to achieve the local desires. Indisputably, these actions could reduce the asymmetry that sometimes characterizes the relationship between the researcher and the community. We understand that new elements may be revealed by the local reality, providing a glimpse of richer and more complex and particular possibilities when they are associated with the local knowledge. In some situations, our academic and professional training (not just as ethnobiologists) and the association of this training with that of researchers in intellectual propriety could provide benefits to the community, for example, foster care that

Chart 1
Forms of data return

There are many ways to translate data into valuable forms for local groups; the selected method will depend both on the groups and the objectives of those that will be served. The following are some ideas that could be considered
Oral
Workshops and interactive seminars For many industrial, technical, and government officials groups, this form of exchange could prove to be useful; a flexible structure, which may include field trips, could help to foster dialogue and awareness
Theater and itinerant exhibitions Rural and urban groups respond better to stories, portrayed by fellow community members or visitors, that offer lessons learned in a more attractive format than a reading or seminar. In the realm of theater, people are able to transmit information that they might normally be embarrassed to share. In some cases, exhibitions and itinerant theaters have proven to be invaluable to the community and neighboring groups in sharing the lessons learned
Exchanges Exchange between groups with similar needs and histories, but different geographical regions, could be an extremely effective means of information trade. When neighbors share information among one another in culturally appropriate manners, there are many benefits. First, the language, expressions and means of communication are clearly understood. Second, confidence is created on a more solid foundation; the information is better accepted when there is no motivation to "sell" ideas. Third, the individuals of a region may have personally experienced the positive or negative effects of specific decisions on land use and are thus able to report on the consequences in a more convincing manner than an extension agent that is not from the communities
Music Songs are a powerful method of cultural expression. New or popular songs that integrate the findings of an investigation in their lyrics may not only transmit relevant scientific facts but also incorporate feelings of cultural loss that surround the impoverishment of an ecosystem. Music has the additional benefit of migrating from one culture to another by itself, taking the message along with it across geographical borders
Readings In some cases, the presentation of scientific results in a standard academic format may be appropriate and useful, particularly in research institutions, universities, and even local government agencies. Even if it is still important to integrate some of the style aspects previously mentioned in these forums, readings presented in a standard format are much less capable of being assimilated than those that encourage participation by the audience
Written
Manuals and illustrated booklets The content of the manuals could vary greatly. Some may be illustrated guides for the use and management of local species, targeting largely illiterate audiences concerned with the loss of cultural knowledge and who are consequently in search of a means to record and validate this knowledge. Others could be in the form of field guides that local groups can sell to tourists or use in resource management; they could also be technical manuals that help local groups better manage resources or negotiate with commercial or governmental representatives that are interested in the local lands. When these guides are aimed at semi-illiterate or illiterate populations, it is critical to test the illustrated materials with the local inhabitants; rural inhabitants, in particular, have an acute perception of the size and shape of fruits, leaves, and wildlife and have well-developed opinions about the techniques of how to process and manage local resources

(continued)

Chart 1 (continued)

Study plans
Teachers from developing countries with high biodiversity usually have limited access to teaching resources concerning the local environment, the traditional use of resources and wider environmental concerns; researchers could offer a valuable service by translating their results into formats that can be easily adopted by teachers in their classrooms
Coloring books
Children respond very well to coloring books in a local or vernacular language; the coloring book allows them to become involved in the topic, and they are important tools for education and learning
Books
The publication of books could help better advertise the information, reaching a greater portion of society than any other method. However, many groups are not able to pay for books; thus, researchers must be clear about who this form of investigation is targeting—typically, books are much more focused toward academics or the government than the poor urban and rural inhabitants

Translated from Shanley and Laird [1]

optimizes health and hygiene; indicating more pest-resistant species or those with better agricultural performance; collaborating in the development of strategies for management and sustainable use; or identifying and helping in the protection and valorization of local resources through intangible assets of intellectual property.

Schmidt and Santos [4] define intangible assets as incorporeal resources controlled by a specific entity that are able to produce future cash flows. Intellectual property encompasses the areas of industrial property and copyrights.

In Brazil, according to Law 9.279 of 15/04/1996, industrial property protects patents, brands, industrial designs, geographical indications, computer programs, and integrated circuits. Copyrights are protected by Law 9.610 of 19/02/1998, covering literary, artistic or scientific works; conferences, speeches, sermons, and other works of this nature; dramatic or dramatic-musical works; choreographed or pantomimed works, whose stage performance is set in writing or any other form; musical compositions, with or without lyrics; audiovisual works, with or without sound, including cinematography; photographic works and those produced by any process analogous to photography; works of drawing, painting, engraving, sculpting, lithography, and kinetic art; illustrations, maps, and other works of this nature; projects, sketches, and plastic works concerning geography, engineering, topography, architecture, landscaping, scenography, and science; adaptations, translations, and other transformations of original works presented as new intellectual creation; computer programs; and collections or compilations, anthologies, encyclopedias, dictionaries, databases, and other works that, by their selection, organization, or content disposition, constitute an intellectual creation.

Programming actions that can help the target communities of ethnobiological studies with the "transfer of knowledge and experiences" often require only a small amount of time, the willingness to articulate to local leaderships and to organize the management of this information through courses, lectures, and/or workshops, taking advantage of the local infrastructure and conditions and, in some cases, the participation of other institutions that can help add value to the community. Scientific arrogance, which often characterizes relationships between technicians and the community, should be replaced with a legitimate relationship of respect for differences.

Finally, this possibility of contribution should be constructed within an egalitarian and dynamic perspective that is guided above all by the horizontalness of the relationship. As an example of this proposal, consider the case of a community that wishes to implement a community garden to cultivate medicinal plants, making use of all of their means and lacking only in technical support. In this case, within our understanding, the entire process should be discussed with the community such that the community may become self-sufficient and independent after the process. This idea is part of the perspective that technical knowledge should not be introduced locally as a form of domination.

2.3 Political–Organizational Support

Another form of contribution is through actions that are not directly related to our academic training. These actions might include, for example, supporting the development of organizational and bureaucratic groups in which the communities are involved, such as associations and co-ops. These organizations, which allow access to resources and a better-structured foundation, end up not taking shape because of the lack of local knowledge about the necessary steps required. In this manner, the researcher could act as a privileged mediator and collaborate in the structuring and/or strengthening of these institutions. The legitimization of an organization through building and registering the status as a collective organization, along with interventions with public authorities, can be used as an example. Normative Instruction N 12 of the National Institute of Industrial Property in Brazil (Instituto Nacional da Propriedade Industrial—INPI) [5], which is responsible for granting industrial property records in Brazil, states that one of the conditions for the protection of Geographic Indication (GI) is that the representation of the collectivity be through these previously mentioned legal entities. Thus, in the majority of cases, Geographic Indication is an intangible asset of collective use. Article 2 of the Normative Instruction cites that Geographic Indication could be either Indication of Source (IS) or Designation of Origin (DO). Article 2 § 1 defines IS as the geographical name of the country, city, region, or location of a territory that has become known as a center of extraction, production, or manufacturing of a specific product or the providing of a specific service, and Article § 2 conceptualized DO as the geographical name of a country, city, region, or location of a territory that designates a product or service whose qualities or characteristics originate exclusively or essentially from a geographic means, including natural and human factors.

2.4 Public Valorization of Local Knowledge

Another form of contribution to the community could be the dissemination of the practices and knowledge to the greater society through the media. This strategy fulfills two basic objectives that are highly related: (1) the recognition and valorization of local experiences to raise awareness in society to this knowledge and, in some cases, it is the only way of preserving these communities; and (2) the dissemination of local knowledge. However, these strategies must be permeated, again, by the notion of ethical commitment. Information that could compromise intellectual property should be heavily screened before its release, such as topics that are secret affairs of the community or that could have economic implications. The text below provides an example of a type of advertising strategy that could be centered in a greater or lesser degree on the story of the experiences of the community and may vary with the profile or requirements of the communication means used. When the information that we wish to convey will be communicated by third parties, we should attempt to communicate our ideas clearly to avoid the publication of distorted, equivocal or scientifically incorrect information.

Threatened Medicinal Plants
By Verônica Falcão
Jornal do Commercio [Journal of Commerce]
Recife—17.09.2000, Sunday.

Five species of semiarid trees are threatened because of the use of the bark of these trees as homemade medicine. The alert is from the botanist Ulysses Paulino de Albuquerque, who studies this topic with funding from the WWF—World Wildlife Fund. According to him, "aroeira-do-sertão" (*Schinus terebinthifolius* Raddi), "baraúna" (*Schinopsis brasiliensis*), "quixaba" (*Sideroxylon obtusifolium*), "imburana" (*Commiphora leptophloeos*), and "angico" (*Anadenanthera peregrina* (L.) Speng) are at risk of extinction because the manner in which the pieces of bark are removed could lead to the death of the plant (...).

(continued)

(continued)

According to the researcher's evaluation, these species are the five most exploited species of medicinal plants in the semiarid region. He estimates, however, that the local populations of this region use 200 plant species, from herbs to large trees, especially for medicinal purposes. Some plants from the caatinga, such as umbu plants (*Spondias tuberosa*), are still used as food.

Albuquerque has studied the relationship between the people and plants of the semiarid region since 1998. Called ethnobiology, this branch of science considers economic, social, cultural, and biological aspects, with the objective of the conservation and management of biodiversity. The work is performed in a rural area of Alagoinha, 232 km from Recife, with 30 families.

In Albuquerque's opinion, the inhabitants of the semiarid region are not the ones that threaten these five species of plants. In a survey conducted in Alagoinha, he observed that families reasonably exploit the trees. "They know the amount of bark that can be taken from the tree, which contributes to the maintenance of the species in the environment," (...).

The researcher defends conservation policies for medicinal plants. To him, the trade of these plants must be controlled to avoid their extinction. "In addition to the plants being used intensively, these plants are not cultivated, and their populations are being reduced."

Another example of the valorization of local knowledge through "returning" the ethnobotanical study is that of the Praia do Sono Community, located in Paraty, on the coast of Rio de Janeiro, which was investigated by Brito [6]. After 2 years of conducting his study, in the beginning of 2012, the researcher returned to the community with the collaboration of researchers from the INPI Academy to identify potential means of protection of the collective intangible assets, especially the geographical indication. The "caiçaras" (coastal inhabitants) of Brazil do not presently have the same legal protection as "quilombola" (African slave descendants) and indigenous peoples. The Praia do Sono caiçaras suffer from harassment by real estate speculators, and those that resist this harassment are pressured by the arrival of luxury condominiums constructed in areas that were previously inhabited by the caiçaras. The possibility for the community to be considered a Fitovida Network Community Group of the state of Rio de Janeiro to strengthen the transmission of local knowledge related to medicinal plants and related practices was identified. This identification would increase the visibility of these caiçaras and help to preserve the local knowledge and help the survival of the community. In this same year, for the first time, the caiçaras participated in the Fitovida Network Shareholder's Meeting at the Bracuí Quilombo, and this year, they participated in the Annual Meeting, which demonstrated the interest of the group in participating in the network. The potential and viability of the protection of a geographical indication for caiçara "Praia do Sono" tourism services is being studied. An ethnobotanical study by [7] in Rio de Cima, Friburgo, Rio de Janeiro identified the potential protection of a geographical indication (GI) of a species of identification of origin

IS "Rio de Cima" for yams planted in that community and commercialized in fairs with a better price than in other nearby regions. The study documented the Yam Festival that occurs annually, and [8] had already mentioned the reputation of this yam. Two other potential collective brands are being verified by researchers at the Fluminense Federal University (Universidade Federal Fluminense—UFF) and the INPI Academy in that community. The exception is that according to Art. 123 item III of Law 9.279, a collective brand is that which is used to identify products or services that originate from members of a specific entity. The appropriate protection option among the various assets is also an aim of ethnobotanical studies.

2.5 Training Multipliers

Another return action concerns the training of multipliers, such as community health agents, teachers, and people, connected with neighborhood associations and unions. This return alternative could be quite effective when the aim is to reach more people because the information can be disseminated more rapidly. Moreover, people who participate in this type of activity are more familiar with these moments, which facilitates the job of the researcher. Another advantage is that the person who will pass on this information in the next step is the multiplier and already has a command of the codes and signals to transmit information to the social group that he/she is a part of.

Similar to the knowledge and experience transfer activities, for this activity, we need some time to articulate ourselves with the local representatives and for preparation of the support structure for these actions. In moments such as these, explanations can be given that illustrate, for example, how the results can help improve the quality of life for people.

3 Initiatives That Depend on Third Parties

Strategies that depend on financial resources from public or private sources (specific announcements by funding agencies) often require the sensitization of these social stakeholders to the problems that the community faces. Considering the possibility of acting in this direction, some paths are possible, including creating guidebooks or manuals and building outreach projects. Printed materials such as booklets or manuals enjoy a privileged position when the ethnobiologists think about the idea of contribution or "return." As a productive space for creativity, these materials can be used to document the practices and plants and/or return the local knowledge that is related to the scientist. Two different focuses are necessary when thinking about the use of these materials: the first focus is ethics because according to the information that we present in these documents, we can facilitate access to the traditional knowledge associated with genetic heritage (see Chap. 11) (e.g., information about the use of medicinal plants and animals) and the second focus results from our interests as researchers and our theoretical and methodological experience. Undoubtedly, these materials can consolidate theoretical discussions and actions concerning the improvement of the quality of life of a community. Considering the focus given to the content of these materials, they could be incorporated into strategies to transmit intercultural knowledge. Visualization of this possible scenario again provides an ethical challenge to researchers.

Concerning written materials, there are various forms of "return" with different purposes (see Chart 1), and we present some examples below. Healers in Belize asked researchers to produce a book that helped teach children about medicinal plants (cf. [1]). Alexiades et al. [9], in turn, developed a bilingual manual (Spanish and in the orthography used by the groups involved in the participatory process) on the medicines of the Esse ejja communities in Peru and Bolivia, with the objective of promoting both the culture and the medicine. There are also some interesting examples in Brazil. Rios et al. [10] designed a manual with information about useful "capoeira" plants (secondary succession) that are considered very important to farmers of the Benjamin Constant Community (Pará, Brazilian Amazon). Fonseca-Kruel et al. [8] followed this same direction and aimed at saving and disseminating information about useful "restinga" plants used by local fishermen in Arraial do Cabo (Rio de Janeiro). In addition to the examples cited, we should also highlight Hanazaki et al. [11], who published an illustrated book about useful plants to the caiçaras of the São Paulo coast.

To conclude the examples of written materials, we would like to highlight two experiences of the LEA (Laboratório de Etnobiologia Aplicada da Universidade Federal Rural de Pernambuco—Laboratory of Applied Ethnobiology of the Federal Rural University of Pernambuco). The first is the product of an interdisciplinary and inter-ethnical project in which LEA was a partner, which aimed to build aid for environmental and cultural sustainability of the medical system of the Fulni-ô indigenous people. As a result, all of the members of the team involved in the project developed a bilingual booklet in Portuguese and Yathê, (the native language) that discusses the importance of traditional practices, the explanatory medical system, and the local specialists in addition to the main medicinal plants and their uses. Finally, we highlight three materials that were produced using the living experience of a few years in a rural community of the municipality of Altinho, in rural Pernambuco: (a) an illustrated catalog of medicinal plants of the Caatinga and the method to prepare them; (b) a comic book that discusses the role of ethnobotanical research in the community and the importance of local knowledge; and (c) an educational book to be used in schools throughout the city that evaluates the biological particularities of the Caatinga.

In turn, outreach projects have two characteristics: they are linked to local needs (which aren't always what the researcher imagines), and they often depend on specific lines of explicit action in the announcements of the funding agencies. The researcher, for example, could believe that a community needs technical information about the correct use of medicinal plants when this community identifies the fact that the "oldest members" are not passing along their experiences and knowledge about those resources as the main problem requiring a solution. The investigations of the LEA identified a community in the semiarid northeast in which many families were unable to attain food self-sufficiency. The researcher could attack the problem from two different sides in an outreach project: attempt to develop partnerships or stimulate public actions that aim at guaranteeing access to these resources for these families; or, as was locally identified, strengthen the network of social relations with the socialization of successful experiences of other families of the same community. In the latter approach, the researchers observed that there were families in the same community that were self-sufficient due to traditional land management techniques and the use of native resources of the region.

An ethnobiology outreach project could have a main objective of developing strategies for community development and improving the quality of life of these people. However, there are many concepts of development and, consequently, methodological references and social implications. Amartya Sen [12], for example, understands development as a practice of individual liberties. According to this author, "*the expansion of freedom is seen (…) as the main goal and the main means of development. Development consists of the elimination of that which deprives of freedom or that limits the choices and opportunities of a person to thoughtfully exercise his/her condition as an agent.*" This concept differs from ideas that relate development with economic growth or the incorporation of new technology. Likewise, a community could have a completely different understanding of quality of life. For some ethnic groups, for example, quality of life is irrevocably linked to health, which in turn is associated with the continuity and perpetuation of specific rituals. Thus, first, while enjoying this freedom, the researcher must respect local interests when developing proposals and, second, must attempt to foresee the effects of the actions on the way of life and local institutions. Accordingly, the researcher must ask himself or herself if his/her actions will lead the community toward a path of autonomy or generate some type of dependence on external elements or goods (without guarantee of a continued and accessible supply to all people).

Normally, the sign of some action in the community creates anxiety and expectations, especially when the community is not informed about the requirements involved in the solicitation of financial resources for these actions. Therefore, the researcher must pay attention to all of these facets, including preparing the community for the proposal to not be accepted. Additionally, at each step, the researcher should discuss and provide clarification to the community about the objectives of the proposal and the utilization of the resources, even if a comprehensive explanation is required to avoid inconvenient situations such as the researcher requesting resources for strictly personal use. There is an entire social dynamic that is involved in the planning and implementation of actions in a community, which could reflect a history of unsuccessful projects and disastrous relationships with other researchers. Without a doubt, these negative experiences could result in the community being fearful and non-trusting.

It is also worth mentioning that the researcher/extension agent must be aware of the structure and social organization of the community because some communities could have a clan organization, and if the intentions of the researcher will benefit the entire community, the projects could be hindered by rivalry between clans or by the researcher being more closely associated with one of the clans.

4 Final Considerations

In this chapter, we intended to reflect on the "return or non-return" and "how to return," which are ideas that are also associated with reflection on our scientific proceedings. This reflection could ideally lead to the building of solid theoretical foundations that should guide the selection and execution of "return" events. In this manner, understanding that the decisions taken concerning the nature of the "return" practices are also within the theoretical–political field of the researcher and that any position is a political position [13], we recognize that the practices of return are also political actions that are therefore able to be tied to questions such as "for whom," "by whom," and "for what."

We believe that there are still many challenges to be overcome when we treat the question of "giving back" or "returning." Without a doubt, there are many lessons to be learned from the experience of others and many problems to be faced (see, for example, the case studies presented by Shanley and Laird [1]), above all, to challenge ourselves to think of this "return" as we reflect and build our projects of scientific investigation. In this sense, the proposal by Bridges [3] to include notes on the subject in our articles could help ethnobiologists to benefit from the experiences constructed in different geographical and cultural contexts to guide their own practices of ways to "benefit," "give back," or "return" the data of their research. The collaboration of intellectual property researchers in ethnobiological studies could greatly help to identify collective intangible assets that should be protected particularly by those associated with local knowledge.

The idea of this chapter was more to encourage a reflection than to actually note paths to follow. However, we cannot escape the intention to view the real possibilities of contribution and, especially, challenge the dilemma that accompanies our investigations while also stimulating the interaction of ethnobotanists and ethnobiologists with researchers in intellectual property. Although we may understand the regularities that govern the relationships of people with plants, we must face the reality of the differences and of the distinct views. "Conhecer um grupo é poder discernir a regra da exceção, a regularidade da excepcionalidade [To know a group is to be able to discern the rule of exception, the regularity of exceptionality]" [14]. If, in principle, we look for what the manuals teach us to investigate in a community—their knowledge, practices, and experience—we cannot forget that, in the effervescent pot of coexistence, we can transform our own experience. Finally, we argue that the practices of return could enrich our ethnobiological work, founded on the idea that to perform science is also a political and ideological action that could be compromised in a new world, a world ruled by human emancipation.

References

1. Shanley P, Laird SA (2002) "Devolviendo": hacienda que los resultados de la investigación sean relevantes para los grupos locales y para la conservación. In: Laird SA (ed) Biodiversidad y conocimiento tradicional: paticipación equitativa en práctica. Editorial Nordan-Comunidad, Uruguay
2. Albuquerque UP (2005) Introdução à etnobotânica, 2ª. ed. Interciência, Rio de Janeiro
3. Bridges KW (2004) Editorial: give and take. Ethnobot Res Appl 2:75
4. Schmidt P, Santos JL (2002) Avaliação de ativos intangíveis. Atlas, São Paulo
5. Brasil INPI Instrução Normativa N 12 de 16/03/2013. Estabelece as condições para o registro das indicações geográficas
6. Brito MR (2010) Comunidade caiçara da praia do Sono, Paraty, RJ, Brasil, uso dos recursos vegetais. Dissertation, Museu Nacional-Universidade Federal do Rio de Janeiro
7. Boscolo OH (2011) Estudos etnobotânicos em área de Mata Atlântica nas comunidades de Galdinópolis e Rio Bonito, Nova Friburgo, RJ, Brasil. Thesis, Museu Nacional-Universidade Federal do Rio de Janeiro
8. Fonseca-Kruel VS, Peixoto AL, Sá CFC, Araújo DSD, Silva WL, Ferreira AJ (2006) Plantas úteis da restinga: o saber dos pescadores artesanais de Arraial do Cabo, Rio de Janeiro. Instituto de Pesquisas Jardim Botânico do Rio de Janeiro, Rio de Janeiro
9. Alexiades M, Huajohuajo G, Huajohuajo M, Mamío R, Peluso D, Quiosche R, Pesha V, Tirina G (2005) Ejjabahuejjaquijji ebiojonesqui shemeño Esse ejjaja esohuijo—Para conocer los remedios del monte. Federación Nativa del Rio Madre de Dios y Afluentes, Puerto Maldonado
10. Rios M, Martins-da-Silva RCV, Sabogal C, Martins J, Silva RN, Brito RR, Brito IM, Brito MFC, Silva JR, Ribeiro RT (2001) Benefícios das plantas da capoeira para a comunidade de Benjamin Constant, Pará, Amazônia brasileira. CIFOR, Belém
11. Hanazaki N, Peroni N, Araújo LG, Toledo BA, Tamashiro JY, Begossi A (2007) Etnobotânica Caiçara no litoral paulista. RiMa, São Carlos
12. Sen A (2000) Desenvolvimento como liberdade. Companhia das Letras, São Paulo
13. Freire P (2004) Pedagogia da Autonomia. Saberes necessários à prática educativa. Paz e Terra, São Paulo
14. Víctora CG, Knauth DR, Hassen MNA (2000) Pesquisa qualitativa em saúde—uma introdução ao tema. Tomo Editorial, Porto Alegre

INDEX

A

B

Ulysses Paulino Albuquerque et al. (eds.), *Methods and Techniques in Ethnobiology and Ethnoecology*,
Springer Protocols Handbooks, DOI 10.1007/978-1-4614-8636-7, © Springer Science+Business Media New York 2014

D

E

F

J

L

M

N